# Contents

*Contents*

# decision mathematics

**Sue de Pomerai**
Seaford Head Community College
Sussex

**John Berry**
Centre for Teaching Mathematics
University of Plymouth

contributory authors
  *Phil Dyke*
  *Mike Herring*

project contributors
  *Steve Dobbs, Bob Francis,*
  *Roger Fentem, Ted Graham,*
  *Howard Hampson, Penny Howe,*
  *Rob Lincoln, Claire Rowland,*
  *Stuart Rowlands, Stewart Townend,*
  *John White*

Collins Educational
*An imprint of HarperCollinsPublishers*

**John Berry** is Professor of Mathematics Education and Director of the Centre for Teaching Mathematics at the University of Plymouth. John Berry worked with SCAA on the development of the new A-level syllabuses, and the applications of new technology on the teaching and assessment of A-level mathematics.

**Sue De Pomerai** is currently Head of Mathematics at Seaford Head Community College in Sussex. She has many years experience of teaching Decision Maths at A-level and is a revisor for Edexel Decision Maths Module.

Published by Collins Educational
An imprint of HarperCollinsPublishers
77–85 Fulham Palace Road
Hammersmith
London W6 8JB

© Sue De Pomerai and John Berry/HarperCollinsPublishers 1998
First published in 1998
ISBN 0 00 322372 8

Designed and illustrated by: Ken Vail Graphic Design, Cambridge, UK
Cover design by: Moondisks, Cambridge, UK
Project editor: Joan Miller
Printed and bound in the UK

# Preface

## discovering advanced mathematics

Mathematics is not just an important subject in its own right, but also a tool for solving problems. Mathematics A-level is changing to reflect this; during the A-level course, you must study at least one area of the *application* of mathematics. This is what we mean by 'mathematical modelling'. Of course, mathematicians have been applying mathematics to problems in mechanics and statistics for years. But now, the process has been formally included throughout A-level maths.

Decision mathematics is a fairly new area of applied mathematics in the A-level curriculum. Until recently most students following a course in advanced mathematics would have studied mechanics and/or statistics. There are many decision-making problems in business and commerce for which a different problem-solving approach is required; one based on an algorithmic strategy instead of the traditional skills of algebra and calculus. The development of cheap and powerful computers has been an incentive for mathematicians to develop efficient algorithms for solving these decision-making problems.

This book in the *discovery advanced mathematics* series is designed as an introduction to Decision Mathematics.

We have written *discovering advanced mathematics* to meet the needs of the new A- and AS-level syllabuses and the Common Core for mathematics. The books provide opportunities to study advanced mathematics while learning about modelling and problem-solving. We show you how to make best use of new technology, including graphics calculators (but you don't need more than a good scientific calculator to work through the book).

In every chapter in this book, you will find:
- an introduction that explains a new idea or technique in a helpful context;
- plenty of worked examples to show you how the techniques are used;
- exercises in two sets, A and B (the B exercises 'mirror' the A set so that you can practise the same work with different questions);
- consolidation exercises that test you in the same way as real exam questions;
- questions from the Exam Boards.

And then, once you have finished the chapter:
- modelling and problem-solving exercises to help you pull together all the ideas in the chapter.

We hope that you will enjoy advanced mathematics by working through this book.

We thank the many people involved in developing *discovering advanced mathematics*. In particular, we thank Mike Herring for many of the B exercises and modelling activities, Karen Eccles for typing the manuscript, and Phil Dyke for helping with the Coding chapter.

John Berry, Sue de Pomerai *March 1998*

# Acknowledgements

We are grateful to the following Examination Boards for permission to reproduce questions from their past examination papers and from specimen papers. Full details are given with each question. The Examination groups accept no responsibility whatsoever for the accuracy or method of working in the answers given, which are solely the responsibility of the author and publishers.

Associated Examining Board (*AEB*)
Northern Examinations and Assessment Board (*NEAB*); also School Mathematics project, 16–19 (*SMP 16–19*)
Oxford and Cambridge Schools Examination Board (*OCSEB*)
Scottish Examination Board (*SEB*)
University of Cambridge Local Examinations Syndicate (*UCLES*)
University of London Examinations and Assessment Council (*ULEAC*)
University of Oxford Delegacy of Local Examinations (*UODLE*); also Nuffield Advanced Mathematics (*Nuffield*)
Welsh Joint Examinations Council (*WJEC*)

We are also grateful to the following for permission to reproduce copyright photographs:

Mary Evans Picture Library 49 (left), 229, 285; By permission of the President and Council of the Royal Society 2, 49 (right), 174.

We should like to thank BBP, PO Box 204, Hatfield, Herts, for the activity on dandelion populations on page 214.

Every effort has been made to contact the holders of copyright material, but if any have been inadvertently overlooked the publishers will be pleased to make the necessary arrangements at the first opportunity.

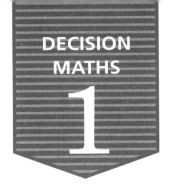

**DECISION MATHS**

**1**

# *Algorithms*

*By the end of this chapter you should be able to:*

■ *understand what is meant by an algorithm*

■ *understand an algorithm presented in a variety of forms*

■ *sort using Bubble sort and Quicksort*

■ *use techniques to assess the speed and efficiency of an algorithm*

■ *use algorithms for packing.*

The best way to start a course in decision mathematics is to look at the sorts of problems that you will be able to solve by the time you reach the end of this book.

■ How do you find the shortest route between two places?

■ How would you design the fire exit routes in a building?

■ How would you plan a job, to make the most efficient use of your time?

■ How can you best match the skills of workers to jobs that need to be done?

■ How can a manufacturer maximise profits?

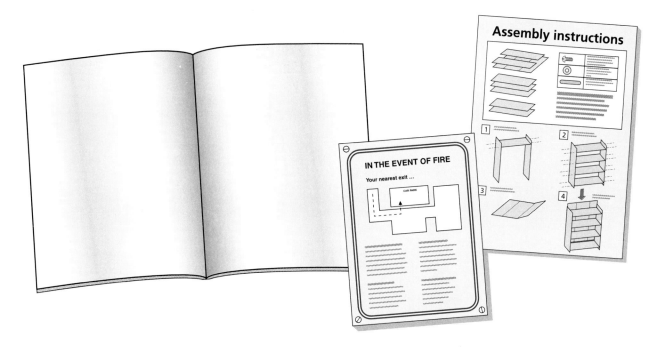

## WHAT IS DECISION MATHS?

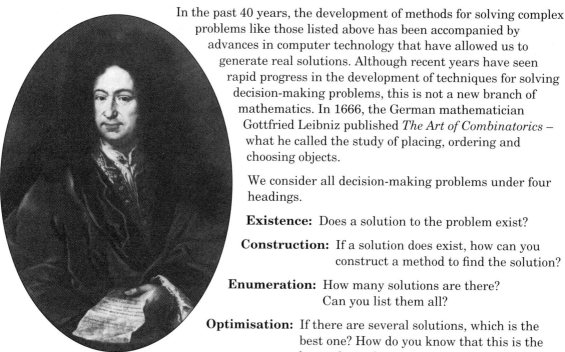

In the past 40 years, the development of methods for solving complex problems like those listed above has been accompanied by advances in computer technology that have allowed us to generate real solutions. Although recent years have seen rapid progress in the development of techniques for solving decision-making problems, this is not a new branch of mathematics. In 1666, the German mathematician Gottfried Leibniz published *The Art of Combinatorics* – what he called the study of placing, ordering and choosing objects.

We consider all decision-making problems under four headings.

**Existence:** Does a solution to the problem exist?

**Construction:** If a solution does exist, how can you construct a method to find the solution?

**Enumeration:** How many solutions are there? Can you list them all?

**Optimisation:** If there are several solutions, which is the best one? How do you know that this is the best solution?

*Gottfried Liebniz (1646–1716)*

## DOES A SOLUTION TO THE PROBLEM EXIST?

Most of the problems that you will have met in your study of mathematics have an answer, so you may not always think to ask if a solution exists. But the following problem shows that this is not always an easy question to answer. It is one of the most famous examples of a decision-making problem.

**Exploration 1.1**

### *The bridges of Konigsberg*

The town of Konigsberg in East Prussia was built on the banks of the River Pregel, with islands that were linked to each other and the river banks by seven bridges.

For many years, the citizens of the town tried to find a route for a walk that would cross each bridge only once and allow them to end their walk where they had started from.

Can you find a suitable route?

**Hint:** Don't spend too long trying to find a solution!

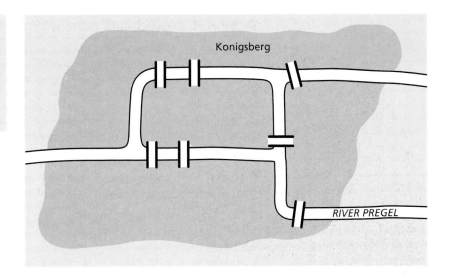

Konigsberg

RIVER PREGEL

*Konigsberg was renamed Kaliningrad and, due to bombing in World War II, only four of the bridges now remain. We shall consider this problem again in Chapter 4, Networks 3: Route inspection problems.*

## Exploration 1.2

### *League positions*

Here's another problem.

We often need to sort things into a particular order, either alphabetically or numerically. This table gives the end of season results for twelve school football teams listed *alphabetically*.

What is the final position of each team in the league?

| Team | P | W | D | L | Points |
|------|---|---|---|---|--------|
| Ashleigh | 11 | 0 | 5 | 6 | 5 |
| Beechwood | 11 | 5 | 3 | 3 | 18 |
| Enford | 11 | 3 | 4 | 4 | 13 |
| Hamworth | 11 | 8 | 2 | 1 | 26 |
| Lyndhurst | 11 | 0 | 3 | 8 | 3 |
| Manor | 11 | 3 | 5 | 3 | 14 |
| Morristone | 11 | 5 | 1 | 5 | 16 |
| Raneham | 11 | 6 | 3 | 2 | 21 |
| St John's | 11 | 4 | 3 | 4 | 15 |
| Springfield | 11 | 3 | 3 | 5 | 12 |
| Tallington | 11 | 9 | 1 | 1 | 28 |
| Yatley | 11 | 2 | 3 | 6 | 9 |

## How is it done?

To find the final positions, we need to sort the list by points. How would you do this?

It is clear that a solution to this problem does exist. The next step is to develop a method by which to *find* the solution. It should become clear that you need to be systematic when sorting the information. Once you have decided on a method, it is important to keep to the same method throughout the problem (although it may be a different method to someone else's).

The name for a systematic process that will produce the required solution is an **algorithm**. An algorithm consists of a set of input data and a list of instructions; to solve a problem, you take the initial data and apply the instructions one at a time until you reach a solution.

*The word algorithm comes from the Persian mathematician Mohammed al-Khowarizmi who lived in the ninth century. His most famous work* Al-jabr wal-muqabalah *(the science of equations) gives us the words algebra and algorithm.*

**Al-Khowarizmi (c. 825 AD)**

All algorithms must have the following properties.

1  Each step must be defined precisely.
2  The algorithm must work for any set of inputs.
3  The answer must depend only on the inputs for a particular problem.
4  It must produce an output and stop after a finite number of steps.

## COMMUNICATING AN ALGORITHM

We use algorithms in many areas of mathematics. You may already be familiar with the one we use for finding the real roots of a quadratic equation, known as **completing the square**.

**Example 1.1**

*List the steps for completing the square as a method for solving a quadratic equation and apply the steps to solve $3x^2 + 10x + 8 = 0$.*

## Solution

| Step 1 | Divide the equation by the coefficient of $x^2$. | $x^2 + \frac{10}{3}x + \frac{8}{3} = 0$ |
|--------|--------------------------------------------------|------------------------------------------|
| Step 2 | Subtract the constant term from both sides. | $x^2 + \frac{10}{3}x = -\frac{8}{3}$ |
| Step 3 | Halve the coefficient of x, square it and add it to both sides. | $x^2 + \frac{10}{3}x + \left(\frac{5}{3}\right)^2 = -\frac{8}{3} + \left(\frac{5}{3}\right)^2$ |
| Step 4 | Rewrite the left-hand side of the equation as a perfect square. | $\left(x + \frac{5}{3}\right)^2 = \frac{1}{9}$ |
| Step 5 | Take the square root of both sides of the equation. | $\left(x + \frac{5}{3}\right) = \pm\frac{1}{3}$ |
| Step 6 | Solve for x. | $x = -\frac{5}{3} \pm \frac{1}{3}$ |

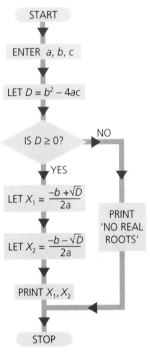

Listing the steps in this way is one way of communicating an algorithm. Another common technique is to use a flowchart. For example, to find the real roots of a quadratic equation, $ax^2 + bx + c = 0$, we can use the algorithm expressed by the formula:

$$x = \frac{\left(-b \pm \sqrt{b^2 - 4ac}\right)}{2a}$$

Each time you complete one loop, you have performed an **iteration** of the algorithm. A simple algorithm such as this is easy and straightforward to apply. Many complex, real-life problems are less easy and we often need to write the algorithm as a computer program.

We can write this algorithm as a set of instructions for a programmable calculator; for example for the TI–80 range of calculators the program is as follows.

```
:clrHome
:Disp "Y = AX² + BX + C"
:Disp "VALUE A"
:Input A
:Disp "VALUE B"
:Input B
:Disp "VALUE C"
:Input C
```

```
:(-B+ √(B²-4AC))/2A STO> D
:(-B- √(B²-4AC))/2A STO> E
:Disp"X ="
:Disp D
:Disp E
```

## EXERCISES

1.1A

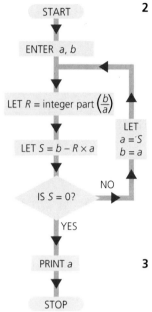

**1** Use the above algorithm for completing the square to find the roots of the equation $2x^2 - 5x + 1 = 0$.

**2** The algorithm in this flowchart finds the highest common factor of two positive integers.

**a)** Use it to work out the highest common factor of a = 105 and b = 180. Fill in the table to show the stages in your working.

| a | 105 | | | |
|---|-----|---|---|---|
| b | 180 | | | |
| R | | | | |
| S | | | | |

**b)** How many iterations does it take to find the HCF of 105 and 180?
**c)** What happens if you enter a = 180 and b = 105?
**d)** Write a program for your calculator based on the flowchart. Check the program with different values of a and b.

**3** The diagram on the right shows a maze. Find a path from the centre (C) to the exit (E). Think of how you might devise an algorithm for escaping from a maze if you had no map.

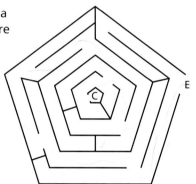

**4** Consider the flowchart on the left.

**a)** Complete this table to show the output.

| A | B | C |
|---|---|---|
| 1 | 1 | 2 |
| 1 | 2 | 3 |
| 2 | 3 | 5 |
| 3 | 5 | 8 |
| 5 | 8 | 13 |

**b)** What does the algorithm produce?

**5** Draw a flowchart or write a short program to set out an algorithm of your choice (for example: multiplication of two 2-digit numbers, the sine rule or cosine rule).

## EXERCISES

1 Use the above algorithm for completing the square to find the roots of equation $5x^2 + 2x - 12 = 0$.

2 a) Use the algorithm from Exercises 1.1A, question 2, to work out the highest common factor of 135 and 220.

b) How many iterations does it take?

3 The 'interval bisection method' is a numerical means of finding the root of a non-linear equation $f(x) = 0$. The idea is to find two values of $x$, $x = a$ and $x = b$ say, such that $f(a) < 0$ and $f(b) > 0$. We then choose $c$ to be the mid-point between $a$ and $b$ i.e. we define $c = \frac{1}{2}(a + b)$.

If $f(c) < 0$ then the root lies between $c$ and $b$; if $f(c) > 0$ then the root lies between $a$ and $c$.

The process is repeated until the interval $[a, b]$ containing the root is narrow enough to give the desired accuracy.

a) Apply the interval bisection method to find a root of the cubic equation $x^3 - 7x + 3 = 0$.
b) Write the interval bisection method as an algorithm.
c) Write a program for your calculator based on your algorithm. Check the program by solving $x^3 - 7x + 3 = 0$.

4 A unique code for referencing, called the International Standard Book Number (ISBN), is used to identify published books. For example, the ISBN of this book is 000 322372 8. It has ten digits, the last being a check digit which is introduced to reduce the likelihood of copying down the number incorrectly.

The check digit is found like this.

- Multiply the first digit by 1, multiply the second digit by 2, the third digit by 3, ... , the ninth digit by 9.
- Add all the results of the multiplications.
- Divide the total sum by 11 and note the remainder.
- If the remainder is 0 to 9 then this is the check digit, if it is 10 then the check digit is X.

Write a program for your calculator that evaluates the check digit. Use your program to check the ISBNs of several more of your books.

5 Write an algorithm to find the smallest prime number greater than a given positive integer $N$.

## ALGORITHMS FOR SORTING

We can use the results for the school football teams, on page 3, to investigate some algorithms for sorting. Since we are only interested in their total number of points we can simplify the original information.

| Team | Points |
|------|--------|
| Ashleigh | 5 |
| Beechwood | 18 |
| Enford | 13 |
| Hamworth | 26 |
| Lyndhurst | 3 |
| Manor | 14 |
| Morristone | 16 |
| Raneham | 21 |
| St John's | 15 |
| Springfield | 12 |
| Tallington | 28 |
| Yatley | 9 |

## Bubble sort

**Bubble sorts** work by placing one number in its correct position in the list each time we run through the algorithm. (Remember, this is called an iteration or **pass**). Numbers that are below their correct position rise up to their proper place like bubbles in a fizzy drink, hence the name.

For our example we want to use a **descending bubble sort** as we want the team with the greatest number of points at the top. The algorithm starts by comparing the first two numbers in the list; if the second is larger than the first, the numbers are exchanged. It then compares the second and third numbers and exchanges them if necessary, and so on through the list. After one iteration we obtain the following table of results.

| | | | | | | | | | | | |
|---|---|---|---|---|---|---|---|---|---|---|---|
| 5 | 18 | 18 | 18 | 18 | 18 | 18 | 18 | 18 | 18 | 18 | 18 |
| 18 | 5 | 13 | 13 | 13 | 13 | 13 | 13 | 13 | 13 | 13 | 13 |
| 13 | 13 | 5 | 26 | 26 | 26 | 26 | 26 | 26 | 26 | 26 | 26 |
| 26 | 26 | 26 | 5 | 5 | 5 | 5 | 5 | 5 | 5 | 5 | 5 |
| 3 | 3 | 3 | 3 | 3 | 14 | 14 | 14 | 14 | 14 | 14 | 14 |
| 14 | 14 | 14 | 14 | 14 | 3 | 16 | 16 | 16 | 16 | 16 | 16 |
| 16 | 16 | 16 | 16 | 16 | 16 | 3 | 21 | 21 | 21 | 21 | 21 |
| 21 | 21 | 21 | 21 | 21 | 21 | 21 | 3 | 15 | 15 | 15 | 15 |
| 15 | 15 | 15 | 15 | 15 | 15 | 15 | 15 | 3 | 12 | 12 | 12 |
| 12 | 12 | 12 | 12 | 12 | 12 | 12 | 12 | 12 | 3 | 28 | 28 |
| 28 | 28 | 28 | 28 | 28 | 28 | 28 | 28 | 28 | 28 | 3 | 9 |
| 9 | 9 | 9 | 9 | 9 | 9 | 9 | 9 | 9 | 9 | 9 | 3 |

First pass

From the list of twelve numbers, eleven comparisons and ten exchanges have been made. By the end of this first pass the smallest number, 3, is in position at the bottom of the list.

The process is repeated in the second pass, but there is no need to consider the last number, since it is in the correct position. This time there are only ten comparisons to be made, leaving the last two numbers in their correct positions. On the third pass nine comparisons are made, and so on until all the numbers are in their correct places.

| 18 | 26 | 26 | 26 | 26 | 26 | 26 | 26 | 28 | 28 |
|----|----|----|----|----|----|----|----|----|----|
| 26 | 18 | 18 | 18 | 21 | 21 | 21 | 28 | 26 | 26 |
| 13 | 14 | 16 | 21 | 18 | 18 | 28 | 21 | 21 | 21 |
| 14 | 16 | 21 | 16 | 16 | 28 | 18 | 18 | 18 | 18 |
| 16 | 21 | 15 | 15 | 28 | 16 | 16 | 16 | 16 | 16 |
| 21 | 15 | 14 | 28 | 15 | 15 | 15 | 15 | 15 | 15 |
| 15 | 13 | 28 | 14 | 14 | 14 | 14 | 14 | 14 | 14 |
| 12 | 28 | 13 | 13 | 13 | 13 | 13 | 13 | 13 | 13 |
| 28 | 12 | 12 | 12 | 12 | 12 | 12 | 12 | 12 | 12 |
| 9  | 9  | 9  | 9  | 9  | 9  | 9  | 9  | 9  | 9  |
| 5  | 5  | 5  | 5  | 5  | 5  | 5  | 5  | 5  | 5  |
| 3  | 3  | 3  | 3  | 3  | 3  | 3  | 3  | 3  | 3  |
| 2nd | 3rd | 4th | 5th | 6th | 7th | 8th | 9th | 10th | 11th pass number |

The final solution to the problem looks like this.

| Team | Points |
|------|--------|
| Tallington | 28 |
| Hamworth | 26 |
| Raneham | 21 |
| Beechwood | 18 |
| Morristone | 16 |
| St John's | 15 |
| Manor | 14 |
| Enford | 13 |
| Springfield | 12 |
| Yatley | 9 |
| Ashleigh | 5 |
| Lyndhurst | 3 |

The algorithm can be expressed like this.

**Step 1:** Compare the first two numbers.
**Step 2:** If the second number is larger than the first, exchange the numbers.
**Step 3:** Repeat steps 1 and 2 for all pairs of numbers until you reach the end of the list.
**Step 4:** Repeat steps 1 to 3 until no more exchanges are made.

In all we made eleven passes, although the last pass did not involve any exchanges. It simply confirmed that all the entries were in the correct position. We made a total of 66 comparisons and 36 exchanges. Work through the sort again and check this.

## Quicksort

**Quicksort** is another algorithm for sorting. It takes the first number in the list as a **pivot**, then separates the rest of the numbers into two subsets:

- those numbers that are smaller than the pivot
- those that are larger than the pivot.

Don't re-order these subsets.

To illustrate the Quicksort algorithm, consider the football teams once again: the first pass of the algorithm will look like this.

pivot →

| 5 | 18 | 18 | 18 | 18 | 18 | 18 | 18 | 18 | 18 | 18 |
|---|----|----|----|----|----|----|----|----|----|----|
| 18 | 5 | 13 | 13 | 13 | 13 | 13 | 13 | 13 | 13 | 13 |
| 13 | 13 | 5 | 26 | 26 | 26 | 26 | 26 | 26 | 26 | 26 |
| 26 | 26 | 26 | 5 | 14 | 14 | 14 | 14 | 14 | 14 | 14 |
| 3 | 3 | 3 | 3 | 5 | 16 | 16 | 16 | 16 | 16 | 16 |
| 14 | 14 | 14 | 14 | 3 | 5 | 21 | 21 | 21 | 21 | 21 |
| 16 | 16 | 16 | 16 | 16 | 3 | 5 | 15 | 15 | 15 | 15 |
| 21 | 21 | 21 | 21 | 21 | 21 | 3 | 5 | 12 | 12 | 12 |
| 15 | 15 | 15 | 15 | 15 | 15 | 15 | 3 | 5 | 28 | 28 |
| 12 | 12 | 12 | 12 | 12 | 12 | 12 | 12 | 3 | 5 | 9 |
| 28 | 28 | 28 | 28 | 28 | 28 | 28 | 28 | 28 | 3 | 5 |
| 9 | 9 | 9 | 9 | 9 | 9 | 9 | 9 | 9 | 9 | 3 |

At the end of this pass the algorithm has separated the numbers into two subsets:

- numbers greater than 5
- numbers less than 5.

We have made eleven comparisons and ten exchanges.

We now repeat the process on each of the two subsets and then on all subsequent subsets created, until we have the list in the desired order.

| 2nd | 3rd | 4th | 5th | 6th |
|-----|-----|-----|-----|-----|
| 18 | 26 | 28 | 28 | 28 |
| 13 | 21 | 26 | 26 | 26 |
| 26 | 28 | 21 | 21 | 21 |
| 14 | 18 | 18 | 18 | 18 |
| 16 | 13 | 14 | 16 | 16 |
| 21 | 14 | 16 | 15 | 15 |
| 15 | 16 | 15 | 14 | 14 |
| 12 | 15 | 13 | 13 | 13 |
| 28 | 12 | 12 | 12 | 12 |
| 9 | 9 | 9 | 9 | 9 |
| 5 | 5 | 5 | 5 | 5 |
| 3 | 3 | 3 | 3 | 3 |

The darker shading shows the numbers which were fixed after each pass and the lighter shading shows the pivots for the next pass. Work through the sort again and find the total number of comparisons and exchanges made.

Other versions of Quicksort use different numbers as the pivot point. One way of doing this is to select the middle number in the list. If we consider the example, we could select either 14 or 16 as the pivot.

## Exploration 1.3

### *Changing the pivot*

Repeat the sort using 14 as the initial pivot. Consider the number of comparisons and exchanges you make, and say whether you think this method may have some advantage over the method above, which pivoted about the top number in the list.

The following figures show the first pass in full and the effects of each subsequent pass.

**Quicksort – alternative method**

| | | | | |
|---|---|---|---|---|
| 5 | 28 | 28 | 28 | 28 |
| 18 | 18 | 18 | 18 | 18 |
| 13 | 13 | 15 | 15 | 15 |
| 26 | 26 | 26 | 26 | 26 |
| 3 | 3 | 3 | 21 | 21 |
| 14 | 14 | 14 | 14 | 16 |
| 16 | 16 | 16 | 16 | 14 |
| 21 | 21 | 21 | 3 | 3 |
| 15 | 15 | 13 | 13 | 13 |
| 12 | 12 | 12 | 12 | 12 |
| 28 | 5 | 5 | 5 | 5 |
| 9 | 9 | 9 | 9 | 9 |

First pass

| 2nd | 3rd | 4th | 5th | 6th |
|---|---|---|---|---|
| 28 | 28 | 28 | 28 | 28 |
| 18 | 18 | 18 | 26 | 26 |
| 15 | 16 | 21 | 21 | 21 |
| 26 | 26 | 26 | 18 | 18 |
| 21 | 21 | 16 | 16 | 16 |
| 16 | 15 | 15 | 15 | 15 |
| 14 | 14 | 14 | 14 | 14 |
| 3 | 13 | 13 | 13 | 13 |
| 13 | 3 | 12 | 12 | 12 |
| 12 | 12 | 3 | 9 | 9 |
| 5 | 5 | 5 | 5 | 5 |
| 9 | 9 | 9 | 3 | 3 |

Pass number                                                                Final result

## Speed and efficiency of an algorithm

When we have the choice of more than one algorithm for a task, we want to choose the best, most **efficient** one for the job. We may define the efficiency in terms of how easy the calculations are, or how long it takes to find the solution. If we are running an algorithm on a computer, we may also consider the memory needed to run the program.

We can compare the two sorting algorithms to see which performed the sort more efficiently, using the total number of comparisons and exchanges made in each case, for the same set of numbers. In this table, $c$ is the number of comparisons and $e$ is the number of exchanges.

| Bubble sort | $c = 66$ | $e = 36$ |
|-------------|----------|----------|
| Quicksort   | $c = 31$ | $e = 19$ |

Quicksort seems to be more efficient in this case.

In general terms, if the bubble sort takes the maximum number of passes, then the total number of comparisons needed for a list of 12 numbers is:

$$11 + 10 + 9 + 8 + 7 + 6 + 5 + 4 + 3 + 2 + 1 = 66$$

and the maximum number of exchanges will also be 66.

For a list of $N$ items there will be a maximum of $N - 1$ comparisons on the first pass, $N - 2$ comparisons on the second pass and so on until we get to zero. Thus the total number of comparisons ($c$) can be expressed by:
$$c \leq 0 + 1 + 2 + 3 + \ldots + (N - 2) + (N - 1)$$

This is an arithmetic progression, with $a = 0$ and $d = 1$.

The sum of an arithmetic progression is given by:

$$S_n = \frac{n}{2}\left[2a + (n - 1)d\right]$$

which gives us:

$$c \leq \frac{N}{2}\left[0 + (N - 1)1\right] \Rightarrow c \leq \frac{N}{2}(N - 1)$$

and the maximum number of exchanges ($e$) will be:

$$e \leq \frac{N}{2}(N - 1)$$

Thus for $N = 12$, $c \leq 6 \times 11 = 66$ and $e \leq 66$.

It turns out that for Quicksort the largest number of comparisons and exchanges is the same as for Bubble sort, but this only occurs when the initial list is in reverse order from what is required. A list that is randomly arranged will be more efficient on average, especially if the initial pivot chosen is close to the median value.

## EXERCISES

1  Sort these sets of data in ascending order using:

**a)** Bubble sort      **b)** Quicksort.
  set 1  8, 7, 6, 5, 4, 3, 2, 1
  set 2  1, 3, 5, 7, 2, 4, 6, 8
  set 3  5, 1, 4, 7, 3, 2, 8, 6
  In each case compare the efficiency of the sorts.

2  Sort this list in *ascending* order, using both Bubble sort and Quicksort.

  50, 48, 76, 30, 12, 4, 28, 56, 63, 77, 21, 57, 29, 41

3  Sort this list of marks in *descending* order.

  14, 25, 12, 31, 18, 13, 20, 28

4  For a list of $n$ numbers {L(1), L(2), L(3), .... L($n$)}, the exchange sort can be stated like this.

  **Step 1:** Look for the smallest number in L($i$).
  **Step 2:** Exchange with number in L(1).
  **Step 3:** Look for smallest L($i$) remaining.
  **Step 4:** Exchange with next L($i$).
  **Step 5:** Repeat steps 3 and 4 until the list is sorted.
  **a)** Use the exchange sort to complete the sort on this list (the first pass is done for you).

| 28 | 20 | 13 | 18 | 31 | 12 | 25 | 14 |
|----|----|----|----|----|----|----|----|

L ↓

| 12 | 20 | 13 | 18 | 31 | 28 | 25 | 14 |
|----|----|----|----|----|----|----|----|

first pass

  **b)** Complete the sort.
  **c)** How many comparisons and exchanges were made?
  **d)** Comment on the efficiency of the exchange sort.

## EXERCISES

1  Sort these data sets in *descending* order using:

**a)** Bubble sort      **b)** Quicksort.
  set 1   4, 5, 6, 7, 8, 9, 10
  set 2   9, 7, 5, 3, 8, 6, 4, 2
  set 3   6, 8, 2, 3, 7, 4, 1, 5

2  Sort this list in *descending* order using:

**a)** Bubble sort   **b)**   Quicksort.
  41, 29, 57, 21, 77, 63, 56, 28, 4, 12, 30, 76, 48, 50

3  Sort this list in *ascending* order.

  28, 20, 13, 18, 31, 12, 25, 14

4  There are several other sorting algorithms, such as Shuttle sort, Shell sort and Merge sort. Find out about these algorithms and compare them to the two we have studied here.

## BIN-PACKING

### Packing a tape

Over a bank holiday weekend you want to record four programmes on your video recorder. You have four 180-minute tapes. The lengths of the programmes are given in minutes.

| A | 90 | E | 120 |
|---|----|---|-----|
| B | 35 | F | 45 |
| C | 105 | G | 60 |
| D | 90 | H | 100 |

Can you record all the programmes? If so, how would you do it?

## Efficient packing

We are all familiar with the problem of packing too many things into too little time or space. You may have difficulty fitting all the activities you want to do into the time available, or you may have trouble fitting things into your suitcase when you go on holiday. Efficient packing is very important in business and industry, whether it is fitting a number of jobs into the minimum time or packing goods into a warehouse to waste the least amount of space.

As with sorting, these problems need a clear strategy but, unlike sorting, there is no known algorithm which will always produce the best, or **optimal**, solution. **Heuristic algorithms** set out to find good solutions Heuristic methods use logical and often intuitive steps to find a good solution for a particular problem.

*The word* heuristic *is from the Greek* heuriskein *meaning find.*

In the exploration above, we were trying to pack the programmes into the tapes, which are called the **bins**. We start by establishing the most and least tapes we will need; these are called the **upper and lower bounds**. Since there are eight programmes, we will not use more than eight tapes, so 8 will be an upper bound.

The total time for the programmes is 645 minutes, each tape lasts 180 minutes, so the least amount of tape we could possibly need is $645 \div 180 = 3.58$, so 4 will be a lower bound.

We know that we shall need at least four tapes and no more than eight tapes. We can now consider three algorithms that will give us a better solution. Problems like this are called **bin-packing** problems.

## Full-bin algorithm

When there is a relatively small amount of information, we may look for combinations that will fill one bin, then the remaining items are packed in the first available space which is big enough. In the example:

A + D = 180

E + G = 180

B + F + H = 180

leaving C (105) to be placed on a separate tape. This gives us a solution using four tapes.

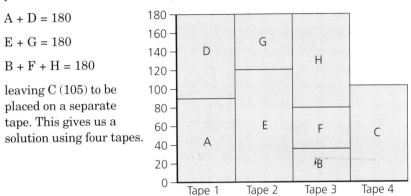

## First-fit algorithm

Full-bin combinations work well for simple problems, but if there is a large amount of information or if no full-bin combinations exist, we need a more general method. The first-fit algorithm places each item in the first available bin that has enough room. Then the solution to the tape problem looks like this.

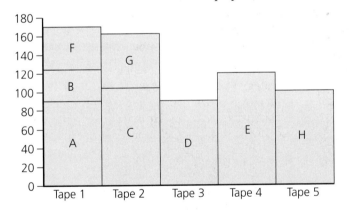

We put programme A on the first tape, using up 90 minutes. Programme B needs 35 minutes and since there is enough space on tape 1 we record it there. Tape 1 now has 55 minutes of space. Programme C needs 105 minutes so we begin Tape 2 and so on.

This algorithm gives a solution requiring five tapes, with more tape left unused that the previous solution. It is less efficient in terms of filling up space, and in fact would not enable us to record all the programmes, if only four tapes are available.

## First-fit decreasing algorithm

This algorithm re-orders the information in descending order of size, then follows the pattern of the first-fit algorithm. This puts the largest items first, so that a more efficient solution can be found. For the programmes above the first step of re-ordering gives:

E 120, C 105, H 100, A 90, D 90, G 60, F 45, B 35

The first-fit algorithm then gives the following solution.

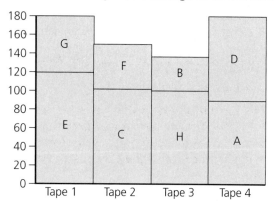

Again we get a solution using four tapes, so there is less space wasted. However, as we must re-order the list first, the algorithm takes rather longer than the first-fit algorithm.

Bin-packing techniques are used to model many practical situations for example:

- How many workers are needed to complete a job in a given amount of time?
- What is the best way to load a ferry, to fit in the maximum number of cars and lorries?
- How does a plumber minimise the number of standard lengths of pipe needed to cut a given number of lengths for a job?
- How many computer disks will be needed to store a selection of programs?
- How does a TV company allocate advertising slots, to minimise the number of breaks in its programmes?

The choice of which 'bin-packing' method to use depends upon the type of problem.

## EXERCISES

**1.3 A**

In this exercise you should try using all three 'bin-packing' methods on each problem you do. Compare the results and see if you can reach any conclusions as to when a particular algorithm will be most effective.

1 Vertical racks in a storage room are designed to take boxes with standard bases but different heights. If the racks are 1 metre high what is the minimum number of racks needed to store the boxes shown?

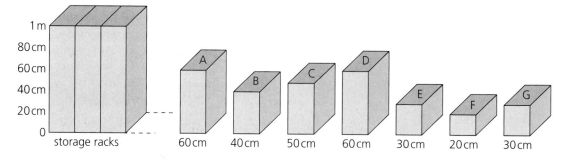

2 A plumber has some 12-foot lengths of copper pipe. For a particular job he needs to cut the sections listed below. What is the best way of cutting the lengths so that he wastes as little pipe as possible?

| Section | A | B | C | D | E | F | G | H | I | J | K | L |
|---|---|---|---|---|---|---|---|---|---|---|---|---|
| Length (feet) | 2 | 2 | 2 | 3 | 3 | 3 | 4 | 4 | 5 | 6 | 6 | 7 |

**3** A computer programmer wishes to fit the following programs onto no more than four 800 kbyte disks.

| Program | A | B | C | D | E | F | G | H | I | J | K | L | M | N | O | P |
|---|---|---|---|---|---|---|---|---|---|---|---|---|---|---|---|---|
| Size (kbytes) | 180 | 150 | 140 | 145 | 200 | 50 | 100 | 130 | 160 | 300 | 170 | 250 | 165 | 200 | 150 | 350 |

Which solution would you recommend? Why?

**4** Celestial Television wishes to schedule the following advertisements during the showing of its Saturday night theatre presentation. However, the producer does not want to incorporate more than three breaks, so as not to disturb the artistic flow of the play. A standard length break is 180 seconds.

Is it possible to schedule all the advertisements?

| | Product | Time (seconds) |
|---|---|---|
| **A** | Doggibrek | 40 |
| **B** | Zap detergent | 30 |
| **C** | Stukup glue | 25 |
| **D** | Deeliteful dessert | 46 |
| **E** | Carl's Cars | 52 |
| **F** | Mammoth meals | 46 |
| **G** | Trudi dolls | 33 |
| **H** | Deals on Wheels | 38 |
| **I** | Dinosnax | 20 |
| **J** | Bodgit builders | 25 |
| **K** | Soppybabe nappies | 40 |
| **L** | Rekkit DIY | 18 |
| **M** | Miss Ellie's cakes | 30 |
| **N** | Pasta la Vista | 20 |
| **O** | Venus beauty products | 40 |
| **P** | Aardvark carphones | 37 |

State, with reasons, the solution that you would recommend to the producer.

**5** A project consists of several activities, as shown in the table below, with their expected duration.

| Activity | A | B | C | D | E | F | G | H | I | J | K | L |
|---|---|---|---|---|---|---|---|---|---|---|---|---|
| Duration (days) | 8 | 7 | 5 | 6 | 5 | 6 | 9 | 4 | 7 | 9 | 8 | 6 |

Use a suitable packing algorithm to determine the number of workers required to complete the job in twelve days.

## EXERCISES

In this exercise you should try using all three 'bin-packing' methods on each problem you do. Compare the results and see if you can reach any conclusions as to when a particular algorithm will be most effective.

1   A storage warehouse has storage racks 2.4 m high. They provide their clients with boxes of three different heights: 60 cm, 80 cm and 120 cm. A client fills three 60 cm boxes, A, B and C, four 80 boxes D, E, F and G and one 120 cm box H. What is the minimum number of racks he needs to store his boxes?

2   Bodgit Brothers are installing a central heating system. They need the following lengths of pipes.

| Length (m) | 2 | 2.5 | 3.0 | 3.5 |
|---|---|---|---|---|
| No. | 4 | 3 | 2 | 3 |

The piping comes in 6 m lengths. What is the best way to cut the lengths, to minimise wastage?

3   Wendy is backing up the files on the hard disk of her computer onto 1.4 Mbyte floppy disks. Here is a printout of her file details.

| | File | Size (kbytes) |
|---|---|---|
| A | bridge.tif | 409 |
| B | cbl.tif | 819 |
| C | chapt1.doc | 114 |
| D | chapt2.doc | 127 |
| E | chapt3.doc | 225 |
| F | chapt4.doc | 356 |
| G | chapt5.doc | 142 |
| H | ctm.tif | 414 |
| I | frame1.dir | 48 |
| J | frame2.dir | 109 |
| K | frame3.dir | 72 |
| L | graph.doc | 271 |
| M | intro.doc | 134 |
| N | motion.doc | 231 |
| O | pipe.doc | 130 |
| P | ted.tif | 672 |
| Q | wave.tif | 532 |

What is the minimum number of disks she needs?
(1 Mbyte = 1000 kbytes)

4   The following acts are all keen to take part in a variety show. The show is organised in three half-hourly parts. Is it possible for all acts to take part? What schedule would you recommend to the producer?

| | Act | Time (minutes) |
|---|---|---|
| A | Daves Duo | 4 |
| B | Jeremy's Jugglers | 5 |
| C | Twinkle Toes dance troupe | 7 |
| D | The Bodmin Clog dancers | 6 |
| E | The Plymstock Brass Band | 10 |
| F | Marvo the Conjurer | 6 |
| G | Victor the Ventriloquist Fiddler | 5 |
| H | Coral Sheen (Comedienne) | 5 |
| I | Maureen Bright's Startling Origami | 4 |
| J | Stuart Stardust – the fire eater | 7 |
| K | Mike the Magnificent (illusionist) | 8 |
| L | Lamplight Lucy (vocalist) | 4 |
| M | Rocky Wilde (female impersonator) | 4 |
| N | Woofers and Tweeters Jazz Band | 7 |
| O | The Caterpillars (band) | 5 |

5   On a trip home from Finland, John and his family want to pack as few suitcases as possible. Each suitcase must have a mass of less than 22 kg. They have parcelled up the things they want to take into twelve bundles.

| Bundle | A | B | C | D | E | F | G | H | I | J | K | L |
|---|---|---|---|---|---|---|---|---|---|---|---|---|
| Mass (kg) | 8 | 12 | 14 | 6 | 6 | 6 | 6 | 14 | 8 | 4 | 8 | 4 |

How many suitcases are needed?

## COURSEWORK INVESTIGATION

**Investigation**

*Newspaper layout*

If you look at the way a newspaper is set out you will see that the text is organised in columns, with pictures and headlines often taking up more than one column width. Consider ways in which bin-packing algorithms could be adapted to help with the layout of a newspaper.

## CONSOLIDATION EXERCISES FOR CHAPTER 1

1   Look again at the flowchart for finding the highest common factor of two numbers. It is claimed that the number of iterations needed to find the HCF of two numbers using this algorithm is approximately given by:

$$\frac{\log\left[\dfrac{b}{1.17}\right]}{\log\left[\dfrac{1+\sqrt{5}}{2}\right]}$$

where $b$ is the larger of the two numbers.

Use your answer to question 2 of Exercises 1A to compare the *predicted* number of iterations with the *actual* number.

Predict the number of iterations with the inputs $a = 233$ and $b = 377$.

Perform the algorithm on these inputs and compare the number of iterations with the predictions.

Can you suggest a reason for the results?

2   Describe the Bubble sort algorithm and use it to sort, in ascending order, the list:
    7, 5, 2, 8, 6, 1.

*(ULEAC Specimen Paper)*

3   The following flowchart defines an algorithm which operates on two inputs, $x$ and $y$.
    a) Run the algorithm with inputs of $x = 3$ and $y = 41$, counting how many times the instructions in box number 3 are repeated.

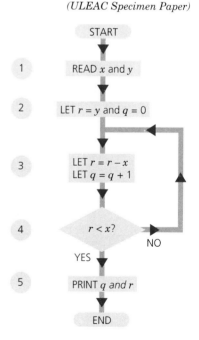

**b)** Say what the algorithm achieves.

The following flowchart defines an algorithm which operates on three inputs, $x$, $y_1$ and $y_2$.

**c)** Work through the algorithm with $x = 3$, $y_1 = 4$ and $y_2 = 1$. Record your work by completing the table below. Keep a count in the spaces provided of how many times the instructions in boxes numbers 3 and 6 are repeated.

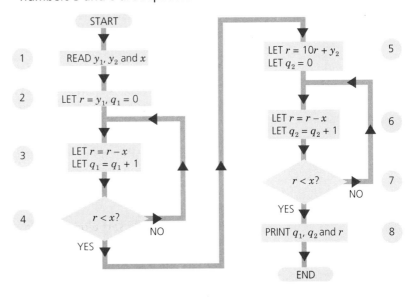

| $r$ | | | | | | | | | | | | | | | | | | | | | | | | | |
|---|---|---|---|---|---|---|---|---|---|---|---|---|---|---|---|---|---|---|---|---|---|---|---|---|---|
| $q_1$ | | | | | | | | | | | | | | | | | | | | | | | | | |
| $q_2$ | | | | | | | | | | | | | | | | | | | | | | | | | |
| No. of repetitions of box number 3 | | | | | | | | | | | | | | | | | | | | | | | | | |
| No. of repetitions of box number 6 | | | | | | | | | | | | | | | | | | | | | | | | | |

**d)** The second algorithm achieves the same result as the first. Say what you think are the advantages and disadvantages of each.

*(O & C MEI Question 3, Decision & Discrete Mathematics Paper 19, January 1995)*

**4   a)** The instructions labelled 1 to 7 overleaf describe the steps of a bubble sort algorithm. Apply the bubble sort algorithm to the list of numbers shown below the instructions. Record in the table provided the state of the list after each pass, and record the number of comparisons and the number of swaps that you make in each pass. (The result of the first pass has already been recorded.)

**Instructions**
1 Let $i$ be equal to 1.
2 Repeat lines 3 to 7, stopping when $i$ becomes 6.
3 Let $j$ equal 1.
4 Repeat lines 5 and 6, stopping when $j$ becomes $7 - i$.
5 If the $j$th number in the list is bigger than the $(j + 1)$th, then swap them.
6 Let the new value of $j$ be $j + 1$.
7 Let the new value of $i$ be $i + 1$.

**List:**   7   9   5   1   11   3

**Table:**

| $i$th pass | $i = 1$ | $i = 2$ | $i = 3$ | $i = 4$ | $i = 5$ |
|---|---|---|---|---|---|
| | 7 | | | | |
| | 5 | | | | |
| | 1 | | | | |
| | 9 | | | | |
| | 3 | | | | |
| | 11 | | | | |
| Comparisons | 5 | | | | |
| Swaps | 3 | | | | |

**b)** Suppose now that the list is split into two halves, {7, 9, 5} and {1, 11, 3}, and that the algorithm is applied to each half separately, giving {5, 7, 9}, and {1, 3, 11}. How many comparisons and swaps does this entail altogether?

**c)** A modified version of the bubble sort algorithm, in which lines 2 and 3 have been changed, is now used to sort the revised list {5, 7, 9, 1, 3, 11}.

**Revised instructions**
1 Let $i$ equal 1.
2 Repeat lines 3 to 7, stopping when $i$ becomes 4.
3 Let $j$ equal $4 - i$.
4 Repeat lines 5 and 6, stopping when $j$ becomes $7 - i$.
5 If the $j$th number in the list is bigger than the $(j + 1)$th, then swap them.
6 Let the new value of $j$ be $j + 1$.
7 Let the new value of $i$ be $i + 1$.

Apply the revised algorithm to the list {5, 7, 9, 1, 3, 11}, recording your results in the table provided.

| *i*th pass | $i = 1$ | $i = 2$ | $i = 3$ |
|---|---|---|---|
| | | | |
| | | | |
| | | | |
| | | | |
| | | | |
| | | | |
| Comparisons | | | |
| Swaps | | | |

Does splitting the list, applying the bubble sort algorithm to each half separately, and then applying the revised algorithm to the revised list, appear to offer any advantages? Justify your answer.

**d)** Explain why the changes to statements 2 and 3 of the original algorithm were made to produce an algorithm to sort the revised list.

*(O & C MEI Question 3, Decision & Discrete Mathematics Paper 19, June 1995)*

**5** **a)** The following eight lengths of pipe are to be cut from three pieces, each of length 2 m.
1.1 m, 1.2 m, 0.3 m, 0.4 m, 0.3 m, 0.7 m, 0.4 m, 0.2 m

   **i)** Use the first-fit decreasing algorithm to find a plan for cutting the lengths of pipe. Show your plan by marking cuts on the diagram below.

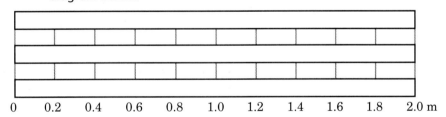

0    0.2    0.4    0.6    0.8    1.0    1.2    1.4    1.6    1.8    2.0 m

   **ii)** Suppose that it is preferable to be left with a small number of longer pieces of pipe, rather than a larger number of shorter lengths. Use any method to produce an improved cutting plan.

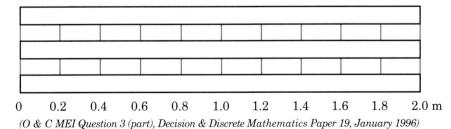

0    0.2    0.4    0.6    0.8    1.0    1.2    1.4    1.6    1.8    2.0 m

*(O & C MEI Question 3 (part), Decision & Discrete Mathematics Paper 19, January 1996)*

**6**    In the following algorithm $f(x) = x^2 + 4x - 60$.

**Step 0:**  Set $L = 0$, $U = 16$.
**Step 1:**  Let $FL = f(L)$, $FU = f(U)$.
**Step 2:**  Let $M = (L + U)/2$.
**Step 3:**  Let $FM = f(M)$.
**Step 4:**  IF $(FM < 0)$ Then let $L = M$, $FL = FM$, go to Step 2.
**Step 5:**  IF $(FM > 0)$ Then let $U = M$, $FU = FM$, go to Step 2.
**Step 6:**  IF $(FM = 0)$ Then stop.

Demonstrate the working of this algorithm by running through it until either you have executed Step 2 five times or until you have stopped at Step 6.
Explain briefly what you think the algorithm is seeking to do. Why might it be called a binary search algorithm?

*(UODLE Question 6, Paper 4 ,1989)*

**7**    Streets in New York are arranged in a grid pattern, so that all routes between points involve moving only 'horizontally' or 'vertically'. Thus the shortest distance between (1, 2) and (4, 6) is $[(4 - 1) + (6 - 2)] = 7$.

A company has deliveries to make in New York. A van can deliver from the depot to two points on each trip, returning to the depot afterwards.

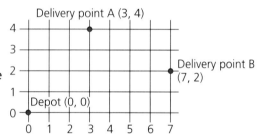
Delivery point A (3, 4)
Delivery point B (7, 2)
Depot (0, 0)

Shortest distance from depot to A and B return:
$$= [(3 - 0) + (4 - 0)] + [(7 - 3) + (4 - 2)] + [(7 - 0) + (2 - 0)]$$
$$= 22$$

**a)** The following delivery points need to be serviced.

| Delivery point | A | B | C | D | E | F |
|---|---|---|---|---|---|---|
| Location | (3, 4) | (7, 2) | (1, 2) | (6, 5) | (5, 1) | (5, 3) |

Copy the following matrix, which gives the shortest distance each pairing. Complete the last row and column.

$$\begin{pmatrix} & A & B & C & D & E & F \\ A & * & 22 & 14 & 22 & 18 & \\ B & 22 & * & 18 & 24 & 18 & \\ C & 14 & 18 & * & 22 & 14 & \\ D & 22 & 24 & 22 & * & 22 & \\ E & 18 & 18 & 14 & 22 & * & \\ F & & & & & & \end{pmatrix}$$

*indicates that the pairing is not possible.

**b)** What is the minimum delivery distance?

*(UODLE Question 6 (part), Decision Mathematics AS, 1991)*

8  A class is given a task to perform by its mathematics teacher. Each pupil is to design an algorithm to achieve what is required. One pupil's attempt is as follows.

```
LET J = 3
REPEAT
LET J = J + 1
LET K = 2
REPEAT
    IF J = [J/K] × K THEN cross out cell J
    LET K = K + 1
UNTIL K ≥ ⌈√J⌉

UNTIL J = 49
```

[x] stands for the largest whole number less than or equal to x; e.g. [3.2] = 3; [5] = 5.

a)  Make a copy of the following table representing all the values of J considered by the algorithm. Follow through the algorithm crossing out cells as directed. (You do not need to show more than one cross in any cell.)

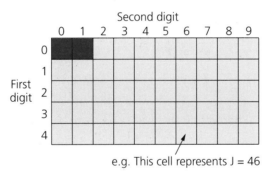

e.g. This cell represents J = 46

What do you think the task set by the teacher was? Does the algorithm produce the correct result?

A second pupil had the idea of looking at each number in turn, starting with 2, and then crossing out all cells representing a *proper* multiple of the number (i.e. 2 × the number, 3 × the number, 4 × the number, etc.).

b)  Make another copy of the table and follow through the second pupil's approach.
    Indicate the order in which you cross out cells by putting a number instead of a cross when you *first* eliminate a cell.

c)  Produce an algorithm, in similar form to that given for the first pupil, for the second pupil's idea.

d)  Describe how the efficiency of the second algorithm could be assessed.

*(UODLE Question 10, Decision Mathematics AS, 1994)*

**9 a)** In a two-person game one player thinks of a word and invites the other to guess what that word is. In response to an incorrect guess the first player indicates whether the mystery word is before or after the guess in alphabetical (dictionary) order.

    **i)** Given that a dictionary is available, describe a strategy for the second player to follow in trying to find the word in as few guesses as possible.

    **ii)** Given that the dictionary has approximately 100 000 words, find the maximum number of guesses which might be required to find the mystery word using your method from part **i)**. Explain your reasoning in arriving at this number.

**b)** In a game of 'Battleships' a player chooses a square on a grid to represent a battleship. A second player tries to find that square by making a sequence of guesses. In response to an incorrect guess, the second player is told *either* whether the battleship is to the east or west of the guess (i.e. to the left or the right), *or* whether it is to the north or south (above or below).

The game is being played on a 15 × 20 grid and a first guess has been made at square K8, as indicated on the diagram.

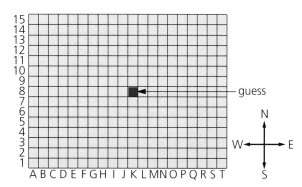

The information provided is that the battleship is to the south of the guess.

    **i)** In a binary search the set of possibilities is, as nearly as possible, halved at each iteration. Given the guess and the information, what would be the next guess?

    **ii)** If the information following that next guess is that the battleship is to the north of the guess, what would be the next guess in a binary search?

    **iii)** What difficulty will be encountered when the information is that the battleship is to the west?

    **iv)** Starting a game afresh, what is the minimum number of guesses that will be needed to be sure of locating the battleship?

**c)** 'Spaceships' is similar to 'Battleships', but is played on a 3-dimensional grid. Playing a binary search strategy on a 15 × 15 × 15 grid, how many guesses will be needed to be sure of locating the spaceship?

*(O & C MEI Question 3, Decision & Discrete Mathematics Paper 19, January 1997)*

**10**  A DIY enthusiast decides to build a lean-to greenhouse on the side of her garden shed. The dimensions of the wooden frame are shown below.

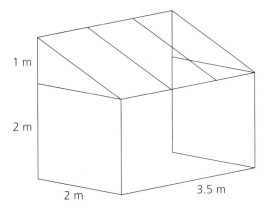

1 m

2 m

2 m

3.5 m

**a)** Calculate the number of lengths of the pieces of wood needed to make the frame.

**b)** Use a suitable packing algorithm to find the cheapest way to build the frame if:

    **i)**  wood is only available in 5 m lengths costing £28 per metre

    **ii)** wood is available in 5 m lengths costing £28 per metre, and also 2 m lengths costing £12.00.

## Summary

*After working through this chapter you should:*

■ understand that an algorithm is a systematic process that can be used to give the solution to a problem

■ know how an algorithm can be stated or described and be able to follow the steps of a given algorithm

■ be able to use Bubble sort and Quicksort algorithms

■ know how you can assess the speed and efficiency of these algorithms, by considering the number of comparisons and exchanges made at each iteration

■ model situations using bin-packing algorithms and apply full-bin, first-fit and first-fit decreasing algorithms to a variety of problems.

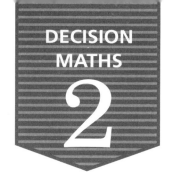

# Networks 1: Shortest path

*By the end of this chapter you will:*

■ *know that we can use networks to model many real-life situations*

■ *find that networks are an application of an area of mathematics called graph theory*

■ *explore Dijkstra's algorithm which finds the shortest distance between any two points or **vertices** in a network.*

## NETWORK PROBLEMS

**Exploration 2.1**  *Route maps*

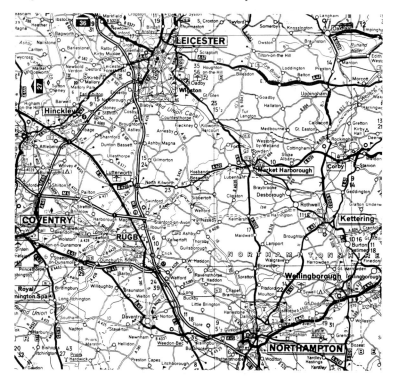

This is a map of the area south of Leicester. It represents many (though not all) of the roads and towns in the area. A conference is being held in Wellingborough. Design a simpler road map of the area to give information on how to get to Wellingborough from nearby towns.

### Simplified diagrams

The original map gives far more detail than is needed just for travelling through the region in a car. It would be easier if we could represent it in a simpler form. Such a simplified plan, showing only the major roads, is called a **route map**.

This is a diagrammatic way of showing how places are linked by roads. It does not necessarily give information of the type of road or places of interest in the area. This type of simplified diagram is useful in many other areas.

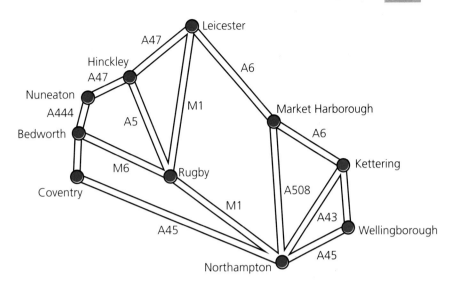

### Electrical circuits

This diagram represents an electrical circuit made up of a battery, four ammeters, a resistor and three lamps. This type of diagram shows how components are connected together, although, as with the route map, it does not give much detail about features such as how long or how thick the individual lengths of wire are.

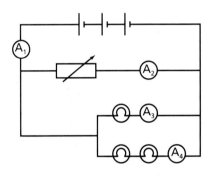

### Circulation diagrams

This is a plan of the ground floor of a house. A plan is useful for showing the layout of a small building, but if you were designing a large public building to be used by lots of people, such as a school, a simpler diagram showing the connections between rooms would be useful.

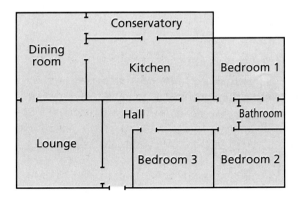

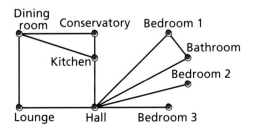

This is a circulation diagram, of the type used by architects. Once again, it gives only limited information; it does not tell us the size or shape of the rooms or the length and width of the corridors.

### Structural diagrams

If you are studying Chemistry, you will probably be familiar with these diagrams. They show how atoms link together to give molecules. These diagrams give no information about how atoms are aligned in space: they are a two-dimensional representation of a three-dimensional object.

Methyl alcohol
(Methylated spirit)   $CH_3OH$

## Introduction to graph theory

All the examples above involve the way in which certain objects are related to each other. In each diagram the objects are represented by **points** and the connections are represented by **lines**. This type of diagram is called a **graph**. Each point representing an object is called a **vertex** (plural: **vertices**). Each line connecting two vertices is called an **edge** if it has no direction or an **arc** when it has direction.

*In some books vertices are called **nodes**.*

The **degree** of a vertex is defined as the number of arcs incident with the vertex.

A graph is **connected** if there is at least one route through the graph connecting any given pair of vertices (nodes). In the following diagram the first graph is connected because there is a route along the edges between any pair of vertices e.g. ACDE connects A to E. In the second graph there is no route to vertex E.

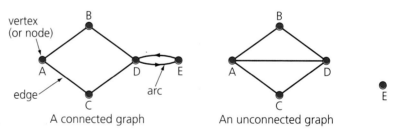

A connected graph          An unconnected graph

A graph where each vertex is connected to every other vertex is **complete**. A complete graph with $n$ vertices is denoted by $K_n$.

$K_3$ It has three vertices and three edges.

$K_4$ It has four vertices and six edges.

$K_5$ It has five vertices and ten edges

### Subgraphs

A **subgraph** is part of a graph which is itself a graph. A graph can have several subgraphs, not all of which are connected. For example, these are all subgraphs of $K_5$.

## Adjacency matrices

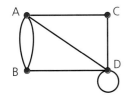

So far we have represented all the graphs by diagrams that are clear and easy to understand. However it is not always practical to draw this kind of graph, especially if there are large numbers of vertices and edges which would create a very complicated picture, or if we wish to store the information in a computer. Another way in which graphs can be represented is by using **matrices**.

|   | A | B | C | D |
|---|---|---|---|---|
| A | 0 | 2 | 1 | 1 |
| B | 2 | 0 | 0 | 1 |
| C | 1 | 0 | 0 | 1 |
| D | 1 | 1 | 1 | 2 |

Above, left, is a graph with four vertices. The matrix on the left here describes the graph. Each row and each column represents a vertex of the graph and the numbers in the matrix give the number of edges joining each pair of vertices, so there are two edges from A to B, one edge from A to C and so on. This is the **adjacency** matrix for the graph.

## Paths

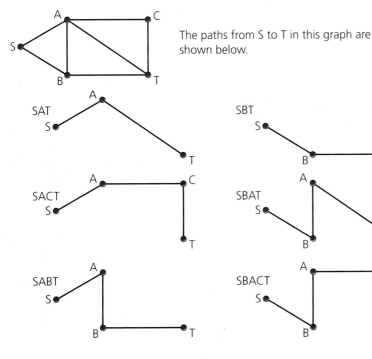

The paths from S to T in this graph are shown below.

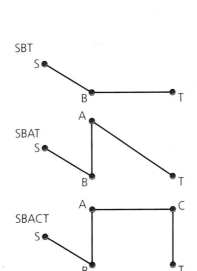

Many of the applications of graph theory are about getting from one vertex to another. If we make a route through a graph which does not go along any edge more than once or visit any vertex more than once, it is called a **path**.

# Networks

A **network** is a graph in which each edge (or arc) is given a value called its **weight**. The weight on an edge may represent a distance, a time or a cost: for example, if we add distances to our route map the resulting network would look like this.

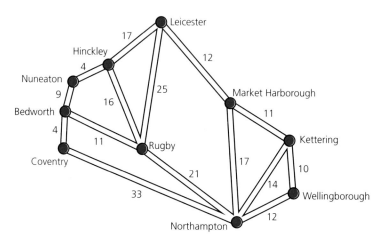

## Representing networks using matrices

Just as we can use an adjacency matrix to represent a graph, we can also use matrices to represent networks. These can show:

- ■ the distance between vertices
- ■ the time it takes to travel between vertices
- ■ the cost of moving between vertices, called **cost matrices**.

Here is an example of a distance matrix and its associated network.

|   | A  | B | C | D  | E |
|---|----|---|---|----|---|
| A | –  | 8 | 5 | 10 | – |
| B | 8  | – | – | 3  | 5 |
| C | 5  | – | – | 4  | 7 |
| D | 10 | 3 | 4 | –  | 4 |
| E | –  | 5 | 7 | 4  | – |

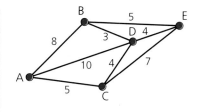

# EXERCISES

**2.1 A**

1 For each of the following graphs give the number of vertices, the number of edges and say whether or not the graph is connected.

In each case, suggest a real situation that the graph might represent.

a)

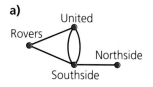

b)

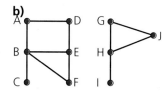

c)

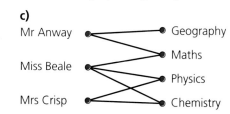

**2** **a)** Draw the graph for this adjacency matrix.

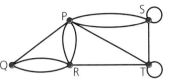

$$\begin{array}{c c c c c c} & A & B & C & D & E \\ A & 0 & 1 & 2 & 1 & 1 \\ B & 1 & 0 & 1 & 2 & 1 \\ C & 2 & 1 & 0 & 0 & 1 \\ D & 1 & 2 & 0 & 0 & 0 \\ E & 1 & 1 & 1 & 0 & 2 \end{array}$$

What is the significance of the 2 here?

**b)** Draw the adjacency matrix for this graph.

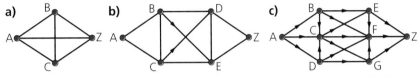

**3** List all the paths from A to Z in these graphs.

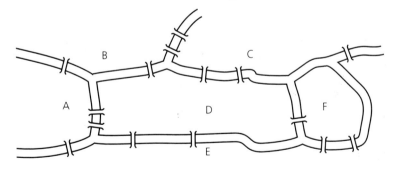

**a)** **b)** **c)**

**4** Draw a circulation diagram for your house or part of your school. Can you suggest any improvements to the design?

**5** Draw the graphs $K_6$ and $K_7$. By considering the number of arcs find a general formula for the number of arcs in $K_n$.

## EXERCISES

**2.1 B**

**1** The figure shows different regions of a city with a river running through it. There are six areas of the city labelled A, B, C, D, E and F and bridges connect the areas as shown.

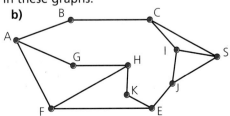

**a)** Draw a graph to represent this city. Each edge should represent a possible route over a bridge between two areas of the city.

**b)** Can you find a path starting from any area of the city which passes through each other part and returns to the starting area?

**2** Construct the adjacency matrix for the bridge crossings for the city in question **1**.

**3** List all the paths from A to S in these graphs.

**a)** **b)**

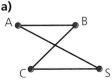

33

4    What is the degree of each vertex in:
**a)** $K_2$      **b)** $K_5$      **c)** $K_n$?
We shall meet this again in Chapter 4, *Networks 3: Route inspection problems.*

5    Use a map of the motorway system and main trunk roads of England to find the shortest distances between each pair of teams in the top half of the Premier Soccer league. Represent this information a graph. Is this graph complete?

## SHORTEST PATH PROBLEMS

**Exploration 2.2**

### *Find the shortest route*

Study this network again. Suppose you live in Nuneaton and are attending the conference in Wellingborough. What is the shortest route to take?

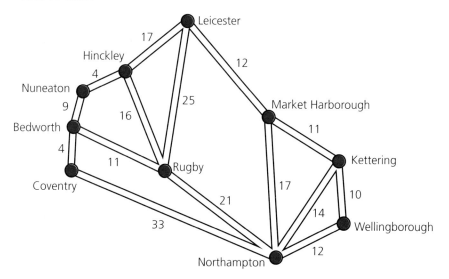

### Finding paths

In Exploration 2.2 you probably investigated different routes through the network by means of a trial and error method. Shortest path problems involve finding paths with optimal properties, not necessarily to do with length. They may be **cost paths** (the cheapest way) or **risk paths** (the safest way). They are very important in industry where the shortest or least-cost route is required.

Now suppose that there are many vertices in your network; in question 3 of Exercises 2.1A you will have discovered that as the number of vertices increases, the number of possible routes through a network increases faster. Looking for every possible path and finding its length – a method called **exhaustive search** – will certainly provide the correct solution (provided you do not make any mistakes or overlook any path) but it will be very time-consuming.

It seems reasonable to suppose that it would be possible to devise an algorithm to find the shortest path through a network more efficiently and which could be programmed into a computer to solve large problems. Such an algorithm was developed by E. W. Dijkstra in 1959.

## Dijkstra's algorithm

Dijkstra's algorithm is an example of a **greedy** algorithm. Greedy algorithms are generally used to solve optimisation problems. They proceed in steps; at every step the algorithm chooses the best option *at that stage*. Once an item is included in the solution, it will not be changed; it never reconsiders its path, but it always provides an optimal solution if the type of problem is appropriate.

Dijkstra's algorithm finds the shortest route between two vertices in the network. At each stage a label is assigned to a vertex, showing the shortest distance from the starting point to that vertex. The label is either temporary or permanent, depending on whether we have considered all the ways of getting to that vertex. When the label is temporary, we simply write the value next to the vertex like this.

When we have decided that the label is permanent, we put a box round it giving the value and its position in the sequence of permanent labelling, which is also the number of iterations performed so far.

Recording the order in which we assign permanent labels to the vertices is an essential part of the algorithm.

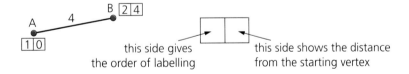

this side gives the order of labelling

this side shows the distance from the starting vertex

The algorithm gradually changes all the temporary labels into permanent ones. This can be demonstrated on this simple network.

## Example 2.1

*Find the shortest path from A to F.*

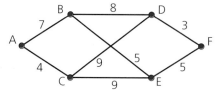

**Solution**

*We start by giving a permanent label 0 to our starting vertex A. Next we consider the vertices which can be reached directly from A and give them temporary labels equal to their distances from A: so B is labelled 7 and C is labelled 4.*

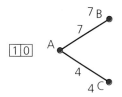

*We now look at the vertices that have temporary labels and select the smallest. We assign a permanent label to it. In this case we assign a permanent label 4 to C.*

*Now we consider all the vertices that are connected directly to our newly-labelled vertex C; they are D and E. We calculate the shortest distance from A to D via C by adding the permanent label at C to the distance from C to D.*

*A → C → D = 4 + 9 = 13 so we assign the temporary label 13 to D.*

*Similarly A → C → E is found to be 4 + 9 = 13 so we assign the temporary label 13 to E.*

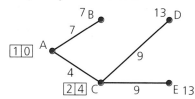

*We now look again at the vertices with temporary labels and select the smallest, to assign a permanent label. We assign permanent label 7 to B.*

*Now we repeat the process, considering all the vertices that can be reached directly from B and give temporary labels to these vertices.*

*A → B → D = 7 + 8 = 15*    *This is larger than the current temporary label at D so we do not change it.*

*A → B → E = 7 + 5 = 12*    *This is smaller than the current temporary label at E so we replace the 13 with 12.*

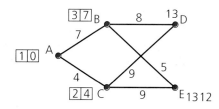

*Our smallest temporary label is now 12 at E so we make this a permanent label and we give F a temporary label 17.*

The fifth permanent label is then assigned to D with a value of 13. This will lead to the replacement of the temporary labels at F, since DF has weight 3, giving a total of 13 + 3 = 16 which is smaller than the current temporary label.

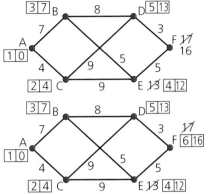

*Finally we assign permanent label 16 to F. Now all the vertices have permanent labels and we can see that the shortest distance from A to F is 16.*

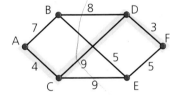

*We find the shortest path by working backwards from F. The final iteration brought us to F from D. The value of 13 at D cannot have come from B (since 13 – 8 = 5, not 7), so it must have come from C. In a similar way, we reached C from A so tracing back through the network: F → D → C → A.*

*Hence the shortest path from A to F is: A → C → D → F with length 16.*

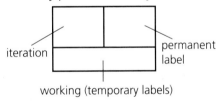

## Note
You may prefer to draw your boxes like this.

```
          ┌──────┬──────┐
         /│      │      │\
iteration │      │      │  permanent
          └──────┼──────┘  label
                 │
     working (temporary labels)
```

We can now express Dijkstra's algorithm as a set of steps.

### Dijkstra's algorithm
**Step 1:** Assign the permanent label 0 to the starting vertex.

**Step 2:** Assign temporary labels to all the vertices that are connected directly to the most-recently permanently labelled vertex.

**Step 3:** Choose the vertex with the smallest temporary label and assign a permanent label to that vertex.

**Step 4:** Repeat steps 2 and 3 until all vertices have permanent labels.

**Step 5:** Find the shortest path by tracing back through the network.

We are now able to use Dijkstra's algorithm to solve the problem given in Exploration 2.2. Here is a solution.

*It is now possible to get computer programs for a home PC which will calculate the shortest route between any 2 towns and will print out the route for you. Alternatively they will calculate the quickest route.*

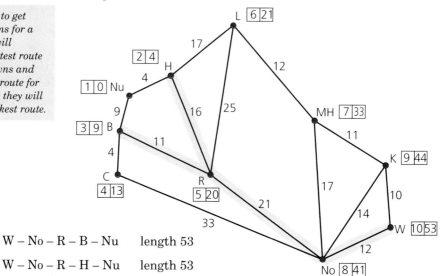

| Tracing back | W – No – R – B – Nu | length 53 |
|---|---|---|
| two paths | W – No – R – H – Nu | length 53 |

Notice that there are two possible solutions, both with length 53 miles:
- Nuneaton, Bedworth, Rugby, Northampton, Wellingborough
- Nuneaton, Hinckley, Rugby, Northampton, Wellingborough.

## EXERCISES

**2.2 A**

**1** Find the shortest path from S to T through these networks.

**a)**

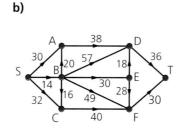

**b)**

**c)**

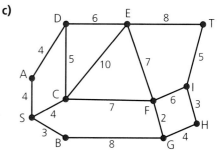

**2** Maria travels from her home in Bickleigh to work in Exeter every day. What is her shortest route to work?

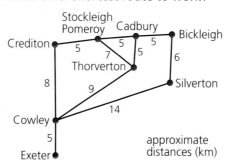

Roadworks cause the road between Thorveton and Cowley to be closed. What is her shortest route to work now?

approximate distances (km)

**3** A delivery company operates a service between certain towns in Kent. The charges per kilogram (in £s) for delivery are shown in this network.

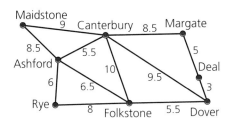

Use Dijkstra's algorithm to find the cheapest way of getting a parcel weighing 4 kg from:

a) Ashford to Dover
b) Deal to Maidstone.

**4** Three fire stations are situated at the points marked P, Q and R on this network. A fire breaks out in a factory at X. Which of the fire stations should send their engines to attend the blaze?

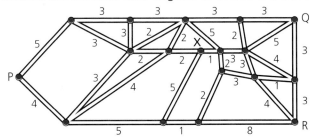

**5** A sports centre is to be built that will be used mainly by people in several different villages. It is decided to site the centre in one of the villages. Use the network below to decide in which village it should be sited so as to minimise the average distance travelled by visitors to the centre.

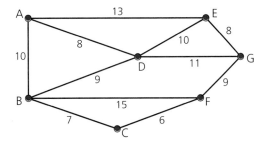

## EXERCISES

2.2 B

**1** Find the shortest path from S to T through the following networks.

a)

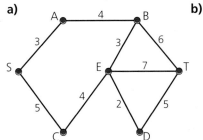

b)

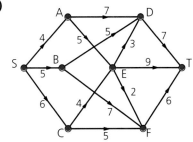

**2**    Each day an Open University secretary cycles from her home in Stony Stratford to Walton Hall along the cycle-ways of Milton Keynes. Use Dijkstra's algorithm to find her shortest path.

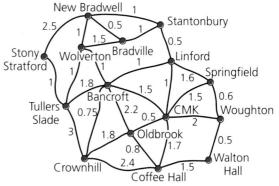

**3**    The table shows the distances in miles along major routes between cities where schools are using the same mathematics A-level scheme.

|            | Bristol | Oxford | London | Birmingham | Nottingham | Cambridge |
|------------|---------|--------|--------|------------|------------|-----------|
| **Bristol**    | —    | 70  | 114 | 89  | —   | —   |
| **Oxford**     | 70   | —   | 56  | 63  | 95  | 55  |
| **London**     | 114  | 56  | —   | 112 | 160 | 55  |
| **Birmingham** | 89   | 63  | 112 | —   | 49  | 102 |
| **Nottingham** | —    | 95  | 160 | 49  | —   | 84  |
| **Cambridge**  | —    | 55  | 55  | 102 | 84  | —   |

**a)**  Represent this information on a network graph.
**b)**  Books for the scheme are being printed in Bristol. The school in Cambridge finds that it needs an extra set of books urgently. The printing company decide to send books by van.
   **i)**  Find the shortest path for the van to travel between Bristol and Cambridge.
   **ii)** Just before leaving the driver is warned of a four-hour delay on the Oxford to Cambridge road. Does the driver need to change the route?

**4**    A delivery company operates a service between cities in Finland. The charges (in Finnish marks per kilogram) for delivery are shown in the following network.

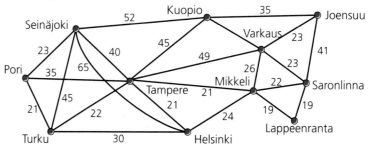

Use Dijkstra's algorithm to find the cheapest way of getting a parcel weighing 5 kg from:
**a)**  Helsinki to Pori        **b)**  Turku to Joensuu.

**5** A salesman in Gothenburg flies regularly using the same airline to the seven cities shown in the table, which also shows the fares (in pounds) for direct flights of the chosen airline between the cities.

| | Gothenburg | Helsinki | Amsterdam | Copenhagen | Stockholm | Berlin | Frankfurt | Warsaw |
|---|---|---|---|---|---|---|---|---|
| **Gothenburg** | — | — | 68 | 238 | 136 | — | 136 | — |
| **Helsinki** | — | — | — | 102 | 204 | — | — | 102 |
| **Amsterdam** | 68 | — | — | — | — | — | 34 | 408 |
| **Copenhagen** | 238 | 102 | — | — | 68 | 136 | — | — |
| **Stockholm** | 136 | 204 | — | 68 | — | — | — | — |
| **Berlin** | — | — | — | 136 | — | — | 170 | 136 |
| **Frankfurt** | 136 | — | 34 | — | — | 170 | — | — |
| **Warsaw** | — | 102 | 408 | — | — | 136 | — | — |

Represent this data by a network. Use Dijkstra's algorithm to find the cheapest route between Gothenburg and the other cities.

## DYNAMIC PROGRAMMING

### Working backwards

We can also solve this type of problem by working backwards through the network. This is known as **dynamic programming**. If we consider the network used in Example 2.1, we begin by looking at the last stage first.

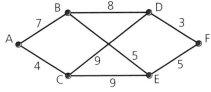

We could reach F from either D or E.

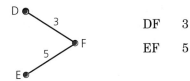

DF    3

EF    5

Considering the previous stage gives us the following possibilities.

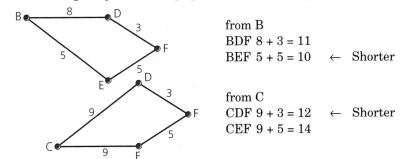

from B
BDF 8 + 3 = 11
BEF 5 + 5 = 10    ← Shorter

from C
CDF 9 + 3 = 12    ← Shorter
CEF 9 + 5 = 14

Then the final stage is like this.

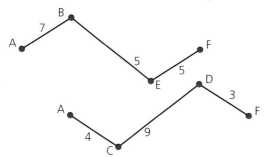

Carry forward the shorter route from each vertex at every step:

from A
ABEF    $7 + 10 = 17$
ACDF    $4 + 12 = 16$    Shortest path

Dynamic programming methods can be applied to many 'shortest path' problems and are frequently used when designing computer programs.

## EXERCISES

**2.3 A**

Repeat Exercises 2.2A, using the dynamic programming method.

## EXERCISES

**2.3 B**

Repeat Exercises 2.2B, using the dynamic programming method.

## COURSEWORK INVESTIGATION

**Investigation**

### Fire escapes

In a school, or other public building, there are notices in all the rooms directing people to the nearest fire exit. By considering the plan of such a building, decide on the fire exits which should be used by people in various parts of the building. What other factors need to be considered (width of corridors, other obstructions etc.)? Is shortest distance the most appropriate way to devise a building evacuation plan?

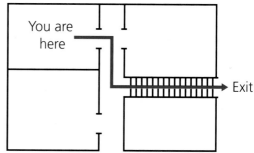

## CONSOLIDATION EXERCISES FOR CHAPTER 2

1   By considering the number of calculations made in finding the shortest path through the network in Example 2.1, by both Dijkstra's algorithm and by dynamic programming, compare the efficiency of the two methods.

**2** This network illustrates the time (in minutes) of rail journeys between the given cities.

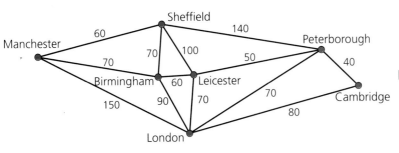

**a)** Assuming that any necessary changes of train do not add any time to the journey, find the quickest rail route from Manchester to Cambridge.

**b)** Assuming that at each city shown a change of train is always necessary, and that each change adds 20 minutes to the journey time, find the quickest rail route from Manchester to Cambridge.

*(AEB Question 1, Discrete Mathematics P4 Specimen Paper, 1996)*

**3** This network models possible air routes between S and T via certain intermediate cities. The numbers on the arcs indicate the fares in units of £100. The journeys can only be made in the direction indicated. Use the shortest path algorithm to find the route between S and T for

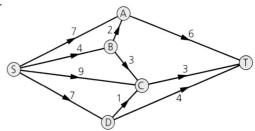

which the total fare is a minimum. Your solution should indicate clearly the order in which the nodes (cities) receive their permanent labels.

*(ULEAC Question 5, Decision Mathematics D1 Specimen Paper)*

**4** The network shows the road connections between a group of villages. The numbers on the arcs represent the times in minutes that a driver takes to travel along the roads represented by those arcs. The network is set up ready for an application of Dijkstra's algorithm.

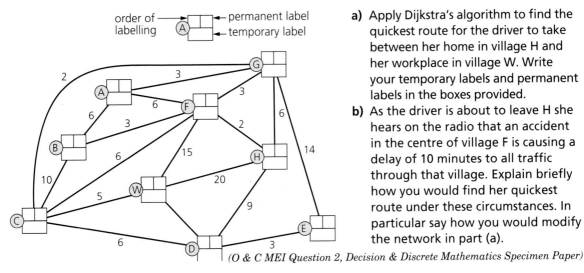

**a)** Apply Dijkstra's algorithm to find the quickest route for the driver to take between her home in village H and her workplace in village W. Write your temporary labels and permanent labels in the boxes provided.

**b)** As the driver is about to leave H she hears on the radio that an accident in the centre of village F is causing a delay of 10 minutes to all traffic through that village. Explain briefly how you would find her quickest route under these circumstances. In particular say how you would modify the network in part (a).

*(O & C MEI Question 2, Decision & Discrete Mathematics Specimen Paper)*

**5**   The diagram represents the roads joining 10 villages, labelled A to J. The numbers give distances in km.

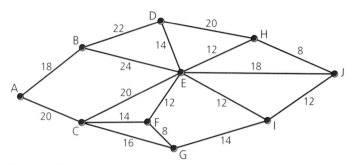

**a)** Use Dijkstra's algorithm to find a shortest route from A to J. Explain the method carefully, and show all of your working. Give your shortest route and its length.

The driver usually completes this journey driving at 60 km/h. The local radio reports a serious fire at village E, and warns drivers of a delay of 10 minutes.

**b)** Describe how to modify your approach to **a)** to find the quickest route, explaining how to take account of this information. What is the quickest route, and how long will it take?

*(UODLE Question 7, Paper 4, June 1991)*

**6**   A salesman plans to travel between two towns S and T. Along certain routes he can conduct some business and make an effective profit

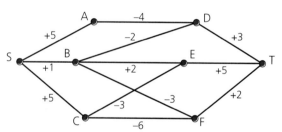

on his journey whereas on others he cannot. The positive weights on the edges represent the cost of the journey while those with negative weights represent the profit he can make by using particular routes.

**a)** Is it possible to solve this problems using Dijkstra's algorithm?

**b)** Solve the problem using dynamic programming.

**7**   **a)** An express delivery company operates its service between certain towns in Kent. The charges per kilogram (in £s) are shown on the network.

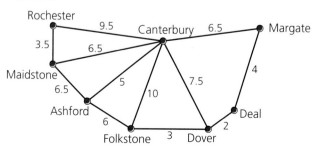

Use Dijkstra's algorithm to find the cheapest way of sending a 4 kg package from:
**i)** Maidstone to Dover **ii)** Deal to Rochester.

**b)** A second company sets up in opposition. It can undercut the first company by £1 per kg on the routes between Rochester and Canterbury, Canterbury and Dover, Maidstone and Ashford, and Ashford and Folkestone, but charges 50p more on all the others. Draw the network for the new company. Would it be cheaper to use this company for the packages in part **a)** and if so, would the routes be the same?

**8** The diagram shows a small network in which the number on each arc represents the cost of travelling along that arc in the directions indicated.

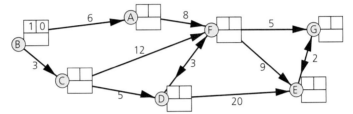

**a)** Apply Dijkstra's algorithm to find the smallest cost, and the associated route, to get from B to E. (Record the order in which you permanently label nodes, and write your temporary labels and permanent labels, in the boxes provided.)

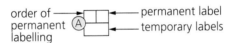

order of ──── permanent label
permanent ──── temporary labels
labelling

Smallest cost from B to E ...... Cheapest route from B to E ......
Show briefly how the route is found in Dijkstra's algorithm, once the destination vertex has been permanently labelled.

**b)** Suppose now that a new route is introduced from A to C, and that a profit of 4 may be made by making a delivery from A to C when using that route. This is shown as a negative cost in part of the network reproduced below.

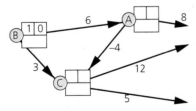

Work through Dijkstra's algorithm for this part of the network with the extra route. Say what is the best route from B to C, and explain why Dijkstra's algorithm fails to find it.

*(O & C MEI Question 2, Decision Mathematics, January 1995)*

**9** The network represents an electronic information network. The vertices represent computer installations. The arcs represent direct links between installations. The weights on the arcs are the rates at which the links can transfer data (in units of 100 000 bits per second).

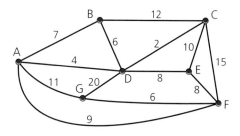

A chain of links may be established between any two installations via any other installations. For such links the rate at which the chain can transfer data is given by the speed of the slowest link in the chain.

**a) i)** The fastest chain linking G to B is needed. Devise a way of adapting Dijkstra's algorithm to produce an algorithm suited to this task. Give your adaptations and explain how your algorithm works.

**ii)** Apply your algorithm, showing all of your steps on a copy of the network.

**iii)** State the fastest chain and its speed of transferring data.

**b)** How could you find a slowest chain linking G to B? Give such a chain.

*(UODLE Question 22, Paper 4, June 1995)*

**10** A skier finds she can travel across a section of mountain country by using a system of drag lifts. She has to queue for each lift in order that she can be towed up a hillside. She can then ski to the bottom of another lift. The bottoms of the lifts are labelled A to H, and the average queue plus tow time for each lift is given in this table.

| Lift | A | B | C | D | E | F | G | H |
|---|---|---|---|---|---|---|---|---|
| Time for queue plus two (minutes) | 5 | 10 | 7 | 25 | 14 | 15 | 7 | 10 |

The time in minutes to ski from the top of a lift to the bottom of a lift, where a route exists, is given in this table.

| From top of: to bottom of: | A | B | C | D | E | F | G | H |
|---|---|---|---|---|---|---|---|---|
| A | | 6 | 8 | — | 10 | — | — | — | — |
| B | 10 | — | — | 5 | — | — | — | — |
| C | 8 | — | — | — | — | — | — | 7 |
| D | 11 | 5 | — | 5 | — | 15 | — | — |
| E | — | 12 | — | — | 4 | — | — | — |
| F | — | — | — | 17 | 5 | — | 4 | — |
| G | — | — | — | — | — | 6 | — | 7 |
| H | — | — | 7 | — | — | — | — | — |

a) Explain why the total time from the bottom of B to the bottom of A is 18 minutes. Draw a network incorporating total times, its vertices representing the bottoms of lifts.
b) Use Dijkstra's algorithm to find the shortest time for the skier to get from the bottom of lift A to the bottom of lift F. Show all of the steps of your method, including the order in which you deal with vertices. Find the fastest route, and explain how you found it.

*(UODLE Question 20, Paper 4, 1993)*

## Summary

*After working through this chapter you should:*

■ know that:
a graph is a diagram showing how objects are related to each other
a vertex or node is a point representing an object in a graph
an edge is a line joining two vertices, without direction
an arc is a line joining two vertices in a given direction
a network is a graph in which each edge (or arc) is given a value or weight

■ know that:
a path is a route through the graph
a connected graph is one in which there is at least one route through
a complete graph is one in which each vertex is connected to every other vertex

■ be able to draw a graph from its adjacency matrix and a network from its cost matrix

■ understand how networks can be used to model problems

■ use Dijkstra's algorithm to find the shortest path through a network, and recognise that algorithms that are used to find optimal paths are called greedy algorithms

■ be aware of dynamic programming techniques as an alternative method for solving this type of problem.

# Networks 2: Minimum-connector and the travelling salesman problem

*By the end of this chapter you should be familiar with:*

■ *a cycle, which is a loop which returns to its starting point*

■ *a spanning tree, which connects all the vertices in a graph with no cycles*

■ *the minimum connector, which is the spanning tree of minimum length*

■ *the travelling salesman problem, which attempts to find a cycle of minimum length that visits all the vertices in a network.*

## Exploration 3.1

### Cable TV: a minimum connector problem

A cable TV company is installing a system of cables to connect all the towns in a region. The numbers in the network show distances in miles.

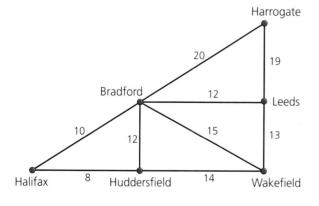

■ Find a layout that will connect all the towns in the region.

■ Find a different layout which uses less cable.

■ What is the least amount of cable needed?

Once the cable is installed an inspector, based in Leeds, visits all the towns to ensure that the connection has been successful. Find a route for her to visit every town.

Is this the shortest possible route?

The first part of this chapter will introduce techniques to help solve this type of problem.

# GRAPH THEORY – CYCLES AND TREES

In Chapter 2, *Networks 1: Shortest path* we discussed the idea of a path through a graph or network. A **cycle** is a path that completes a loop and returns to its starting point.

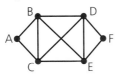  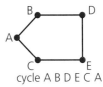

cycle A B D C A   cycle A B D E C A

There are two types of cycle which are of interest in solving real-life problems.

**Hamiltonian cycles** are cycles that pass through every vertex of the graph exactly once, and only once, and return to the starting point. This is the type of cycle that would be needed to obtain a solution to the second part of Exploration 3.1. The path ABDFECA is an example of a Hamiltonian cycle.

**Eulerian cycles** are cycles that include every edge of a graph exactly once. The path ABCDBEDFECA is an example of an Eulerian cycle. These cycles are often simply called **Euler cycles**.

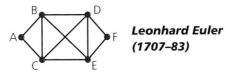

**Leonhard Euler (1707–83)**

We shall meet this kind of cycle again in Chapter 4, *Networks 3: Route inspection problems*.

A connected graph which contains no cycles is called a **tree**.

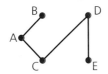

A **spanning tree** is a subgraph that includes all the vertices in the original graph, and is also a tree. A graph will have several trees.

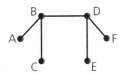 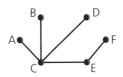 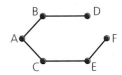

These are all spanning trees of the path ABCDECA above. There are others, can you find them? Confirm that, for a network of $n$ vertices, there will be $n - 1$ edges in the minimum spanning tree.

# EXERCISES

**3.1A**

1

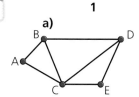

a)

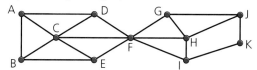

b)

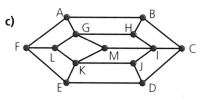

c)

i)   Find three cycles in each of the graphs above.
ii)  Find a spanning tree for each of the graphs.
iii) Which of your cycles in (i) are Hamiltonian?

2   By considering the complete graphs $K_3$, $K_4$, $K_5$ and $K_6$, deduce that the set of complete graphs $K_n$ are Hamiltonian for $n \geq 3$. Why is it necessary to have the condition $n \geq 3$?

3   a)   Explain why this graph is not a Hamiltonian cycle.

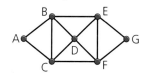

b)  By adding one edge to the graph, find a Hamiltonian cycle beginning and ending at A.

4   There is a theorem which states that an Eulerian graph can be split into cycles, no two of which have any edges in common. Show that the graph below can be split into four cycles, no two of which have edges in common. How can the cycles be combined to form an Eulerian cycle?

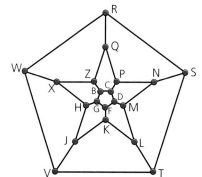

5   William Hamilton invented the Icosian game, which is played on a board like the one below. The idea of the game is to find a Hamiltonian cycle starting with five given letters. Find two Hamiltonian cycles beginning with the letters BGHJK.

*Hamilton sold the game to a dealer in 1859 for £25.00 It was not a commercial success, so the dealer got himself a poor bargain!*

## EXERCISES

1   Sketch a graph of the following.

a) Four vertices, one adjacent to all the others but no others adjacent. Is this a tree?
b) Five vertices, with cycles of length 2, 3, 4 and 5.

2   Study graph $G_4$ below. Find cycles of lengths, 5, 6, 8 and 9. Do cycles of length 7 exist?

3   Consider the graphs $G_1$, $G_2$ and $G_3$ below.

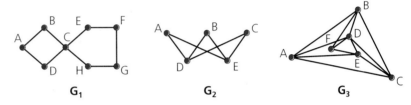

In each case, determine whether the graph is:
a) Hamiltonian
b) Eulerian.
For those graphs which exhibit the appropriate property, identify a cycle.

4   Give an example, with at most seven vertices, which you have not already encountered in this chapter, of:

a) a cycle that is both Hamiltonian and Eulerian
b) an Eulerian cycle that is not Hamiltonian
c) a Hamiltonian cycle that is not Eulerian
d) a cycle that is neither Hamiltonian nor Eulerian.
In each case:
e) find a spanning tree for your graph
f) find cycles for your graph.

5   Suggest conjectures for:
a) the number of edges of any tree with $n$ vertices
b) the number of edges in a complete graph of $n$ vertices.
    Justify your conjectures.

**Note:**
The graph in question 2 in Exercises 3.1B is known as the **Petersen graph**.

## THE MINIMUM-CONNECTOR PROBLEM

Let us return to the problem in the first part of Exploration 3.1. We can now see that to solve it, we need to find the spanning tree of shortest length for the network. This type of problem is a **minimum connector problem**. There are several algorithms for finding the minimum connector for a network; we shall look at two of these.

## Kruskal's algorithm

*J.B. Kruskal Jr published the algorithm in 1956 in the **American Mathematical Society proceedings**.*

Kruskal's algorithm is a greedy algorithm, it chooses the smallest available edges, not worrying about any connection to edges that have already been chosen, except that it is careful not to form a cycle. The algorithm can be stated like this.

**Step 1:** Rank the edges in order of length.
**Step 2:** Select the shortest edge in the network.
**Step 3:** Select, from the edges which are not in the solution, the shortest edge which does not form a cycle. (Where two edges have the same weight, select at random.)
**Step 4:** Repeat Step 3 until all the vertices are in the solution.

## Example 3.1

*Use Kruskal's algorithm to find the least amount of cable needed to solve the problem in Exploration 3.1.*

### Solution
*First, rank the edges in order of length.*

| Order of connection | | | |
|---|---|---|---|
| | 1 | Halifax to Huddersfield | 8 |
| | 2 | Halifax to Bradford | 10 |
| | 3 | Bradford to Leeds | 12 } (As these two edges have the same |
| This edge forms a loop | → | Bradford to Huddersfield | 12 } length, the order is arbitrary.) |
| | 4 | Leeds to Wakefield | 13 |
| This edge forms a loop | → | Huddersfield to Wakefield | 14 |
| This edge forms a loop | → | Bradford to Wakefield | 15 |
| | 5 | Leeds to Harrogate | 19 |
| This edge forms a loop | → | Bradford to Harrogate | 20 |

*Now we begin to select edges, starting with the smallest.*
*We have now connected all the vertices into the spanning tree.*
*The length of our minimum spanning tree is 8 + 10 + 12 + 13 + 19 = 62 miles.*

*It is not necessary to redraw the network at every iteration. The solution can be shown on one diagram provided you note the order in which the edges were connected into the spanning tree.*

## Prim's algorithm

When programming a network into a computer, it is generally useful to adopt a matrix format. In Chapter 2, *Networks 1: Shortest path*, we used the idea of an adjacency matrix to describe graphs. It is also possible to describe a network using **distance matrices**. The distance matrix for example 3.1 would look like this.

|     | L  | H  | W  | B  | Ha | Hu |
| --- | -- | -- | -- | -- | -- | -- |
| L   | ∞  | 19 | 13 | 12 | ∞  | ∞  |
| H   | 19 | ∞  | ∞  | 20 | ∞  | ∞  |
| W   | 13 | ∞  | ∞  | 15 | ∞  | 14 |
| B   | 12 | 20 | 15 | ∞  | 10 | 12 |
| Ha  | ∞  | ∞  | ∞  | 10 | ∞  | 8  |
| Hu  | ∞  | ∞  | 14 | 12 | 8  | ∞  |

The symbol ∞ means that there is no direct connection between these vertices. If we use ∞ (of infinite lengths) it ensures these edges will not feature in the solution.

*R.C. Prim published this algorithm in the **Bell System Technical Journal** in 1957.*

Prim's algorithm works from a starting point and builds up the spanning tree step by step, connecting edges into the existing solution. It can be applied directly to the distance matrix, as well as to the network itself, so it is more suitable for using with a computer if the network is large. The algorithm can be stated follows.

**Step 1:** Choose a starting vertex.
**Step 2:** Join this vertex to the nearest vertex directly connected to it.
**Step 3:** Join the next nearest vertex, not already in the solution, to any vertex in the solution, provided it does not form a cycle.
**Step 4:** Repeat until all vertices have been included.

We now apply the algorithm to the distance matrix above, with the solution taken straight from the graph shown alongside.

Choose a starting vertex, say L. Delete row L. Look for the smallest entry in column L.

1
↓

|     | L    | H  | W  | B  | Ha | Hu |
| --- | ---- | -- | -- | -- | -- | -- |
| ~~L~~ | ~~∞~~ | ~~19~~ | ~~13~~ | ~~12~~ | ~~∞~~ | ~~∞~~ |
| H   | 19   | ∞  | ∞  | 20 | ∞  | ∞  |
| W   | 13   | ∞  | ∞  | 15 | ∞  | 14 |
| B   | (12) | 20 | 15 | ∞  | 10 | 12 |
| Ha  | ∞    | ∞  | ∞  | 10 | ∞  | 8  |
| Hu  | ∞    | ∞  | 14 | 12 | 8  | ∞  |

● L

LB is the smallest edge joining L to the other vertices. Put edge LB into the solution. Delete row B. Look for the smallest entry in columns L and B.

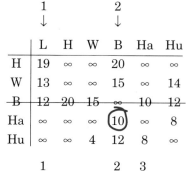

|    | L  | H  | W  | B   | Ha   | Hu |
|----|----|----|----|-----|------|----|
| H  | 19 | ∞  | ∞  | 20  | ∞    | ∞  |
| W  | 13 | ∞  | ∞  | 15  | ∞    | 14 |
| ~~B~~ | ~~12~~ | ~~20~~ | ~~15~~ | ∞ | ~~10~~ | ~~12~~ |
| Ha | ∞  | ∞  | ∞  | (10) | ∞   | 8  |
| Hu | ∞  | ∞  | 4  | 12  | 8    | ∞  |

1↓ (L)  2↓ (B)

BHa is the smallest edge joining L and B to the other vertices. Put edge BHa into the solution. Delete row Ha. Look for the smallest entry in columns L, B and Ha.

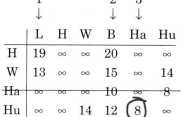

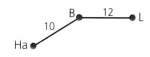

|    | L  | H  | W  | B   | Ha  | Hu |
|----|----|----|----|-----|-----|----|
| H  | 19 | ∞  | ∞  | 20  | ∞   | ∞  |
| W  | 13 | ∞  | ∞  | 15  | ∞   | 14 |
| ~~Ha~~ | ∞ | ∞ | ∞ | ~~10~~ | ∞ | ~~8~~ |
| Hu | ∞  | ∞  | 14 | 12  | (8) | ∞  |

1↓ (L)  2↓ (B)  3↓ (Ha)

HaHu is the smallest edge joining L, B and Ha to the other vertices. Put edge HaHu into the solution. Delete row Hu. Look for the smallest entry in columns L, B, Ha and Hu.

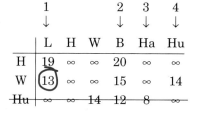

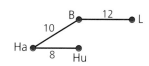

|    | L    | H  | W  | B   | Ha  | Hu |
|----|------|----|----|-----|-----|----|
| H  | 19   | ∞  | ∞  | 20  | ∞   | ∞  |
| W  | (13) | ∞  | ∞  | 15  | ∞   | 14 |
| ~~Hu~~ | ∞ | ∞ | ~~14~~ | ~~12~~ | ~~8~~ | ∞ |

1↓ (L)  2↓ (B)  3↓ (Ha)  4↓ (Hu)

LW is the smallest edge joining L, B, Ha and Hu to the other vertices. Put edge LW into the solution. Delete row W. Look for the smallest entry in columns L, B, Ha, Hu and W.

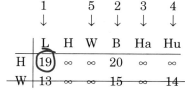

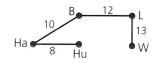

|    | L    | H  | W  | B   | Ha  | Hu |
|----|------|----|----|-----|-----|----|
| H  | (19) | ∞  | ∞  | 20  | ∞   | ∞  |
| ~~W~~ | ~~13~~ | ∞ | ∞ | ~~15~~ | ∞ | ~~14~~ |

1↓ (L)  5↓ (W)  2↓ (B)  3↓ (Ha)  4↓ (Hu)

LH is the smallest edge joining L, B, Ha, Hu and W to the other vertices. Put LH into the solution.

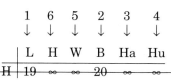

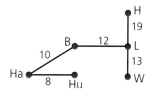

|    | L    | H  | W  | B   | Ha  | Hu |
|----|------|----|----|-----|-----|----|
| ~~H~~ | ~~19~~ | ∞ | ∞ | ~~20~~ | ∞ | ∞ |

1↓ (L)  6↓ (H)  5↓ (W)  2↓ (B)  3↓ (Ha)  4↓ (Hu)

We have now connected all the vertices into the spanning tree. Can you see that we have arrived at the same solution as with Kruskal's algorithm?

In reality, we would not rewrite the matrix at every iteration, but would work from one matrix, deleting the lines as we progress through the algorithm, and making a record of the order in which we connected the vertices along the top of the matrix, so the solution would look like this.

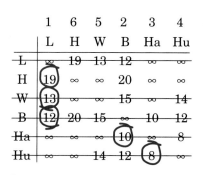

|    | L    | H  | W  | B   | Ha  | Hu |
|----|------|----|----|-----|-----|----|
| ~~L~~ | ∞ | ~~19~~ | ~~13~~ | ~~12~~ | ∞ | ∞ |
| H  | (19) | ∞  | ∞  | 20  | ∞   | ∞  |
| ~~W~~ | (13) | ∞ | ∞ | ~~15~~ | ∞ | ~~14~~ |
| ~~B~~ | (12) | ~~20~~ | ~~15~~ | ∞ | ~~10~~ | ~~12~~ |
| ~~Ha~~ | ∞ | ∞ | ∞ | (10) | ∞ | ~~8~~ |
| ~~Hu~~ | ∞ | ∞ | ~~14~~ | ~~12~~ | (8) | ∞ |

1 (L)  6 (H)  5 (W)  2 (B)  3 (Ha)  4 (Hu)

length 19 + 13 + 12 + 10 + 8 = 62 miles

**EXERCISES**

**3.2 A**

1   Use Kruskal's algorithm to find the minimum spanning tree for the network on the right.

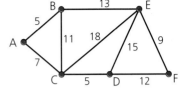

2   Use Prim's algorithm to find the minimum spanning tree for the network on the right. (You will need to put the information into a table first.)

3   The table below gives the distances between some cities in the north of England and Scotland. Use Prim's algorithm to find the shortest route connecting them.

|  | **Aberdeen** | **Carlisle** | **Edinburgh** | **Glasgow** | **Inverness** |
|---|---|---|---|---|---|
| **Aberdeen** | ∞ | 221 | 125 | 145 | 105 |
| **Carlisle** | 221 | ∞ | 100 | 95 | 263 |
| **Edinburgh** | 125 | 100 | ∞ | 40 | 160 |
| **Glasgow** | 145 | 96 | 40 | ∞ | 168 |
| **Inverness** | 105 | 263 | 160 | 168 | ∞ |

From the information in the table, draw a network and use Kruskal's algorithm to find the shortest route connecting the cities. Comment on the relative efficiency of the two algorithms for a network of this size. Do you think your conclusion would be different if the network had twice as many vertices?

4   This distance chart gives the distances, in miles, between towns in Ireland. Adapt the matrix so that you can use Prim's algorithm to find a minimum spanning tree for these towns.

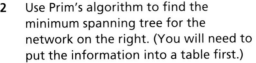

| Belfast | | | | | | | |
|---|---|---|---|---|---|---|---|
| 205 | Cork | | | | | | |
| 103 | 154 | Dublin | | | | | |
| 196 | 122 | 135 | Galway | | | | |
| 204 | 58 | 120 | 64 | Limerick | | | |
| 70 | 281 | 146 | 176 | 231 | Londonderry | | |
| 126 | 200 | 133 | 90 | 147 | 86 | Sligo | |
| 197 | 73 | 96 | 141 | 77 | 242 | 176 | Waterford |

5   Using the information in the distance chart above, construct a network and use it to find the minimum spanning tree using Kruskal's algorithm.

## EXERCISES

**3.2 B**

1  a) Use Kruskal's algorithm to find and draw the minimum spanning tree for this network.
   b) Put the information in the network in a table and use Prim's algorithm to verify your answer to part a).
   c) Discuss the differences between Kruskal's algorithm and Prim's algorithm. Which method do you think would be faster to use on a computer?

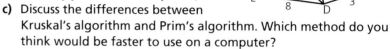

2  To generate a maximum spanning tree i.e. spanning trees with the greatest possible weight, we generate a new weighted graph with the same vertices and edges as the one given but where each weight $W_L$ is replaced by $M - W_L$ where $M$ is any number greater than maximum weight $w_i$, and proceed as before. Determine the maximum spanning tree for the network above and verify your answer using Prim's algorithm.

3  The table below gives the distances between some cities in the north of England and Scotland.

|  | Aberdeen | Carlisle | Edinburgh | Glasgow | Inverness | Perth |
|---|---|---|---|---|---|---|
| **Aberdeen** | ∞ | 221 | 125 | 145 | 105 | 81 |
| **Carlisle** | 221 | ∞ | 100 | 95 | 263 | 144 |
| **Edinburgh** | 125 | 100 | ∞ | 40 | 160 | 42 |
| **Glasgow** | 145 | 96 | 40 | ∞ | 168 | 60 |
| **Inverness** | 105 | 263 | 160 | 168 | ∞ | 115 |
| **Perth** | 81 | 144 | 42 | 60 | 115 | ∞ |

Determine, by any appropriate method, the shortest route connecting the centres. Compare the answers with that for question 3 of Exercises 3.2A and comment.

4  Find *all* the minimum spanning trees for this weighted graph.

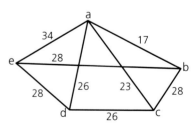

5  a) The table at the top of page 57 gives the distances in miles, between the first eight centres, listed alphabetically, in my road atlas. Construct a graph to illustrate the data. How many distances would be given if
      i) 10 centres   ii) $n$ centres were to be used?

By considering a completed rectangle for the figure justify your conjecture in part **ii)**.
b) Use the information in the table to construct a matrix and hence find the minimum spanning tree by applying Kruskal's algorithm.
c) Verify your result in part **ii)** by adapting the matrix so that Prim's algorithm may be applied to the data.

| Aberdeen | | | | | | | |
|---|---|---|---|---|---|---|---|
| 470 | Aberystwyth | | | | | | |
| 602 | 218 | Barnstaple | | | | | |
| 430 | 119 | 176 | Birmingham | | | | |
| 599 | 268 | 202 | 178 | Brighton | | | |
| 511 | 126 | 98 | 85 | 151 | Bristol | | |
| 468 | 221 | 250 | 101 | 122 | 156 | Cambridge | |
| 533 | 110 | 137 | 107 | 186 | 45 | 191 | Cardiff |

## THE 'TRAVELLING SALESMAN' PROBLEM

In the second part of Exploration 3.1 you were asked to find a route for the inspector who visits every town, in other words, every vertex in the network, and returns to the starting point, such a cycle being referred to as a **tour**.

*TSP is the accepted abbreviation for the travelling salesman problem.*

This type of problem, known as the 'travelling salesman problem' (TSP), is well known because it is faced regularly in industry by people who need to minimise distance, time or cost, for example the salesman after whom the problem was named. It is also interesting because there is no exact method for solving it. Mathematicians are trying to find a suitable method that guarantees an optimal solution, but a simple algorithm has not yet been devised. It therefore falls into the group of problems which we classified as **construction problems** in Chapter 1, *Algorithms*. We know that a solution exists, but an algorithm which provides an optimal solution for all networks has not yet been constructed.

**Exploration 3.2**

### *Find the lengths of cycles*

Find the lengths of all the cycles in this network, beginning and ending at A, and including every vertex at least once.
Identify the cycles of minimum length.

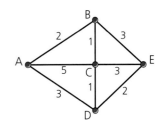

## Finding the optimal solution

Carrying out an exhaustive search will give the optimal solution, but it is tedious and time-consuming. It can be shown that for a complete graph with $n$ vertices, there are $\frac{1}{2}(n-1)!$ possible cycles, so in a network with fifteen vertices there would be roughly $4.36 \times 10^{10}$ cycles; a computer capable of checking 1000 cycles every second would take over a year to check them all! Checking all possible cycles is obviously not a practical proposition, even using a computer. However there is a method of finding a solution that is close to the optimal solution.

In Exploration 3.2 there are many cycles, but the four shown below are of particular importance.

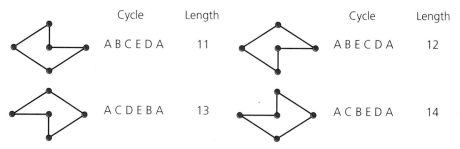

| | Cycle | Length | | Cycle | Length |
|---|---|---|---|---|---|
| | A B C E D A | 11 | | A B E C D A | 12 |
| | A C D E B A | 13 | | A C B E D A | 14 |

**Note:** *The order of listing the vertices could be reversed, but the cycle would be effectively the same e.g. ABCEDA ≡ ADECBA.*

These cycles are important because they visit each vertex once and once only, in other words, they are Hamiltonian cycles. The shortest cycle in the exploration is ABCEDA, length 11. The TSP can be considered as an attempt to find a Hamiltonian cycle of minimum length in the network, if such a cycle exists.

**Exploration 3.3**

## *Is there a Hamiltonian solution?*

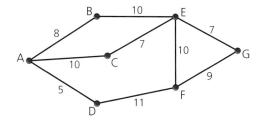

Find a cycle of minimum length, starting and finishing at A.

Is it possible to find a cycle which visits each vertex once, and only once?

## Solving the classical problem

In fact it is impossible to find a Hamiltonian cycle in the network in Exploration 3.3, so the cycle must revisit a vertex, to get home. In a practical problem this would be quite acceptable, but in the 'classical' problem of finding Hamiltonian cycles it is not. From the network preceding the exploration, it can also be seen that even if Hamiltonian cycles exist, there may be several of different lengths. Despite this, we can use mathematical modelling to enable us to use the classical problem as a method for finding a cycle which will give an upper bound to the TSP.

## Upper bounds

Question 2 of Exercises 3.1A established that any complete network is Hamiltonian and will contain cycles which visit every vertex within themselves once and only once before returning to the starting point. In this method, the first step is to convert the network into a complete network. We do this by joining each vertex to every other, by the shortest path in each case.

Consider again the network in Exploration 3.3.

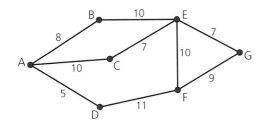

Vertex A is joined directly to vertices B, C and D. To make the network complete, we need to add edges joining A directly to E, F and G. In a small network like this one we can find the shortest distances, (minimum weights) by observation.

To reach E from A, we could take:

| ABE | length 18 |
| ACE | length 17 |

As ACE has the minimum weight, edge AE would have weight 17, giving this result. If we repeat the same process for all pairs of vertices we will get this complete network.

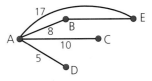

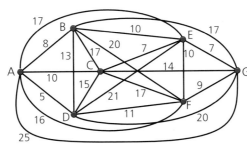

Drawn in this way, the network looks rather cluttered and untidy. It would be difficult to work directly on a diagram like this. To avoid such an untidy network, we could show the information on a distance matrix as we did for Prim's algorithm. This gives the following matrix for the original network.

|   | A | B | C | D | E | F | G |
|---|---|---|---|---|---|---|---|
| A | ∞ | 8 | 10 | 5 | ∞ | ∞ | ∞ |
| B | 8 | ∞ | ∞ | ∞ | 10 | ∞ | ∞ |
| C | 10 | ∞ | ∞ | ∞ | 7 | ∞ | ∞ |
| D | 5 | ∞ | ∞ | ∞ | ∞ | 11 | ∞ |
| E | ∞ | 10 | 7 | ∞ | ∞ | 10 | 7 |
| F | ∞ | ∞ | ∞ | 11 | 10 | ∞ | 9 |
| G | ∞ | ∞ | ∞ | ∞ | 7 | 9 | ∞ |

This matrix shows the complete network above.

|   | A | B | C | D | E | F | G |
|---|---|---|---|---|---|---|---|
| A | ∞ | 8 | 10 | 5 | 17 | 16 | 25 |
| B | 8 | ∞ | 17 | 13 | 10 | 20 | 17 |
| C | 10 | 17 | ∞ | 15 | 7 | 17 | 14 |
| D | 5 | 13 | 15 | ∞ | 21 | 11 | 20 |
| E | 17 | 10 | 7 | 21 | ∞ | 10 | 7 |
| F | 16 | 20 | 17 | 11 | 10 | ∞ | 9 |
| G | 25 | 17 | 14 | 20 | 7 | 9 | ∞ |

As we know that this complete network will contain Hamiltonian cycles, we now set about building a tour by looking for a cycle with a low weight. For this we will use the **Nearest neighbour algorithm**. This is another greedy algorithm which selects the shortest available edge at each stage, provided that the edge does not form a cycle (except for the last edge which completes the tour). The algorithm works by visiting the nearest vertex which has not yet been visited, until it returns to the starting point. Applying it to this problem gives the following matrix.

|  | 1 | 7 | 6 | 2 | 5 | 3 | 4 |
|---|---|---|---|---|---|---|---|
|  | A | B | C | D | E | F | G |
| A | ∞ | ⑧ | 10 | 5 | 17 | 16 | 25 |
| B | 8 | ∞ | ⑰ | 13 | 10 | 20 | 17 |
| C | 10 | 17 | ∞ | 15 | ⑦ | 17 | 14 |
| D | ⑤ | 13 | 15 | ∞ | 21 | 11 | 20 |
| E | 17 | 10 | 7 | 21 | ∞ | 10 | ⑦ |
| F | 16 | 20 | 17 | ⑪ | 10 | ∞ | 9 |
| G | 25 | 17 | 14 | 20 | 7 | ⑨ | ∞ |

Start at vertex A. Delete Row A. Choose the smallest entry in column A (AD, 5) and put edge AD in the solution.

Delete Row D. Choose the smallest entry in column D (DF, 11) and put edge DF in the solution.

Delete Row F. Choose the smallest entry in column F (FG, 9) and put edge FG in the solution.

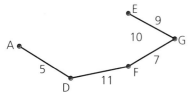

Delete Row G. Choose the smallest entry in column G (GE, 9) and put GE in the solution.

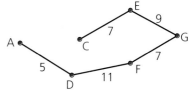

Delete Row E. Choose the smallest entry in column E (EC, 7) and put EC in the solution.

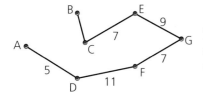

Delete Row C. Choose the smallest entry in column C (the only remaining entry is CB, 17) and put CB in the solution.

We have now connected all the vertices into the solution, so all that remains to do is connect B back to our starting point A, giving the Hamiltonian cycle ADFGECBA, length 64.

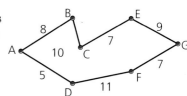

We now need to go back to our original network and interpret this solution in the context of the real problem.

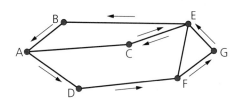

Edges AD, DF, FG, GE, EC and BA were in the network. However, there is no direct edge CB. This was added to make the complete network by taking the path CEB.

Hence our tour is really ADFGECEBA, as shown in the diagram.

This cycle provides us with an upper bound for the optimal length of a tour in the network.

So we can say:

length of optimal tour ≤ 64

The main drawback of the Nearest neighbour algorithm is that it does not consider the weight of the final link until the end. This could prove to be excessively long, although this is not the case in this particular example.

Now we can apply this technique to the problem posed in the second part of Exploration 3.1. First, we transform the network into a complete network, so that we can find a Hamiltonian cycle. This is shown in the matrix below.

|  | Leeds | Harrogate | Wakefield | Bradford | Halifax | Huddersfield |
|---|---|---|---|---|---|---|
| **Leeds** | ∞ | 19 | 13 | 12 | 20 | 24 |
| **Harrogate** | 19 | ∞ | 32 | 20 | 30 | 32 |
| **Wakefield** | 13 | 32 | ∞ | 15 | 22 | 14 |
| **Bradford** | 12 | 20 | 15 | ∞ | 10 | 12 |
| **Halifax** | 20 | 30 | 22 | 10 | ∞ | 8 |
| **Huddersfield** | 24 | 32 | 14 | 12 | 8 | ∞ |

Using the Nearest neighbour algorithm we can find a tour starting and ending at Leeds.

|  | 1 L | 6 H | 5 W | 2 B | 3 Ha | 4 Hu |
|---|---|---|---|---|---|---|
| **L** | ∞ | (19) | 13 | 12 | 20 | 24 |
| **H** | 19 | ∞ | (32) | 20 | 30 | 32 |
| **W** | 13 | 32 | ∞ | 15 | 22 | (14) |
| **B** | (12) | 20 | 15 | ∞ | 10 | 12 |
| **Ha** | 20 | 30 | 22 | (10) | ∞ | 8 |
| **Hu** | 24 | 32 | 14 | 12 | (8) | ∞ |

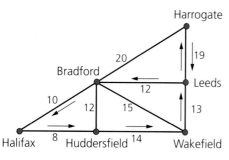

L, B, Ha, Hu, W, H, L:
length = 12 + 10 + 8 + 14 + 32 + 19 = 95 miles

As before, we must now go back to the original problem and interpret our solution in this context.

The tour is:

Leeds, Bradford, Halifax, Huddersfield, Wakefield

Leeds, Harrogate, Leeds

since the edge WH in the complete network goes via Leeds.

## Lower bounds

*The best lower bound is the one closest to the optimal tour – that is the greatest lower bound.*

Having established a tour which provides an upper bound for the solution to the problem, we shall now look at a method for determining a lower bound. We could simply take the minimum connector as being a lower bound, but it would not be a very good one.

A common technique involves:

- deleting a vertex *and* the edges incident on it from the network

- finding a minimum connector for the remaining network

- then reconnecting the deleted vertex by the two shortest edges.

This is acceptable provided the remaining network is connected, but consider the following network.

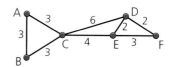

If we delete vertex C, we leave a disconnected network for which no minimum connector can be found.

Once again we return to the classical problem of finding Hamiltonian cycles. If we replace our network by a complete network, the deletion of C does not create a disconnected network and we may proceed with the method.

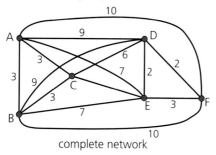

complete network

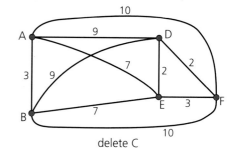

delete C

## Example 3.2

*Use the matrix for the complete network to find a lower bound for the inspector in Exploration 3.1.*

**Solution**

*First, delete vertex L, and find the minimum connector for the remaining network.*

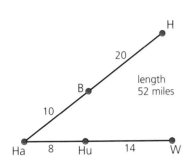

| | 1 | 5 | 2 | 3 | 4 |
|---|---|---|---|---|---|
| | H | W | B | Ha | Hu |
| H | ∞ | 32 | 20 | 30 | 32 |
| W | 32 | ∞ | 15 | 22 | 14 |
| B | 20 | 15 | ∞ | 10 | 12 |
| Ha | 30 | 22 | 10 | ∞ | 8 |
| Hu | 32 | 14 | 12 | 8 | ∞ |

*This gives a lower bound for a TSP on the part of the network not containing Leeds.*

*However, any solution would have to involve travelling out of Leeds and back again, so we take the two shortest edges leading to Leeds and assume that we travel out of town on one of them and back into town on the other. The two shortest edges are LB (12) and LW (13), which provides a lower bound for a TSP on the part of the network containing only Leeds.*

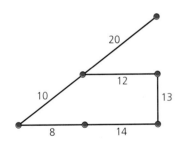

*By adding these two together, we obtain a lower bound for the whole network.*

*Length 25 + 52 = 77 miles*

*We know that this is a lower bound since it is made up of two lower bounds for subdivisions of the network. We can see that it is a better lower bound than simply using the minimum spanning tree for the network, but is it the best we can do? To answer this question we would need to delete every vertex in turn and compare the results. This is shown in the table below.*

| Vertex deleted | Lower bound for deleted vertex | Lower bound for remaining network | Lower bound for TSP |
|---|---|---|---|
| Leeds | 25 | 52 | 77 |
| Bradford | 22 | 54 | 76 |
| Harrogate | 39 | 43 | 82 |
| Wakefield | 27 | 49 | 76 |
| Halifax | 18 | 56 | 74 |
| Huddersfield | 20 | 54 | 74 |

*From the table we can see that the maximum lower bound is 82 miles and is obtained by deleting Harrogate. Thus we can say:*

82 ≤ length of route ≤ 95

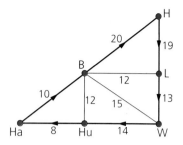

Now we can find a route which will lie within this range. One such is Leeds, Wakefield, Huddersfield, Halifax, Bradford, Harrogate, Leeds with length 84 miles.

Taking into account the practical considerations, this method provides a solution which is good enough. In the inspector's situation, in Exploration 3.1, there would be many other factors which might influence the optimal route on any given day, such as length of journey time, traffic congestion, roadworks and so on, which cannot be accounted for in a mathematical model of this kind. The approximate solution is quite adequate for most TSP problems.

## EXERCISES

**3.3 A**

**1**  A delivery vehicle leaves the warehouse at A to supply goods to the shops marked B, C, D, E and F on the network, before returning to the warehouse.

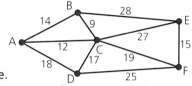

**a)** By transforming the network into a complete network, find an upper bound for the distance the vehicle will have to travel.
**b)** Find the best lower bound by deleting each vertex in turn.

**2**  In Exercises 3.2A, you solved the minimum connector problem for these cities.

|  | Aberdeen | Carlisle | Edinburgh | Glasgow | Inverness |
|---|---|---|---|---|---|
| **Aberdeen** | ∞ | 221 | 125 | 145 | 105 |
| **Carlisle** | 221 | ∞ | 100 | 95 | 263 |
| **Edinburgh** | 125 | 100 | ∞ | 40 | 160 |
| **Glasgow** | 145 | 96 | 40 | ∞ | 168 |
| **Inverness** | 105 | 263 | 160 | 168 | ∞ |

A buyer, based in Glasgow, needs to visit branches of a chainstore in each of these cities. Obtain an estimate for the upper bound and one for the lower bound by deleting Aberdeen.

**3**  Marco sells ice cream each day during the summer season, at the places shown in this network.

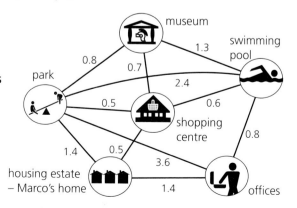

a) Find the shortest route that takes him to each place, starting and finishing at his home, and not visiting any place twice.

b) During the summer holidays, he sells more ice creams at the park than anywhere else. Design a route that enables him to visit the park twice each day.

4   The Island of Guernsey is a popular holiday destination. From the map, draw a network showing the minimum distances between places of interest. Plan a tour of minimum length which visits all the tourist attractions and starts and ends in St Peter Port.

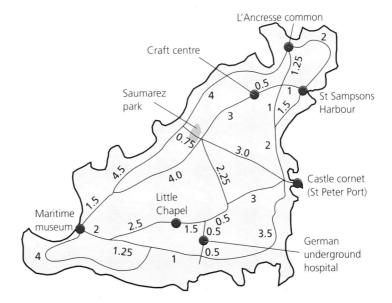

5   In a biscuit factory four types of cookie are made each day by one machine. After each run of one type, the machine must be cleaned, the cleaning time being dependent on the types of cookie being made. The product manageress wishes to find a manufacturing sequence, starting with butter cookies and returning to butter cookies, that will enable each type of cookie to be made whilst minimising the cleaning times.

to be made

|          |           | butter | ginger | chocolate | almond |
|----------|-----------|--------|--------|-----------|--------|
|          | butter    | —      | 15     | 18        | 16     |
| last made| ginger    | 25     | —      | 21        | 23     |
|          | chocolate | 22     | 20     | —         | 24     |
|          | almond    | 18     | 12     | 16        | —      |

a) Explain how this problem can be modelled as a TSP.

b) Use the Nearest neighbour algorithm to establish an upper bound for the cleaning time in one complete cycle.

**EXERCISES**

3.3 B

1   By redrawing this network as a complete network, find a Hamiltonian cycle that will provide an upper bound for a tour starting and ending at vertex A.

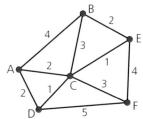

2   Five computer programs have to be run in sequence. Each program needs its own resources, for instance main memory, drives etc. and changing from one set to another uses up valuable time. The table below gives the times for converting resources between programs. Find an ordering of the programs which if run in this order should result in a comparatively small total conversion time.

Program conversion time

|   | a | b | c | d | e |
|---|---|---|---|---|---|
| a | – | 80 | 10 | 30 | 20 |
| b | 80 | – | 60 | 20 | 20 |
| c | 10 | 60 | – | 40 | 40 |
| d | 30 | 20 | 40 | – | 70 |
| e | 20 | 20 | 40 | 70 | – |

3   In question 3 of Exercises 3.2B you solved the minimum connector problems for the augmented Northern centres. A buyer, based in Glasgow, needs to visit branches of a chain store in each of the centres. Obtain an estimate for the upper bound and one for the lower bound by discounting Perth.

4   In question 2 of Exercises 3.3A the buyer must travel directly between Carlisle and Edinburgh. Describe briefly the method of obtaining a set of lower bounds and obtain three of these.

5   The distances between towns in a particular country are given in the graph below. Because a river flows through the country, there is just one road connecting two towns on either side of the river.

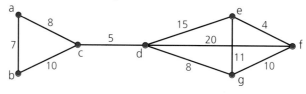

How does this simplify your task of obtaining a good lower bound for the TSP for these towns? Use this fact to obtain two lower bound estimates.

## COURSEWORK INVESTIGATIONS

### Investigation 1 — Cycles

■ The table below shows the number of vertices ($n$), edges ($e$) and 3-cycles ($c_n$) in the complete graphs $K_1$, $K_2$, $K_3$ and $K_4$.

| $n$ | 1 | 2 | 3 | 4 |
|---|---|---|---|---|
| $e_n$ | 0 | 1 | 3 | 6 |
| $c_n$ | 0 | 0 | 1 | 4 |

Extend this table to include $K_5$ and $K_6$.

■ How many 3-cycles has $K_n$? Can you complete the row for 4-cycles?

■ A graph $G_n$ is obtained from the complete graph $K_n$ by deleting just one 3-cycle, i.e. deleting the three edges which make up the 3-cycle.

Draw examples of $G_4$ and $G_5$. Explain why the choice of 3-cycle from $K_n$ would not affect the resulting graph $G_n$ if the vertices were unlabelled.

■ If $e_n^*$ and $c_n^*$ denote the number of edges and 3-cycles respectively in $G_n$ complete this table.

| $n$ | 3 | 4 | 5 | 6 | 7 |
|---|---|---|---|---|---|
| $e_n^*$ | 0 | 3 | 7 | | |
| $c_n^*$ | 0 | 0 | | | |

■ Obtain a formula for $e_n^*$ in terms of $n$.

Determine and explain a relationship between $e_n^*$ and $c_n^*$. Obtain a formula for $c_n^*$ in terms of $n$ and $c_n$.

## COURSEWORK INVESTIGATIONS

### Investigation 2 — Tour of the building

Plan a tour for a visitor to your school, ensuring that they visit all the communal areas such as the sports facilities, and at least some of the subject areas, as well as a stop for coffee in the staff room. Suppose the visitor is in a wheelchair. How would you amend the tour? Would it be necessary to adapt the algorithms to take account of ramps and lifts?

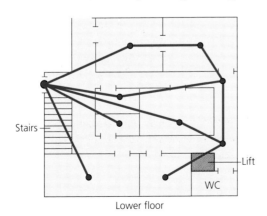

Stairs

Lift

WC

Lower floor

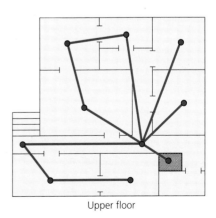

Upper floor

## CONSOLIDATION EXERCISES FOR CHAPTER 3

**1**   The figure shows some places in the south of England. The numbers on the arcs are the distances, in miles, between them. A cable TV company based in Salisbury wishes to link all the places using a minimum length of cable.

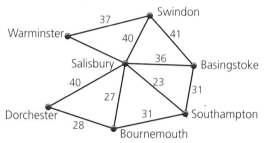

Use the greedy algorithm (Kruskal's algorithm) to find this minimum length of cable. In your solution you should indicate clearly the order in which the places are added to the minimum spanning tree.

*(ULEAC Question 2, Specimen Paper, 1996)*

**2**   Five new houses labelled A, B, C, D and E on the diagram are to be connected to a drainage system. Each house is to be connected to the sewer (point S on the diagram), either directly or via another house. Alternatively houses may be connected to an intermediate manhole at M, directly or via another house. This manhole must in turn be connected to S, either directly or via another house.

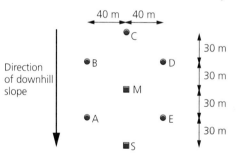

All connecting pipes must be such that water can drain downhill. The direction of the downhill slope is shown on the diagram.

**a)**   Copy and complete the matrix at the top of page 69, showing the lengths of all 16 possible connecting pipes.

**b)**   Starting from S, use Prim's algorithm to find a minimum connector for A, B, C, D, E, M and S. Show which pipes are in your connector, indicating the order in which they were included, and give their total length.

Explain whether or not your connector represents a system which drains correctly.

*(UODLE Question 9, Paper 4, 1993)*

| | To | | | | | | |
|---|---|---|---|---|---|---|---|
| | **A** | **B** | **C** | **D** | **E** | **M** | **S** |
| **A** | – | – | – | – | – | – | 50 |
| **B** | 60 | – | – | – | – | 50 | |
| **C** | | | | | | | |
| **D** | | | | | | | |
| **E** | | | | | | | |
| **M** | | | | | | | |
| **S** | | | | | | | |

From

*(UODLE Question 9, Paper 4, 1993)*

**3** **a)** The five points on this graph may be connected in pairs by straight lines. Use a greedy algorithm to find a connector of minimum length for the five points. Show your answer on a copy of the diagram, explain your method, and give the length of your minimum connector.

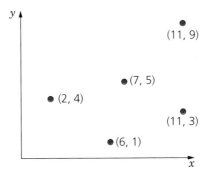

**b)** The points on this graph may also be connected in pairs by straight lines.

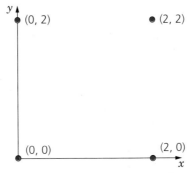

**i)** Find a minimum connector for the four points.
**ii)** Find a fifth point so that the minimum connector for the five points is shorter than the minimum connector for the original four points. Show your minimum connector and give its length.

*(UODLE Question 7, Paper 4, 1996)*

69

**4** The following matrix gives the costs of flight tickets, in £, for direct flights between six connected cities.

**To**

|  | | A | B | C | D | E | F |
|---|---|---|---|---|---|---|---|
| | A | — | 45 | 60 | 58 | 90 | 145 |
| | B | 45 | — | 67 | 25 | 83 | 100 |
| **From** | C | 60 | 70 | — | 50 | 70 | 320 |
| | D | 50 | 25 | 50 | — | 35 | 210 |
| | E | 100 | 80 | 70 | 35 | — | 72 |
| | F | 145 | 110 | 300 | 175 | 80 | — |

a) A tour is a journey from a city, visiting each other city once and only once, and returning to the starting city. Use a greedy algorithm to find a tour, starting and finishing at A, with a low associated cost. Show that the algorithm has not produced the minimum cost tour.

b) How many different tours which start and finish at A are there altogether?

c) Suppose that, in addition to the cost of tickets, airport taxes must be paid on leaving an airport according to the following table.

| A | B | C | D | E | F |
|---|---|---|---|---|---|
| 20 | 30 | 20 | 40 | 10 | 20 |

Thus the flight from A to B will cost £45 for the ticket and £20 tax, a total of £65. Complete the following matrix of costs, i.e. the total of fares and taxes.

|  | | A | B | C | D | E | F |
|---|---|---|---|---|---|---|---|
| | A | — | 65 | | | | |
| | B | | | | | | |
| | C | | | | | | |
| | D | | | | | | |
| | E | | | | | | |
| | F | | | | | | |

d) Give an example where the cheapest route from one city to another differs when taxes are taken into account from that when taxes are not taken into account. Show your two routes and give their costs.

e) Do airport taxes have an effect on the problem of finding the cheapest tour, starting and finishing at A? Why?

*(O & C MEI Question 1, Decision & Discrete Mathematics Paper 19, June 1995)*

**5** The following matrix represents the distances (in miles) between the six nodes of a network.

|   | A | B | C | D | E | F |
|---|---|---|---|---|---|---|
| **A** | 5 | 11 | 21 | 16 | 9 | 10 |
| **B** | 11 | 0.5 | 10 | 31 | 7 | 23 |
| **C** | 21 | 10 | 57 | 14 | 12 | 20 |
| **D** | 16 | 31 | 14 | 11 | 11 | 9 |
| **E** | 9 | 7 | 12 | 11 | 18 | 12 |
| **F** | 10 | 23 | 20 | 9 | 12 | 43 |

a) Delete vertex A from the network and use Prim's algorithm to find the minimum connector for the remaining five vertices – start from B and indicate the order in which you include vertices. State which arcs are in your minimum connector and give its total length.

b) By considering the connections between A and the other five vertices, construct a lower bound for the optimal solution of the travelling salesperson problem for the network.

c) Explain why your answer to b) is a lower bound for the length of the optimal solution of the travelling salesperson problem for the network.

d) Use the greedy algorithm (that is, go to the nearest vertex not yet visited) to find a solution to the travelling salesperson problem which starts and finishes at A. Calculate its length and compare it with the lower bound which you found in b).

*(UODLE Question 10, Paper 4, 1994)*

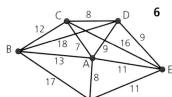

**6** This network represents the major routes joining six towns. The distances beside the arcs are the distances in kilometres between the towns.

The council wish to keep the towns connected during cold weather. To this end they wish to determine a minimum connector tree (minimum spanning tree) in order to decide which roads to salt and grit. Use a greedy algorithm to determine a suitable tree for them. Begin building your tree from town A. You should state the algorithm that you use carefully, especially the criteria that you use to include arcs as you build your tree. What is the minimum length of road that the council need to salt and grit?

In summer a council employee must leave town A, visit every other town, checking the chlorination of swimming pools there, and return to town A. Find an initial upper bound on the minimum distance that he must travel and improve on this.

By deleting town A from the network, also determine a lower bound on the minimum distance.

*(UODLE Question 20, Paper 4, 1989)*

**7** The map below shows six youth hostels, A, B, C, D, E and F, in an area of rough country, together with paths connecting them. The numbers by the paths show lengths, in miles, of sections of path.

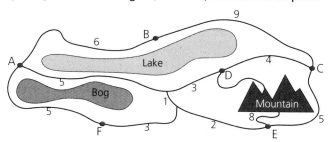

**a)** Draw a planar network showing the shortest direct distances between youth hostels. For example, the shortest direct distance between E and F is 5 miles. However, between B and D there is no direct link. (You may do this by inspection. The application of an algorithm is not required.)

**b)** Use Prim's algorithm to find a minimum spanning tree for your network. Describe each step in your use of the algorithm, and draw your minimum spanning tree.

**c)** A walker wishes to spend a night at each hostel during her stay in the district, starting and ending at hostel A. Construct an upper bound for the distance which she must walk.

**d)** If she requires to traverse the mountain pass between D and E, suggest how you might modify your answer to part **c)** in order to construct an upper bound for the distance she must cover.

*(UODLE Question 7, Paper 4, 1990)*

**8** A manufacturing process consists of five stages, which can be done in any order, before returning to the start again. The costs at each stage are dependent on which stage of the process has been completed immediately before the current one and are shown in the matrix below.

Stage completed

|  |  | **A** | **B** | **C** | **D** | **E** | in thousands |
|---|---|---|---|---|---|---|---|
|  | **A** | – | 5 | 6 | 3 | 8 | of pounds |
| Stage to | **B** | 8 | – | 7 | 4 | 4 | |
| be done | **C** | 3 | 7 | – | 8 | 6 | |
| next | **D** | 4 | 5 | 6 | – | 3 | |
|  | **E** | 5 | 6 | 4 | 5 | – | |

**a)** Assuming that the process starts each time with stage A, find the order which will incur the minimum cost.

**b)** If stage C must be done directly after stage B, adapt the method to find a new least cost plan.

**9**   The company accountant for the Quasar group of computer stores is planning an inspection. The locations are shown on this network.

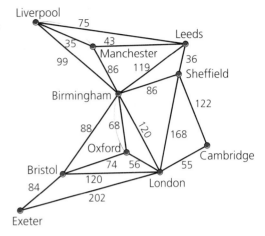

a) Plan a tour for him, starting and finishing at London, which is as short as possible.

b) If he spends 3 hours at each store and allows 1 hour for each 40 miles he travels, how many hours will the trip take?

c) If he can only work at the stores from 8.30 a.m. until 6 p.m., and does not wish to drive between 9 p.m. and 7 a.m., how many days will he need to be away and where should he book overnight hotel accommodation?

**10**   Plan a tour of minimum length around places of interest in your area.

## Summary

*After working through this chapter you should be familiar with:*

■ a cycle, which is a loop which returns to its starting point

■ a spanning tree, which connects all the vertices in a graph with no cycles

■ the minimum connector, which is the spanning tree of minimum length

■ the travelling salesman problem, which attempts to find a cycle of minimum length which visits all the vertices in a network

■ Hamiltonian and Eulerian cycles in a simple graph, and be able to use them to solve the minimum spanning tree, and to solve the minimum connector using Kruskal's and Prim's algorithms

■ finding upper and lower bounds for the travelling salesman problem and appreciate that this method will only provide an approximate solution.

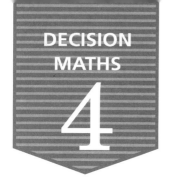
# Networks 3: Route inspection problems

*In this chapter we shall:*

■ *discuss planar graphs, which are graphs that can be drawn in the plane with no edges crossing, and their special properties*

■ *discuss isomorphic graphs, which are graphs that have the same number of vertices and the degrees of corresponding pairs of vertices are the same*

■ *investigate the degree of a vertex, which is the number of edges incident on the vertex*

■ *consider route inspection problems and find the shortest route around a network, travelling along every edge at least once and ending where you started.*

**Exploration 4.1**

### Postal routes

A postman leaves the depot and has to deliver along all the streets in the area shown. What is the shortest distance he must travel?

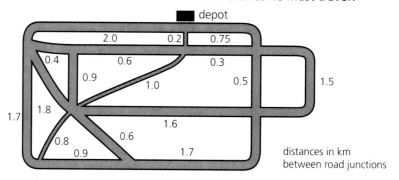

distances in km
between road junctions

## GRAPH THEORY

### Planar graphs

A graph that can be drawn in the plane with no edges crossing is called a **planar graph**.

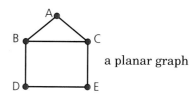

a planar graph

It is not always immediately obvious if a graph is planar, so it may be necessary to redraw it to check. Is this graph planar?

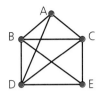

Yes, since it can be drawn with no edges crossing by taking edges AD and CD outside.

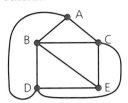

## Euler's formula

A planar graph divides the plane into regions which are called **faces**. If we reconsider the graph above we can see that this graph has six faces, one of which, $f_6$, has no boundary and is called the **infinite** face.

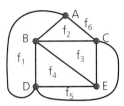

Euler found a simple relationship between the number of vertices ($v$), edges ($e$) and faces ($f$) in a connected planar graph.

$$v - e + f = 2$$

Euler originally developed this formula to apply to the **platonic** solids.

For the graph above this gives:

$$v = 5, e = 9, f = 6 \quad \Rightarrow \quad v - e + f = 5 - 9 + 6 = 2$$

To test whether or not a graph is planar we can use the fact that, if it is, it must satisfy Euler's formula.

**Example 4.1**

*Is the graph $K_5$ planar?*

### Solution
*If $K_5$ is planar it must satisfy Euler's formula: $v - e + f = 2$.*

$$v = 5, e = 10 \quad \Rightarrow \quad 5 - 10 + f = 2 \Rightarrow f = 7$$

*So in order to satisfy Euler's formula, the graph must have seven faces. Now each face must be bounded by at least three edges, and since each edge forms a boundary to two faces we can write:*

$$2e \geq 3f$$

**Note:** *Even the infinite face is bounded by at least three edges.*

*since e = 10, then* $20 \geq 3f$    $\Rightarrow$    $\dfrac{20}{3} \geq f$

*However, this contradicts the fact that f = 7 for the graph to be planar and hence $K_5$ must be non-planar.*

## Degree of a vertex

The **degree** of a vertex is the number of edges incident at that vertex. Consider this graph again.

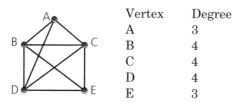

| Vertex | Degree |
|--------|--------|
| A | 3 |
| B | 4 |
| C | 4 |
| D | 4 |
| E | 3 |

If you find the sum of the degrees of the vertices in the graph (18) and compare it with the number of edges in the network, what do you notice?

The sum of the degrees of the vertices for this graph is eighteen and it has nine edges. Draw a few graphs of your own and confirm that the sum of degrees of the vertices is always twice the number of edges in the graph.

More formally , this is known as the **Handshaking theorem** and can be written:

$\Sigma$ *is the Greek letter sigma. It is used in mathematics to indicate 'the sum of'.*

$$\Sigma \deg v = 2e$$

Can you explain why this is the case?

Since the sum of the degrees of the vertices is always even, it follows that there will always be an even number of odd vertices.

## Isomorphic graphs

Two graphs are **isomorphic** if they have the same number of vertices and if the degrees of corresponding pairs of vertices are the same. As an example compare the following two graphs.

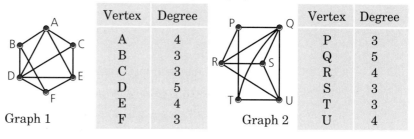

| Vertex | Degree |
|--------|--------|
| A | 4 |
| B | 3 |
| C | 3 |
| D | 5 |
| E | 4 |
| F | 3 |

Graph 1

| Vertex | Degree |
|--------|--------|
| P | 3 |
| Q | 5 |
| R | 4 |
| S | 3 |
| T | 3 |
| U | 4 |

Graph 2

These graphs are isomorphic. Both have six vertices and there is a vertex correspondence A – R, B – P, C – S, D – Q, E – U, F – T.

Do you remember the exploration at the beginning of Chapter 1?

## Exploration 4.2

### *Konigsberg revisited*

The town of Konigsberg in East Prussia was built on the banks of the River Pregel, with islands which were linked to each other and the river banks by seven bridges. The citizens of the town tried for many years to find a route for a walk which would cross each bridge only once and allow them to end their walk where they had started. Can you find a suitable route?

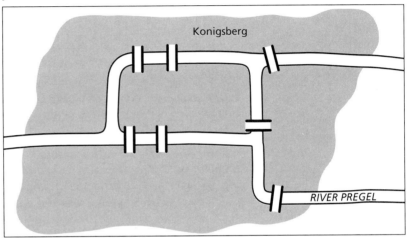

He published his proof in a paper entitled *Solution Problematis ad geometriam situs a pertinentis* (The solution of problem relating to the geometry of position) in 1736.

It is said that the inhabitants of Konigsberg tried unsuccessfully to find a route for many years, but it was not until Euler tackled the problem that it was proved to be impossible. Although his proof was not written in the language of modern graph theory, it contains many of the ideas that have formed the basis of the subject and can reasonably be described as the first paper on the subject.

We can now represent the town as a graph where the land areas are the vertices and the seven bridges are the edges.

The problem becomes that of finding an Euler cycle in this graph. We met the idea of an Euler cycle in Chapter 3, *Networks 2: Minimum-connector and the travelling salesman problem.* We need a path which passes along every edge of the graph exactly once and returns to its starting point. Euler not only showed that it is impossible to find such a path in Konigsberg, but extended his ideas to any graph and hence discovered a rule for finding when a graph contains a cycle which passes along every edge exactly once.

## Exploration 4.3

### *Eulerian graphs*

Consider the following graphs: are they Eulerian?
Can you suggest a method by which you might identify a graph which is Eulerian?

a)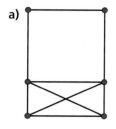
b)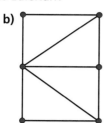
c)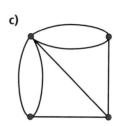
d)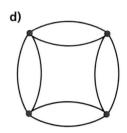

## Solving the Konigsberg bridge problem

Euler discovered that for a connected graph to contain a cycle of this type, the degrees of all the vertices must be even numbers. The third graph (c) in Exploration 4.2 is isomorphic to the one in the Konigsberg problem and is therefore a representation of the Konigsberg bridges. We can see that this is not an Eulerian graph since all the vertices are odd. Thus the problem posed by the residents of Konigsberg has no solution, unless you repeat one or more of the bridges.

## EXERCISES

**4.1 A**

a)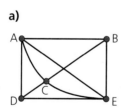
b)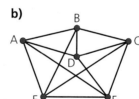
c)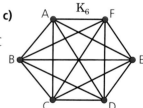

1   Identify which of the graphs in the figure above are planar, redrawing them where necessary. Check your results using Euler's formula.

2   For all the graphs in the figure above, state the degree of every vertex and confirm that the Handshaking theorem holds.

3   State which of the graphs in question 1 are Eulerian, giving an Euler cycle where possible.

4   Identify which of the following graphs are isomorphic.

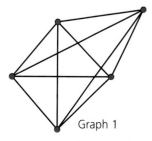
Graph 1

Graph 2

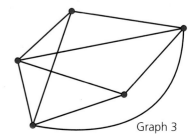
Graph 3

**5**   A traversable graph is one which you can draw without removing your pen from the paper and without going over the same edge twice.

This graph is traversable.          This graph is not traversable.

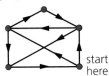   

Are all traversable graphs also Eulerian? Explain your answer.

## EXERCISES

**4.1B**

**1**   Identify which of the following are planar graphs.

a)                                              b)

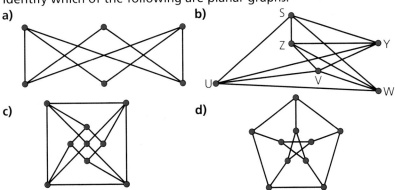

c)                              d)

For those which can be redrawn with no edges crossing, check your result using Euler's formula.

For all the graphs above, determine the degree of each vertex and confirm that the Handshaking theorem holds. For each of them that is not planar, draw a graph that is isomorphic to it.

**2**   State which of the graphs in question 1 are Eulerian. For each that is, construct an Euler cycle.

**Note:** *Edge-disjoint cycles are cycles, in the same graph, that have no edges in common.*

**3**   Identify the edge-disjoint cycles in this Euler graph and use them to construct an Euler cycle.

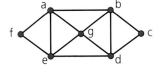

**4**   Use the two results $v - e + f = 2$ and $2e \geq 3f$ to deduce that $e \leq 3v - 6$. Hence confirm that $K_5$ is non-planar.

**5**   Consider this graph.

Show that the Handshaking theorem holds for the graph.
Draw a graph which has the same degree-sequence as G which
**a)** is isomorphic to G       **b)** is not isomorphic to G.

## ROUTE INSPECTION PROBLEMS

### The 'Chinese postman'

The problem in Exploration 3.1 was to find a route through the network that traverses every edge exactly once and returns to its starting point. This type of problem, a route inspection problem, is often called the 'Chinese postman problem' because it was first published by the Chinese mathematician Mei-Ko Kwan in 1962.

If the network contains an Euler cycle then there is no problem. The optimum length is simply the sum of the weights on all the edges. If the network does not contain an Euler cycle, then the postman will have to repeat some of the roads and the problem is then which should he repeat, to keep his journey to a minimum. To solve the problem we first need to draw a network to represent the area that the postman needs to cover.

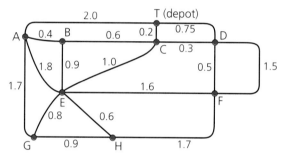

The roads are the edges of the network and the intersections are represented by vertices.

The next step is to list the degrees of all the vertices.

    A 4    B 3    C 4    D 4    E 6    F 4    G 3    H 3    T 3

We can see that there is no Euler cycle since the network contains four odd vertices: B, G, H and T, so we shall need to repeat some of the edges. The next question is which do we repeat?

Consider a vertex which has three edges joining it. If we move to vertex B along AB and leave along BC, we will need to re-enter B along BC and there is no way out, other than by repeating an edge, so a vertex of odd degree will need to have at least one edge at that vertex repeated. We choose to repeat the edges joining odd vertices which will add the minimum distance to the solution. The best way to do this is to consider the distances between all possible pairings of odd vertices like this.

| Pairings of odd vertices | Distance | Route |
|---|---|---|
| TB and GH | 0.8 + 0.9 = 1.7 | (TCB + GH) |
| TG and BH | 2 + 1.5 = 3.5 | (TCEG + BEH) |
| TH and BG | 1.8 + 1.7 = 3.5 | (TCEH + BEG) |

We choose to repeat the edges which add the smallest distance, in this case TB and GH. The length of the shortest route is then simply the sum of all the edge lengths plus the extra length due to the repeated edges.

17.25 + 1.7 = 18.95

The distance the postman will have to travel is 18.95 km and a possible route would be:

T A B E A G̲H̲ E G̲H̲ F D F E C D T̲C̲ B C̲T̲

This is shown in the diagram.

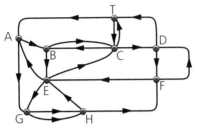

**Note:** *There may be several possible routes of this length.*

The algorithm can be stated like this:

**Step 1:** Identify all the odd vertices in the network.

**Step 2:** Consider all the routes joining pairs of odd vertices and choose the routes of shortest distance.

**Step 3:** Find the sum of the weights on all the edges in the network and add the distance found in step 2.

**Step 4:** Find a possible route round the network which repeats the edges identified in step 2 and is of the correct distance.

## EXERCISES

**4.2 A**

**1** During the school holidays, a group of children decide to plan a route which goes along all the paths in the local park, starting and ending at the main gate. Adrian says it is possible to do this without repeating any of the paths, but Mallika disagrees with him. Who is right? Devise a route they might take, saying how many paths, if any, they need to repeat.

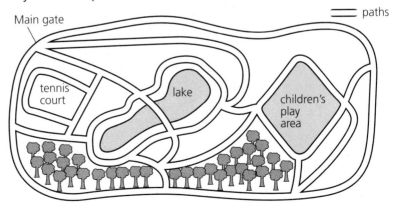

**2**  Mr and Mrs McCreedy do their shopping at Supasavers every week. The layout of the shop is shown in the plan below, the shaded areas are display shelves.
From the plan, draw a graph to represent Supasavers and plan a route which enables them to look at the goods on all the display shelves while covering the minimum distance.

entrance

**3**  Solve the route inspection problem for this network, starting and ending at S, stating clearly which edges you would need to repeat.

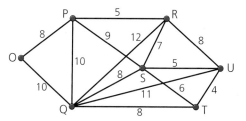

**4**  The morning after a severe gale, the Cambria highways department set off from their depot to inspect all the roads to make sure none of the towns are cut off by fallen trees. Find a route for them which is as short as possible.

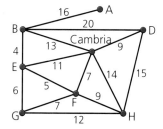

**5**  The distances between eight villages are given in the table below. Use the table to devise a route of minimum length, starting and ending at A, which travels along each route at least once.

|   | A | B | C | D | E | F | G |
|---|---|---|---|---|---|---|---|
| **A** | – | 4 | 7 | 8 | – | – | – |
| **B** | 4 | – | 5 | – | 4 | 5 | – |
| **C** | 7 | 5 | – | 9 | – | – | – |
| **D** | 8 | – | 9 | – | – | – | 12 |
| **E** | – | 4 | – | – | – | 7 | 8 |
| **F** | – | 5 | – | – | 7 | – | 6 |
| **G** | – | – | – | 12 | 8 | 6 | – |

If the road between E and F is closed due to road works, how will this affect your method? What is the new solution?

## EXERCISES

4.2 B

1   The ground floor plan of a house is given below.

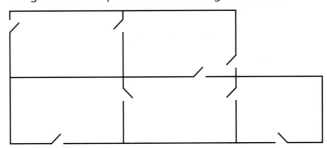

Draw a graph to represent the above. Discuss whether it is possible to walk in and out of the house so that each door of the house is used exactly once.
(**Hint:** Consider the cases whether you are going to return to the starting position or not.)

2   You are given a domino set with the doubles removed. We can represent the pieces as (0, 1), (0, 2), ... , (0, 6), (1, 2), ... , (1, 6), ... , (5, 6). Discuss the possibility of arranging the pieces in a connected series such that one number on a piece (as illustrated here) always touches the same number on its neighbour).

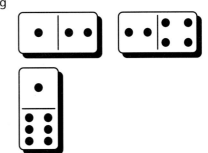

(**Hint:** Use a seven-vertex complete graph and check if it is Eulerian.) Would inclusion of the doubles alter your conclusions?

3   The distances between villages in the Welsh valleys are given in the table below. Devise a route of minimum length, starting and ending at Dowlais, which visits every village. Determine this minimum length if all distances are quoted in miles.

|            | Dowlais | Treharris | Aberdare | Abercynon | Nelson | Bargoed |
|------------|---------|-----------|----------|-----------|--------|---------|
| Dowlais    | –       | 10        | 6        | –         | –      | 8       |
| Treharris  | 10      | –         | –        | 4         | 2      | –       |
| Aberdare   | 6       | –         | –        | 7         | –      | –       |
| Abercynon  | –       | 4         | 7        | –         | 3      | –       |
| Nelson     | –       | 2         | –        | 3         | –      | 8       |
| Bargoed    | 8       | –         | –        | –         | 8      | –       |

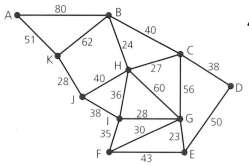

4   A gas pipeline has to be laid alongside some of the existing roads in a region in order to link all the towns in the map. At each town there is a pumping station which ensures the gas can be pumped to or from adjacent towns.

a)  Before the project can go ahead the entire length of the road network has to be inspected. Determine the minimum distance for the inspection route.

b)  The connectivity needs to be increased and one extra pipeline can be built between any pair of towns. The pipeline can go across country but it must not cross an existing pipeline. Draw in the new pipeline and give an estimate of the distance.

5   A clothing company, MH limited, is to make sweaters that will include their logo which is in the form of a network incorporating the company's initials. A diagram of the rectangular shaped logo is shown here, where the numbers indicate the lengths, in millimetres, between vertices.

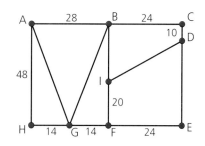

a)  Machinists start and end at vertex A, stitching without a break in the pattern. Why is it important to produce the logo with the minimum of stitching? Determine the order of stitching and state the minimum length.

b)  What will the effect be if the top and bottom edges AB, BC, HG, GF and FE are removed from the logo design?

## COURSEWORK INVESTIGATION

**Investigation**       *Museum visit*

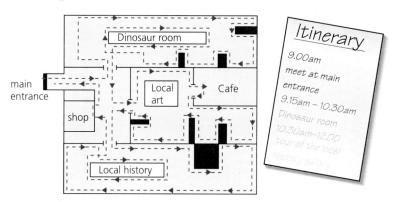

You are organising a trip to a local museum with a party of children. You want them to see all the exhibits, but do not want them to have to walk further than necessary. By modelling the museum as a network, plan a suitable route.

By considering sensible timings, draw up an itinerary for the visit.

## CONSOLIDATION EXERCISES FOR CHAPTER 4

**1** A connected graph G has six vertices and nine edges.

a) State the sum of the degrees of the vertices of G.
b) If G has two vertices with order four, draw G and give the degrees of the other vertices.
c) State, with reasons, whether G is (i) planar (ii) Eulerian.

**2** A simple graph, G, has five vertices, and each of those vertices has the same degree, $d$.

a) State the possible values of $d$.
b) If G is connected, what are the possible values of $d$?
c) If it is Eulerian, what are the possible values of $d$?
d) If G is connected and planar, what is the only possible value of $d$? Give an illustration of such a graph.

*(AEB Question 3, Discrete Mathematics Specimen Paper Pure 4, 1996)*

**3** How many continuous pen strokes are needed to draw the graphs in the diagrams below?

a)
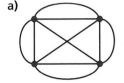

b)
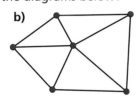

If a graph has $k$ vertices of odd degree, what is the smallest number of continuous pen strokes needed to draw the graph?

**4** Solve the route inspection problem for this network.

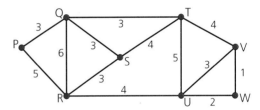

**5** Sarah and Martin decide to spend a few days on a touring holiday in the region shown in this network. They wish to travel along all routes shown, using the local buses. The bus fares between towns are shown in the table. Assuming that they wish to start and finish at S, what route should they take in order to minimise the cost?

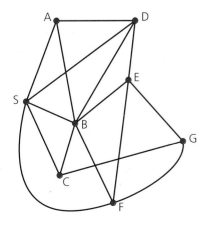

|  | S | A | B | C | D | E | F | G |
|---|---|---|---|---|---|---|---|---|
| **S** | – | 83 | 95 | 100 | 150 | – | 195 | – |
| **A** | 83 | – | 75 | – | 88 | – | – | – |
| **B** | 95 | 75 | – | 60 | 115 | 90 | 120 | – |
| **C** | 100 | – | 60 | – | – | – | – | 135 |
| **D** | 150 | 88 | 115 | – | – | 73 | – | – |
| **E** | – | – | 90 | – | 73 | – | 110 | 65 |
| **F** | 195 | – | 120 | – | – | 110 | – | 76 |
| **G** | – | – | – | 135 | – | 68 | 76 | – |

**6**  The clothing company, New Xperience, has decided to break into the sportswear market. It has been decided to decorate all garments with a simple logo, in the form of a network incorporating the company's initials. A diagram of the logo is shown below. Vertices have been labelled A, B, C, D, E, F, G for convenience. The numbers indicate distances in cm between vertices.

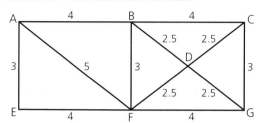

**a)** Plans involve the production of millions of garments, so it is important to produce the logo with the minimum of stitching. Machinists will start and end stitching at A, with no breaks. Explain why it will be necessary to oversew some edges. Use an appropriate algorithm to determine the order of stitching which minimises the length sewn. Explain each step carefully. Give your minimum length, and a corresponding order in which vertices should be stitched.

**b)** Amongst the items produced will be an ice-skating outfit. It is decided that, for this outfit, the standard logo will be decorated by sewing on a sequin at each of the seven vertices. The stitching for this additional feature is to be done without breaks, starting and ending at A.
Construct a minimum spanning tree for the network, explaining your criteria for including each of the arcs. Use it to calculate an upper bound for the length a machinist must sew in order to attach the sequins. Indicate how to reduce the upper bound by introducing circuits. The machinists start at A. They always stitch along arcs to the nearest vertex with no sequin, before returning to A. Give the three orders of vertex stitching that are possible under this policy, and the corresponding lengths of stitching.

*(UODLE Question 12, Decision Mathematics AS, 1991)*

**7**  In a network, when is a vertex said to be
**a)** even    **b)** odd?

In the Chinese postman problem why is it necessary to identify all the odd vertices in the network?

In the schematic network of roads, at the top of page 87, D is the depot, other vertices are road junctions. Distances are in 100 m.
A highways maintenance depot must inspect all the manhole covers within its area. This means that an engineer must leave the depot, D, drive along each of the roads (arcs) in the network above at least once, and return to the depot. What is the minimum distance that he need drive? Give a route which enables him to drive this distance.

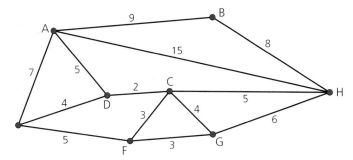

(You may identify the shortest path between any pair of vertices by inspection).

Every morning a road sweeper must leave the depot and sweep the gutters of all the roads. He must, therefore, travel along each road at least twice, once for each gutter. What is the minimum distance that he must travel? Give reasons. (There is no need to give a possible route for him. Gutters may be swept in either direction.)

*(UODLE Question 20, Applied Mathematics Paper 4, 1988)*

8   The map shows a number of roads in a housing estate. Road intersections are labelled with capital letters and the distances in metres between intersections are shown. The total length of all of the roads in the estate is 2300 m.

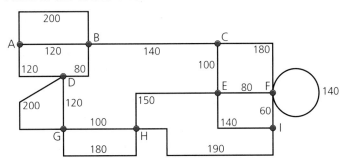

a) A newspaper deliverer has to walk along each road at least once, starting and ending at A. The shortest route to achieve this is required.

i) List those intersections which are of odd order and explain their significance to the problem.

ii) By investigating all possible pairings of odd intersections, find the minimum distance which the newspaper deliverer has to walk. (You are not required to apply a shortest distance algorithm to solve this. You are required to show the distance computations which lead to you choosing a particular pairing of odd intersections.)

iii) For each intersection other than A, give the number of times that the newspaper deliverer must pass through that intersection whilst following the shortest route.

| Intersection | B | C | D | E | F | G | H | I |
|---|---|---|---|---|---|---|---|---|
| No. of visits | | | | | | | | |

**b)** The newspaper deliverer only calls at a proportion of houses. The postwoman has to call at most houses, and since the roads are too wide to cross continually back and forth, she finds it necessary to walk along each road twice, once along each side. She requires a route to achieve this in the minimum distance.

   **i)** Describe how to produce a network to model this problem.

   **ii)** Without drawing such a network, say why it will be traversable and calculate the minimum distance which the postwoman will have to walk.

**c)** The street-cleaner needs to drive his vehicle along both sides of each road. He has to drive on the correct side of the road at all times. He too would like a shortest route.

   Explain how the street-cleaner's problem differs from the postwoman's, and say how the network would have to be modified to model this.

   *(O & C MEI Question 4, Decision & Discrete Mathematics Paper 19, 1996)*

**9**  The matrix below gives fares for direct bus journeys between towns, P, Q, R, S, T, U and V, in a tourist area. Blanks indicate no direct service.

Fares in pence

|   | P | Q | R | S | T | U | V |
|---|---|---|---|---|---|---|---|
| **P** | – | 57 | – | 35 | – | – | 70 |
| **Q** | 57 | – | 53 | 61 | – | 160 | – |
| **R** | – | 53 | – | – | 49 | – | – |
| **S** | 35 | 61 | – | – | 30 | – | 44 |
| **T** | – | – | 49 | 30 | – | 48 | 52 |
| **U** | – | 160 | – | – | 48 | – | 40 |
| **V** | 70 | – | – | 44 | 52 | 40 | – |

**a)** Draw a network to represent this information.

**b)** Using Dijkstra's method, find the minimum cost route from P to U. Describe in full detail all of the steps of the method, specifying the order in which you assign values to vertices.

**c)** A holidaymaker sets out from V to travel over all of the routes in the area, returning to V at the end. Find the minimum cost for the journey and the route that should be taken. Explain the significance of towns P and U in your solution.

**d)** A 'Summertime special' service is to be provided linking R to S at a cost of 35p. What is the new minimum cost and the best route for the holidaymaker in part **c)**? (In this part the necessary minimum connectors may be found by inspection.)

   *(UODLE Question 22, Applied Mathematics Paper 4, 1992)*

**10**  A commercial baker is to produce an iced and decorated cake in large numbers for retail in a supermarket chain. The decoration consists of a pattern of piped icing as shown here.

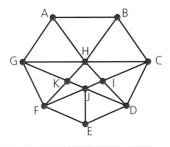

|   | A | B | C | D | E | F | G | H | I | J | K |
|---|---|---|---|---|---|---|---|---|---|---|---|
| **A** | – | 120 | – | – | – | – | 100 | 100 | – | – | – |
| **B** | 120 | – | 100 | – | – | – | – | 100 | – | – | – |
| **C** | – | 100 | – | 90 | – | – | – | 110 | 90 | – | – |
| **D** | – | – | 90 | – | 110 | – | – | – | 80 | 100 | – |
| **E** | – | – | – | 110 | – | 110 | – | – | – | 80 | – |
| **F** | – | – | – | – | 110 | – | 90 | – | – | 100 | 80 |
| **G** | 100 | – | – | – | – | 90 | – | 110 | – | – | 90 |
| **H** | 100 | 100 | 110 | – | – | – | 110 | – | 70 | – | 70 |
| **I** | – | – | 90 | 80 | – | – | – | 70 | – | 60 | – |
| **J** | – | – | – | 100 | 80 | 100 | – | – | 60 | – | 60 |
| **K** | – | – | – | – | – | 80 | 90 | 70 | – | 60 | – |

The piping consists of straight lines of icing connecting the nodes A–K as shown above. The piping nozzle can be programmed to move from a node directly to any other node and to dispense or not to dispense icing *en route*. The approximate distances between nodes are shown in mm in the matrix.

a)  The sum of lengths of the arcs of the network is 2020 mm. By referring to the orders of the vertices explain why it will be necessary for the nozzle to cover more than 2020 mm in order to complete the icing pattern.

b)  Use Dijkstra's algorithm to find the shortest distances from A to J and from A to E. Show all of your working, including the order in which you assign permanent labels to vertices. Give routes associated with your shortest distances, explaining how you found them.

c)  Use the method of solving the route inspection (Chinese postman) problem to find the shortest distances that the nozzle can travel to complete the pattern. (You are not required to use an algorithm to find the shortest distances between vertices other than those required in **b**).)

Give the shortest distance and an appropriate route, indicating when then nozzle should not be dispensing icing.

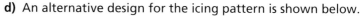

**d)** An alternative design for the icing pattern is shown below.

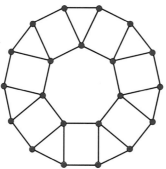

Count how many odd nodes there are in this pattern. In how many ways can they be paired together to construct possible solutions for the corresponding route inspection problem?

*(UODLE Question 9, Applied Mathematics Paper 4, 1994)*

## Summary

*After working through this chapter you should:*

■ know that planar graphs are graphs that can be drawn in the plane with no edges crossing

■ know that isomorphic graphs are graphs that have the same number of vertices and the degrees of corresponding pairs of vertices are the same

■ know that the degree of a vertex is the number of edges incident on the vertex

■ be able to say whether a graph is Eulerian and find an Euler cycle, which travels along every edge of a graph and finishes up at its starting point

■ be able to apply these techniques to solve route inspection (Chinese postman) problems in networks.

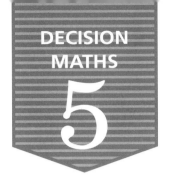

# DECISION MATHS

## 5

# *Network flows*

*By the end of this chapter you should be able to:*

■ *find out how to model flows with networks and use the associated vocabulary*

■ *find the maximum flow using the labelling procedure*

■ *use the maximum flow – minimum cut theorem.*

**Exploration 5.1**

### *Grid lock!*

The centre of a small town suffers from traffic congestion at rush hour. The local council decides to introduce a one-way system to try to solve the problem, so the planning officer develops some plans. Her plan for part of the town centre is shown below; the arrows indicate the direction of traffic flow and the figures give the maximum number of vehicles per hour that can pass along each road.

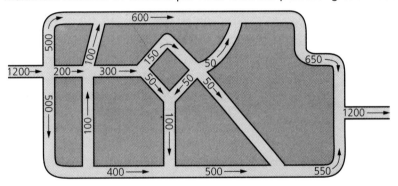

A survey has shown that at peak times this section of the town centre has a flow of 1200 vehicles per hour. Will the planning officer's design cope with this amount of traffic?

In this chapter, we shall consider this problem.

## NETWORK FLOWS

Networks can be used to model situations involving the flow of a substance or single items, such as water or oil in a pipe, traffic in a road system, electricity through a circuit. In these situations the edges of the network represent the routes along which the items can flow.

In most of the networks we have met so far, we have been able to travel in either direction along an edge. When modelling real situations this is not always the case. Sometimes the movement can only be in one direction. This is usually the case when we are dealing

with flows such as water in a pipe or traffic in a one-way system. Then we can use a **directed network** to model the situation.

Consider the following directed network.

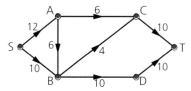

An edge which is directed is called an **arc**; arrows on the arcs indicate the direction of flow. Thus we would refer to the arc AC as being the route along which the substance flows with direction from A to C.

A flow must go from one point to another so a network modelling a flow must have:

■  a starting point – called the **source**, denoted by S

■  a finishing point – called the **sink**, denoted by T

and since there is usually a limit on the magnitude of the flow in any arc, the weights on the arcs show the **maximum capacity** of that arc.

Thus in this network arc SA has a maximum capacity of 12, arc AB has a maximum capacity of 6 and so on.

The flow along an arc does not have to be maximum, but when it is we say the arc is **saturated**.

A maximum flow is the largest possible flow through a network from S to T. An example of a possible flow through the network above is given by this diagram.

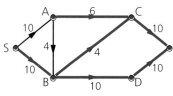

| path | flow |
|------|------|
| S A C T | 6 |
| S B D T | 10 |
| S A B C T | 4 |

The saturated arcs are
SB, AC, BC, BD, CT, DT.

---

## EXERCISES

5.1A

**1**   Suggest what these directed networks could represent.

**a)**

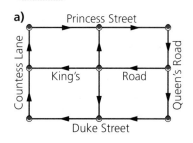

**b)**

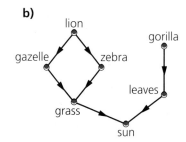

**c)**

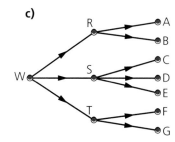

**2** State, with reasons, which two of the following directed graphs are isomorphic.

**a)** 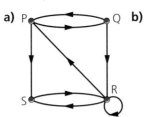 **b)** 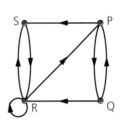 **c)**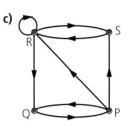

**3** Draw the directed network that is represented by this matrix.

From

|      |   | S | A | B | C | T |
|------|---|---|---|---|---|---|
|      | **S** | — | — | — | — | — |
|      | **A** | 4 | — | — | 2 | — |
| To   | **B** | 3 | — | — | — | — |
|      | **C** | 5 | — | 2 | — | — |
|      | **T** | — | 6 | 2 | 8 | — |

**4** Find a possible flow from S to T through this network. Is it a maximum flow? Justify your answer by considering the arcs which are saturated.

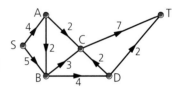

**5** Find, by inspection, the maximum flow from S to T in this network. State the arcs in your solution which are saturated.

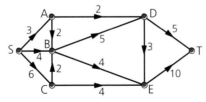

---

**EXERCISES**

**5.1 B**

**1** Draw a directed network for the following situations.

**a)** In a squash ladder, A had beaten B and D, B had beaten C, C had beaten A and D, D had beaten B.

**b)** A talks to B, C and F. C talks to A, E and D. E talks to F, F talks to E, and A and D talk to C.

**c)** Communication links between five centres are: $a$ can pass information to $d$, $b$ to $c$ and $d$, $c$ to $a$, $d$ to $e$ and $e$ to $a$ and $b$.

Express each of the above in a matrix representing an arc between two vertices by 1, and by 0 if no directed edge exists.

**2**   Use a tree diagram to find all the paths from S to T in the directed graph below.

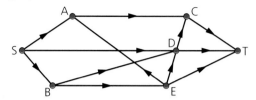

**3**   Draw a directed network represented by this matrix.

From

| To | | S | A | B | C | D | E | J |
|---|---|---|---|---|---|---|---|---|
| | **S** | – | – | – | – | – | – | – |
| | **A** | 5 | – | – | – | – | 4 | – |
| | **B** | 6 | – | – | – | – | – | – |
| | **C** | – | 10 | – | – | 5 | – | – |
| | **D** | 4 | 5 | 4 | – | – | 5 | – |
| | **E** | – | – | 8 | – | – | – | – |
| | **J** | – | – | – | 7 | 3 | 4 | – |

**4**   Which of the following directed graphs are isomorphic? Justify your conclusions.

**a)** 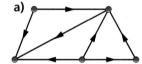    **b)** 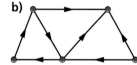    **c)**     **d)**

**5**   Consider the network below in which each arc is labelled with its capacity. Find a flow of value 7 from S to T. Draw a diagram with the flows in each arc.

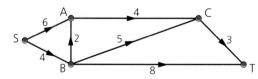

## LABELLING PROCEDURE

To solve a problem such as the one in Exploration 5.1, we need a systematic way of ensuring that the flow we have found is the best one possible. The basic idea when looking for a maximum flow is to find a flow by inspection, then to increase its value step by step until we cannot increase it any further. We shall develop such a method using this directed network.

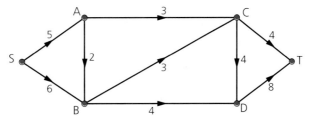

The first step is to look for a **feasible flow** through the network. We do this by finding routes through the network from S to T, for example SACT, and considering the capacities of the arcs along this route.

- ■ SA has capacity 5.
- ■ AC has capacity 3.
- ■ CT has capacity 4.

So the maximum flow along SACT is 3, that is the capacity of the arc with the minimum capacity.

*Arrows in the direction S to T indicate flow is still available in that arc.*

Arc AC is now saturated, but arc SA and CT still have some capacity available, which we call the **excess capacity**. We can show this on the network, by means of arrows alongside the arcs showing their excess capacity.

**Note:**
AC has an excess capacity 0, so AC is now saturated.

It is also useful to record the flow that we are sending down each arc by means of arrows directed in the opposite direction.

*Arrows in the direction T to S indicate flow so far along that arc.*

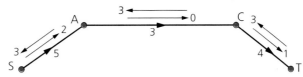

These values are called **artificial backward capacities**. As well as recording the flow in that arc, they also offer a means of altering the flow in that arc if, at some stage, we see a better solution and wish to reduce the flow along that arc and direct it along a different arc.

It is unlikely that our initial feasible flow would consist of just one path; we should normally start with a flow that takes in two or three of the most direct routes from S to T, for example an initial flow in this network might be like this.

**Step 1:**

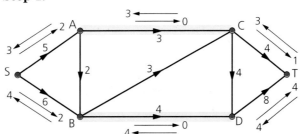

We write this as:

| Path | Flow | Saturated arcs | |
|------|------|----------------|---|
| SACT | 3 | AC | |
| SBDT | 4 | BD | Total flow 7 |

giving an initial flow of 7.

Notice that the saturated arcs have been recorded both on the network, by listing them, and on the diagram, by marking them in a coloured highlighter. It is useful to do this as you perform each step, because it helps to decide when you have an optimal solution.

Now we need to try to improve on this flow if possible. We do this by looking for other paths from S to T which consist entirely of unsaturated arcs; these are called **flow augmenting paths**. One such is SBCT which has a capacity of 1.

**Step 2:**

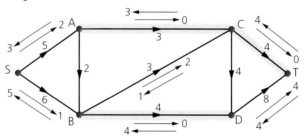

We add this to our network, amending the figures for the artificial backward capacities and excess flows. Arc CT is now saturated.

| Path | Flow | Saturated arcs | |
|------|------|----------------|---|
| SACT | 3 | AC | |
| SBDT | 4 | BD | |
| SBCT | 1 | CT | Total flow 8 |

We repeat this process until there are no possible improvements. The next path could be SBCDT which has a maximum flow of 1.

**Step 3:**

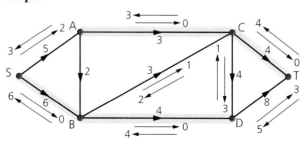

| Path | Flow | Saturated arcs | |
|------|------|----------------|---|
| SACT | 3 | AC | |
| SBDT | 4 | BD | |
| SBCT | 1 | CT | |
| SBCDT | 1 | SB | Total flow 9 |

Then we take SABCDT which also has a flow of 1 available.

**Step 4:**

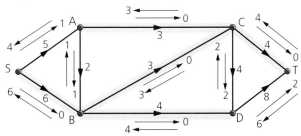

| Path | Flow | Saturated arcs | |
|------|------|----------------|---|
| SACT | 3 | AC | |
| SBDT | 4 | BD | |
| SBCT | 1 | CT | |
| SBCDT | 1 | SB | |
| SABCDT | 1 | BC | Total flow 10 |

There are now no other paths through the network from S to T; every way we try is 'blocked' by saturated arcs, so our final solution is a flow of 10, as illustrated in this diagram.

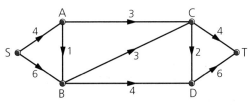

*Hint: It is useful to use different colours to mark the labels on your network.*

With practice, you do not need to draw a separate network at every stage. If you start with a large diagram and work neatly, you can show all your working on that diagram, listing the paths clearly underneath.

We can demonstrate this technique by finding a solution to the problem set in Exploration 5.1. First we need to represent the proposed one-way system as a directed network, where the arcs represent the roads and the vertices the road junctions, as shown in the diagram overleaf. Then we follow the steps as listed above, showing all the flow paths on one network, and listing the working underneath.

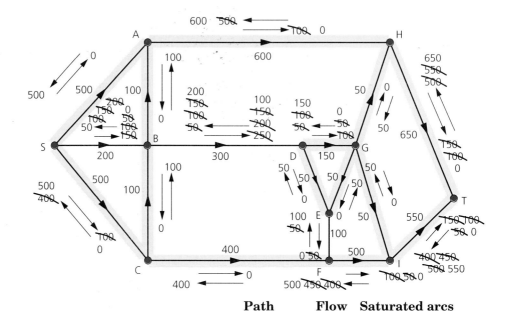

| Path | Flow | Saturated arcs |
|------|------|----------------|
| **Initial feasible flow** SAHT | 500 | SA |
| SCFIT | 400 | CF |
| **Flow augmenting paths** SBDGHT | 50 | GH |
| SBDGIT | 50 | GI |
| SBDEFIT | 50 | DE |
| SBDGEFIT | 50 | SB, DG, GE, EF, FI, IT |
| SCBAHT | 100 | SC, CB, BA, AH, HT |

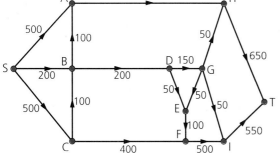

The maximum flow through the road system is 1200 (so the planning officer was correct).

This diagram, which shows the final flows along each arc in our network, is a way of displaying the solution clearly on a diagram.

## Exploration 5.2    *Networks with many sources and sinks*

Two reservoirs, $R_1$ and $R_2$, supply water to two towns, $T_1$ and $T_2$, via three pumping stations, A, B and C as shown in the network. The numbers represent the amount of water which can be pumped from station to station, in thousands of gallons per hour.

How can the labelling procedure be adapted to deal with a situation with more than one source and sink?

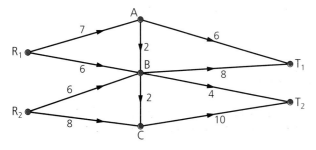

## Supersource and supersink

The solution to the problem in Exploration 5.2 is to introduce a **supersource** S which feeds into $R_1$ and $R_2$. Since there are thirteen units of flow leaving $R_1$, the arc $SR_1$ must have a capacity of 13. Then the arc $SR_2$ will have a capacity of 14 units, to supply the arcs leaving $R_2$. This can be seen in the diagram.

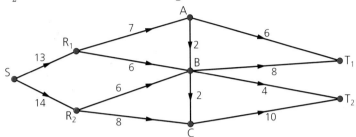

We deal with two sinks, $T_1$ and $T_2$, in a similar way by introducing a **supersink**, T, into which $T_1$ and $T_2$ feed. Again the arcs $T_1T$ and $T_2T$ must have sufficient capacity to allow all the flow entering $T_1$ and $T_2$ to leave.

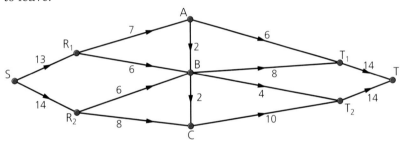

Once the supersource and supersink have been introduced, the flow through the network can be calculated using the labelling procedure.

## EXERCISES

**5.2A**

1   Use the labelling procedure to find the maximum flow from S to T in this network.

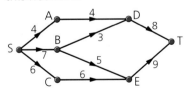

2   Six towns are connected by a local rail network.

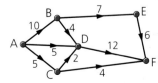

The network shows the maximum volumes of freight (in thousands of tonnes per week) that can be carried between the towns.

a) A large warehouse at A needs to supply 15 tonnes of goods each week to a retail outlet at F. Find a flow through the network that will make this possible.

b) High demand for the product means that the retailer wishes the supply to be as large as possible. Using your answer to part a) as an initial flow find the maximum tonnage of goods which the warehouse can supply each week.

c) The lines from A to C and C to D are upgraded and can now carry 8 thousand and 6 thousand tonnes per week respectively. How does this affect the maximum flow from A to F? Which one other line would you choose to upgrade to increase the flow still further?

3    The network shown in the diagram has a possible 26 units of flow leaving S and a possible 26 units of flow entering T. Explain why it is not possible to achieve a flow of 26 through the network from S to T.

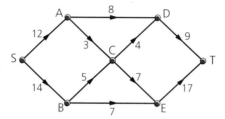

Find the maximum flow through the network.

4    Calculate the maximum amount of water which can be supplied to the two villages in Exploration 5.2.

5    The network below shows the flow rates for traffic in a one-way system; the figures give maximum numbers of vehicles per hour (in hundreds). Find the maximum flow through this network.

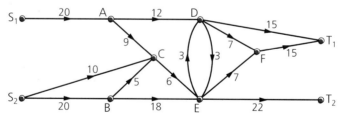

If roadworks at junction F restricted the flow from F to $T_1$ to 800 vehicles per hour, would this affect your maximum flow?

## EXERCISES

1    Use the labelling procedure to find the maximum flow from S to T in the network illustrated in question 5 of Exercises 5.1B. State the saturated arc in this case.

2    Repeat the procedure of the question above to find the maximum flow from S to T in the network you constructed in question 3 of Exercises 5.1B, stating which arcs are saturated.

3 An exporter in Southampton
wishes to ship containers of
a product to an importer in
Taiwan. There are several
channels through which the
containers may be sent, as
shown in this network. The
number on each arc
represents the maximum

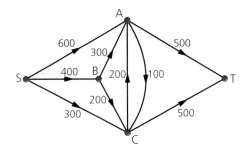

load which each channel can handle. Determine the maximum
number of containers that can be sent through the network.

4 a) Natural gas is produced at two sources $S_1$ and $S_2$ and transported
by a network of underwater pipelines to a refinery T. The
maximum capacity of each pipe, in appropriate units, is given in
the network. Determine the maximum flow in the network.

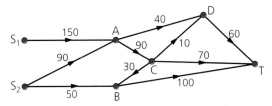

b) Due to the volatile nature of natural gas, there must be a
minimum flow through each pipeline. If the figures above now
represent the minimum capacity that each pipe must carry, devise
a method by which the minimum flow may be found in a such a
way that the given minimum capacities are satisfied.

5 Consider question 5 of Exercises 5.2A. The roadworks now restrict the
flow of traffic at junction E to 1500 vehicles, whereas the roadworks
from F to $T_1$ have now been removed.

Redraw the network so that the vertex E is replaced by an arc of
capacity 15 from $E_1$ to $E_2$ and all arcs directed towards E in the original
network are directed towards $E_1$ and all arcs directed away from E in
the original network are directed way from $E_2$ in the new network.

How is your maximum flow affected in this case?

## THE MAXIMUM FLOW – MINIMUM CUT THEOREM

In the previous section we developed a method for finding the
maximum flow through a directed network. Now we need to know how
we can be sure that the flow we have found is the maximum possible,
particularly if we have a large network. We can get some indication of
the solution to this problem by considering question 3 in Exercises
5.2A. You were asked to give a reason why the maximum flow in a
network is not 26 even though it is possible for 26 units to flow from S
and it is also possible for 26 units to flow into T.

We can start by considering a simple network with a bottleneck.

In this network we can clearly see that although twelve units can flow from S and twelve units can flow into T, the maximum flow through the network is only four because all flow must pass through arc CD.

We can make this idea more precise by introducing the idea of a **cut**, which is a line dividing the network into two parts, one part containing the source, S, and the other containing the sink T. The capacity of the cut is the sum of the capacities of the arcs it crosses, providing those arcs are directed from S to T.

Thus we can take a cut through SA and SB, marked $C_1$ on the diagram below. The capacity of this cut is $7 + 5 = 12$. We can certainly say that the maximum flow cannot exceed the capacity of this cut.

The table below describes some of the other possible cuts in this network.

| Cut | Arcs in cut | Vertices to left of cut | Vertices to right of cut | Capacity of cut |
|---|---|---|---|---|
| $C_1$ | SA, SB | S | A, B, C, D, E, F, T | $7 + 5 = 12$ |
| $C_2$ | SA, BC | S, B | A, C, D, E, F, T | $7 + 5 = 12$ |
| $C_3$ | AC, SB | S, A | B, C, D, E, F, T | $7 + 5 = 12$ |
| $C_4$ | AC, BC | S, A, B | C, D, E, F, T | $7 + 5 = 12$ |
| $C_5$ | CD | S, A, B, C | D, E, F, T | $4 \leftarrow$ minimum |
| $C_6$ | DE, EF | S, A, B, C, D | E, F, T | $8 + 4 = 12$ |
| $C_7$ | DE, FT | S, A, B, C, D, F | E, T | $8 + 4 = 12$ |
| $C_8$ | ET, DF | S, A, B, C, D, E | F, T | $8 + 4 = 12$ |
| $C_9$ | ET, FT | S, A, B, C, D, E, F | T | $8 + 4 = 12$ |

If we consider the other cuts through the network we can see that the cut through CD ($C_5$ in the table) has a capacity of 4. This is obviously the smallest capacity of any cut through this network and is called the **minimum cut**. We can also see that the maximum flow cannot exceed the value of this cut since it separates S from T and so all flow must go through the arcs in the cut. Hence it follows that:

maximum flow ≤ capacity of any cut

Since this is true for any cut, the maximum flow can never be greater than the capacity of the minimum cut.

This rule holds for any network. It is called the **Maximum flow – minimum cut theorem** and can be formally stated as:

maximum flow = value of minimum cut

If we return to the network in question 3 of Exercises 5.2A, we can quickly deduce that, since the total flow out of vertices A and B is less than 26, this cannot be a feasible flow. Then we can use the Maximum flow – minimum cut theorem to find a value for the maximum flow and hence confirm the solution gained, using the labelling procedure.

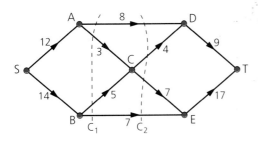

It is very tedious (and difficult) to list every cut through the network, so we use inspection to check the obvious choices for minimum value. Two such cuts are shown on the diagram.

$C_1$: arcs in cut: AD, AC, BC, BE      capacity: $8 + 3 + 5 + 7 = 23$

$C_2$: arcs in cut: AD, CD, CE, BE      capacity: $8 + 4 + 7 + 7 = 26$

*Check: Was this the answer you obtained for question 3 in the exercises?*

A quick inspection of the network will tell us that $C_1$ is the minimum cut with a capacity of 23, thus the maximum flow through the network is 23 units.

One final point to note is that in our original definition of a cut we specified that the arcs must be directed from S to T. We must therefore consider what happens in a case like the one here, where a cut crosses an arc which is directed from T to S.

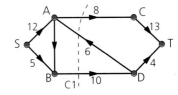

In this case the capacity of the cut through AC, AD and BD in the direction S to T is:

$8 + 0 + 10 = 18$

In other words, an arc directed from T to S does not contribute towards the value of the cut.

## EXERCISES

**5.3 A**

**1 – 4**  Return to questions 1, 2, 4 and 5 in Exercises 5.2A. Use the Maximum flow – minimum cut theorem to confirm that the flows you obtained were maximum flows.

**5**  Use the labelling procedure to find a maximum flow in this network.

Prove that the flow is a maximum by finding a cut of the same value.

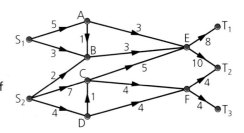

## EXERCISES

**5.3B**

**1  a)** In an analysis of a given basic network, a flow with value 10 is found. What can be deduced about the capacity of a minimum cut?
   **b)** If a cut with capacity 10 is found in a basic network, what can be deduced about the value of the maximum flow?
   **c)** If a cut with capacity 10 and a flow of value 7 are found in a basic network, what can be deduced about the value of the maximum flow?
   **d)** If a cut with capacity 10 and a flow with value 10 is found in a basic network, what can be deduced about the value of the maximum flow?
   **e)** If you calculate a flow with value 10 and a cut with capacity 7, what can you deduce?

**2**  The network below has eight cuts. Draw up a table listing the cuts and hence determine the minimum cut.

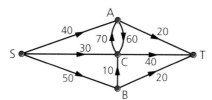

**3**  Use the Maximum flow – minimum cut theorem to verify that your answers to questions 1–4 in Exercises 5.2B were correct.

**4**  A gas pipeline network is shown on the right. The left-hand figure on each arc is the minimum allowable flow and the maximum allowable flow is the right-hand figure. The problem is to determine, if it exists, the maximum flow such that the flow along each arc is not less than the lower capacity.

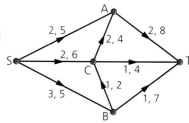

   **a)** The first step is to find a flow which satisfies the maximum and minimum capacity conditions.
   **b)** Find the maximum flow using the flow augmented method.
   **c)** Find a cut through the network which proves your solution to **b)**.

**5**  Study these networks and decide whether a flow can exist in each of them.

   **a)**

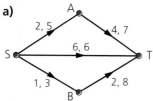

   **b)**

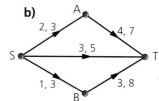

## COURSEWORK INVESTIGATION

**Investigation**

### *Bottlenecks, a flow problem*

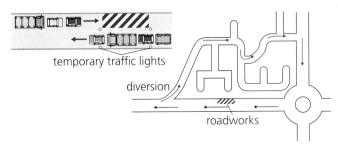

temporary traffic lights

diversion

roadworks

Roadworks often cause bottlenecks by narrowing a road and restricting the flow of traffic. Use Maximum flow algorithms to model the effect of roadworks on roads in your area. Consider whether it may be better to divert some, or all, of the traffic. How could the introduction of temporary one-way systems alleviate the situation?

## CONSOLIDATION EXERCISES FOR CHAPTER 5

**1 a)** If you can find a cut with capacity 9 in a network, what can you deduce about the maximum flow?

**b)** If you can find a flow of value 15 in a network, what can you deduce about the minimum cut?

**c)** If you have a flow of value 28 and a cut of capacity 35, what can you deduce?

**2 a)** Find the minimum cuts for these networks.

**i)** 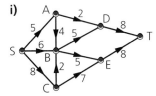  **ii)**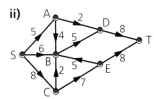

**b)** Confirm your cuts are minimum cuts by finding a flow of the same value in each network.

**c)** 'Every minimum cut in a network consists entirely of saturated arcs.' Is this statement correct? Give reasons for your answer.

**3** Gas is supplied to two factories, $T_1$ and $T_2$, from a storage tank at S,

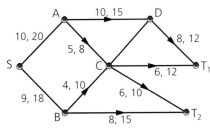

along a series of pipes. As well as a maximum rate of flow, there is also a minimum rate of flow so that the pressure in the pipes does not fall to a dangerously low level. This is shown in the network here, where the first figure shows the minimum flow and the second the maximum flow possible in each pipe (in litres per minute).

**a)** Find a feasible flow which satisfies the restriction on a minimum flow.

**b)** Using this feasible flow as a starting point, find the maximum flow through the pipes.

**c)** Find a cut through the network which proves your solution to **b)**.

**4**    The diagram shows a pipe network. The numbers on the arcs give the maximum capacities of the pipes.

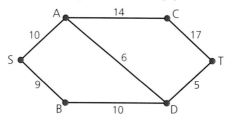

a) Use a labelling algorithm to find the maximum flow from the source S to the sink T. Show and describe the steps of your algorithm.

b) Find a cut to prove that your flow is maximal. Explain why it proves that the flow is maximal.

*(UODLE Question 8, Paper 4, 1995)*

**5**    The diagram shows a capacitated network representing traffic flow. The numbers on each of the arcs indicate the capacity in hundreds of cars per hour of the arcs (roads). The ringed numbers

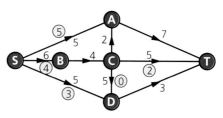

on the arcs indicate the flow along that arc for a particular flow $f$.

a) Copy and complete the figure by finding the flows along the remaining arcs.

b) Hence determine the value of the flow $f$.

c) Show that this flow is a maximum flow by finding a cut whose capacity has the same value.

*(ULEAC Question 1, Specimen Paper, 1996)*

**6**    The diagram shows a directed flow network with two numbers on each arc. The uncircled numbers show the capacity of each arc. The circled numbers are the flows currently passing through the network.

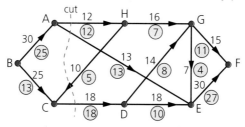

a) Identify the source and the sink.

b) What is meant by a cut? What is the flow across the given cut? Why is this not a maximum flow?

c) Find and indicate one flow-augmenting path, and use it to augment the flow as far as possible.

d) Find a maximal flow.

e) Explain how you know that the flow which you quoted in part **d)** is maximal.

*(UODLE Question 10, Paper 4, 1990)*

**7** The network below shows the maximum rates of flow (in vehicles per hour) between towns S, A, B, C, D and T in the direction from S to T.

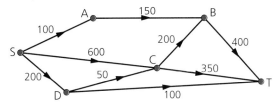

**a)** By choosing a minimum cut, or otherwise, find the maximum traffic flow from S to T. Give the actual rates of flow in each of the edges BT, CT and DT when this maximum flow occurs.

**b)** When a maximum flow occurs from S to T, how many of those vehicles per hour pass through C?

**c)** It is decided to reduce the traffic flow through C (in the direction from S to T) to a maximum of 480 vehicles per hour. In order to maintain the same maximum flow from S to T the capacity of a single edge is to be increased. Which edge should be chosen, and by how much must its capacity be increased?

*(AEB Question 4, Specimen Paper, 1996)*

**8** The diagram shows a gas distribution network consisting of:

- three supply points A, B, C
- three intermediate pumping stations, P, Q, R
- two delivery points, X, Y
- connecting pipes.

The figures on the arcs are measures of the amounts of gas which may be passed through each pipe per day. The figures by A, B and C are measures of the daily availability of gas at the supply points.

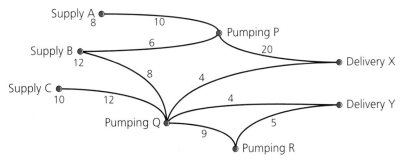

**a)** Copy the network, introducing a single source with links to A, B and C, the capacities on the links reflecting the supply availabilities.

**b)** Introduce a single sink linked to X and Y, the links having large capacities.

**c)** Use inspection to find the maximal daily flow through your network, making a list of the flows through each pipe.

**d)** Find a suitable cut to prove that your flow in part **c)** is maximal.

**e)** Interpret the flows in the new links, from the source and to the sink.

**f)** A new pipeline is proposed with capacity 5 units per day, connecting P and Q. Demonstrate the use of a labelling procedure to augment the flow, and thus find the new maximal flow.

**g)** R is now to become a delivery point. Explain how to adapt the approach of part **b)** and find the maximal daily flow of gas in total that can be delivered to R, X and Y. You do not need to calculate this flow.

*(UODLE Question 19, Paper 4, 1993)*

**9** The two matrices represent a road network connecting seven towns. The first matrix gives the distances between the towns in kilometres. The second matrix gives the capacities of the roads – the maximum numbers of vehicles which can pass between towns in one hour (in thousands of vehicles).

| | A | B | C | D | E | F | G |
|---|---|---|---|---|---|---|---|
| A | – | 10 | 8 | – | – | 30 | – |
| B | 10 | – | – | 8 | 15 | – | – |
| C | 8 | – | – | 5 | – | 7 | – |
| D | – | 8 | 5 | – | – | 15 | 4 |
| E | – | 15 | – | – | – | – | 12 |
| F | 30 | – | 7 | 15 | – | – | 10 |
| G | – | – | – | 4 | 12 | 10 | – |

Distances

| | A | B | C | D | E | F | G |
|---|---|---|---|---|---|---|---|
| A | – | 3 | 3 | – | – | 2 | – |
| B | 3 | – | – | 6 | 2 | – | – |
| C | 3 | – | – | 2 | – | 4 | – |
| D | – | 6 | 2 | – | – | 2 | 1 |
| E | – | 2 | – | – | – | – | 2 |
| F | 2 | – | 4 | 2 | – | – | 2 |
| G | – | – | – | 1 | 2 | 2 | – |

Capacities

**a)** Draw the road network.

**b)** Find the maximum hourly flow of vehicles from B to F, showing how this may be achieved. *Prove* that this is a maximum.

**c)** Use, and demonstrate your use of, Dijkstra's algorithm to find the shortest route between B and F.
What percentage of maximum hourly flow of vehicles use the shortest route?

*(UODLE Question 22, Paper 4, 1993)*

**10** The arcs in the flow network below represent oil pipelines with capacities in thousands of barrels of oil a day. Vertices 1, 2 and 3 are oil wells with production capacities of 150, 300 and 200 thousand barrels a day, respectively. Vertices 4, 5 and 6 are oil refineries which may process up to 200, 250, and 250 thousand barrels of oil a day, respectively. One barrel of crude oil processes to give one barrel of refined oil. Vertices 7, 8 and 9 are centres of demand which require 150, 100 and 300 thousand barrels of refined oil a day, respectively.

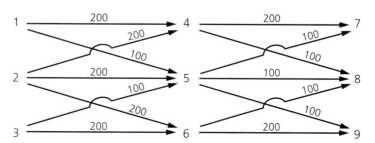

Modify the network, in a manner which you should explain, so that:
a) it has a single source
b) it has a single sink, and
c) capacities are associated only with arcs, not vertices.
Hence, by the application of a formal algorithm, find a feasible flow which satisfies the demands at 7, 8 and 9.

There is an accident at the refinery at 5 which reduces its production capacity to 50 thousand barrels a day. It is possible to increase, temporarily, production capacity at refineries at 4 and 6 to 300 thousand barrels a day. Show, by considering cut-sets, that there is no feasible flow through the network which satisfies all the demand, even with these temporary increases.

*(UODLE Question 22, Paper 4, 1988)*

## Summary

*After working through this chapter you should:*

■ understand that an arc is a directed edge in a network, a source is a starting point, a sink is a finishing point, the capacity is the maximum flow along the arc and an arc is saturated when it is carrying the maximum capacity

■ know that a feasible flow through a network is an allowable path and flow-augmenting paths consist entirely of unsaturated arcs

■ use the labelling procedure to find the maximum flow through a network

■ know the Maximum flow – minimum cut theorem and its application.

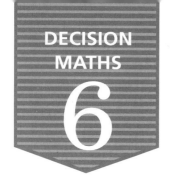

# DECISION MATHS

# 6

# Critical path analysis

*By the end of this chapter you should be able to:*

■ *draw precedence networks, which show the 'order' in which steps have to be carried out, using activity on vertices and activity on arcs procedures*

■ *find the critical path, which is the minimum time needed to complete a project, following the longest path through the precedence network and including all the critical activities which are activities that must be started on time to avoid delaying the whole project*

■ *draw cascade charts showing precedence relations, in the form of bars drawn against a time scale, that help to make best use of all resources*

■ *use resource levelling to balance the amount of work to be done and the resources available for a project.*

**Exploration 6.1**

### *Kate's decorating – 1*

Kate decides to redecorate her room. She buys some wallpaper and paint, and then sits down to plan the job. What tasks will she need to do and in what order? She is hoping to get it done over a weekend. Is she being realistic?

## PRECEDENCE NETWORKS

Many companies use network analysis to help in the planning of large projects in order to ensure efficient use of time and resources. If, like Kate in the exploration, you are planning a project, the first thing to do is make a list of the steps involved in the project from start to completion, then estimate the length of time each step will take. Kate's decorating job might look like this.

| | Activity | Duration (hours) |
|---|---|---|
| A | Tidy the room and move the furniture. | 2 |
| B | Remove the curtains and curtain rail. | 0.25 |
| C | Strip the old wallpaper. | 1.5 |
| D | Wash and sand the paintwork. | 2 |
| E | Apply a first coat of paint to ceiling and woodwork and let it dry. | 6 |
| F | Apply a second coat of paint and let it dry. | 6 |
| G | Hang the wallpaper. | 5 |
| H | Hang the curtains. | 0.25 |
| I | Reinstate the furniture. | 1.5 |

Next, Kate needs to consider which activities need to be completed before others can be started so she can draw up a list of **precedence relations**, showing which activities must immediately precede others.

| | Activity | Duration (hours) | Preceding activity |
|---|---|---|---|
| A | Tidy the room and move the furniture. | 2 | |
| B | Remove the curtains and curtain rail. | 0.25 | |
| C | Strip the old wallpaper. | 1.5 | A, B |
| D | Wash and sand the paintwork. | 2 | C |
| E | Apply a first coat of paint and let it dry. | 6 | D |
| F | Apply a second coat of paint and let it dry. | 6 | E |
| G | Hang the wallpaper. | 5 | C |
| H | Hang the curtains. | 0.25 | F, G |
| I | Reinstate the furniture. | 1.5 | H |

*Precedence networks are sometimes called **activity networks**.*

Having done all of this we can construct a **precedence network**, which is a way of showing this information on a diagram.

## Constructing precedence networks

*Some examination boards specify one particular method for constructing these networks. You should check your syllabus before proceeding.*

There are two methods for constructing precedence networks. We shall consider each in turn.

### Activity on vertex (node)

In this type of diagram, the activities are represented by the vertices of a network. The edges show the order of precedence.

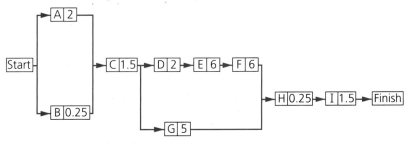

Here is the technique for drawing this type of network.

■ You must have a start vertex and a finish vertex.

■ Each activity is represented by a vertex which shows the activity and the duration.

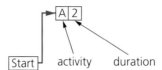

■ Edges in the network show only the order in which activities must be done. They indicate where an activity cannot be started until another has been completed, so we can see that activity H (hanging the curtains) cannot be done until both the paint is dry (F) and the wallpaper has been hung (G).

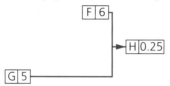

### Activity on arc

In this type of network, the arcs represent the activities, while the vertices represent **events**, by which we mean the finish of one activity and the start of another.

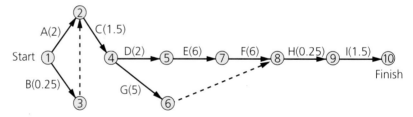

Here is the technique for drawing this type of network.

■ You must have a start vertex and a finish vertex.

■ Activities are represented by arcs, which show the activity and the duration.

■ Vertices represent events which are numbered in such a way that each activity can be described uniquely by a pair of events $(i, j)$ where $i$ is the start event and $j$ is the finish event. For example activity D is represented by $(4, 5)$.

■ **Dummy arcs** may need to be introduced where an activity is directly preceded by more than one other, as in the case of H. Dummy arcs are marked as dotted lines and have zero duration.

## EXERCISES

In this exercise, you may use either type of network.

**1** The two diagrams give alternative – but equivalent – ways of representing the precedence with which eight activities of a project must be performed.

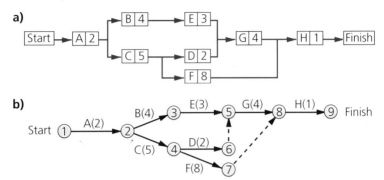

**a)**

**b)**

Use the networks to produce a table showing the duration of each activity and its immediate predecessor(s).

**2** The table gives a series of tasks, the times taken to complete them and the immediately preceding activities.

| Task | A | B | C | D | E | F | G |
|------|---|---|---|---|---|---|---|
| Duration (minutes) | 3 | 5 | 5 | 4 | 5 | 8 | 3 |
| Preceding activity | | | A | A | B | D, E | F, C |

Show this information on a precedence network.

**3** A conservatory is to be built on the back of a house. Details of the jobs to be done and timings are shown below.

| | Activity | Duration (days) | Preceding activity |
|---|----------|-----------------|--------------------|
| A | Lay the foundations. | 3 | |
| B | Build the walls. | 5 | C |
| C | Lay the drains. | 2 | A |
| D | Lay the floor. | 1 | B |
| E | Make the door and window frames. | 2 | |
| F | Erect the roof. | 3 | D |
| G | Fit the door frame and door. | 0.5 | E, F |
| H | Fit the windows. | 2 | G |
| I | Plaster the walls and ceiling. | 8 | H |

Draw a precedence network to show this information.

**4** When Marc returns home from work, he has to prepare and eat his evening meal. His meat pie and dessert are already prepared; the pie needs to be cooked and the dessert has to be removed from the refrigerator just before eating.

The table shows the activities which need to be done.

| | Activity | Duration (minutes) | Preceding activity |
|---|---|---|---|
| **A** | Set the table. | 5 | |
| **B** | Heat the oven. | 3 | |
| **C** | Peel the potatoes. | 5 | |
| **D** | Cook the potatoes. | 20 | |
| **E** | Put the pie in the oven. | 1 | |
| **F** | Cook the pie. | 30 | |
| **G** | Cook the frozen peas. | 10 | |
| **H** | Eat the main course. | 15 | |
| **I** | Take the dessert from fridge. | 1 | |
| **J** | Eat the dessert. | 7 | |
| **K** | Wash up. | 8 | |

Copy the table and fill in the 'preceding activity' column.

Show the information on a precedence network.

**5** You are planning a dinner party for friends and wish to serve a three-course meal. Decide on a menu, then list the activities that will need to be completed. Make a table of times and precedence and draw a network for preparing your meal.

## EXERCISES

**6.1B**

**1** Each of the following represents a part of a complete network. They are not necessarily complete in themselves. Construct networks to represent them, using the activity on arcs method.
a) Activities X and Y depend on activities A and B.
b) Activity X depends on activity A and activity Y depends on activities A and B.
c) Activity X depends on activities A and B, activity Y depends on activities B and C and activity Z depends on activity C only.
d) Activity X depends on activity A only, activity Y depends on activity B only and activity Z depends on activities A, B and C.
Repeat the above exercise using the activity on vertex method.

**2** Redraw the following networks to eliminate any redundant or unnecessary dummy activities.

a)

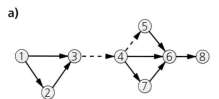

b)

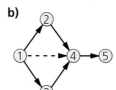

c)

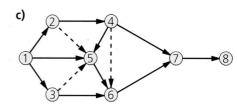

**3** List the activities involved in each of the following situations. Construct the two types of precedence networks in each case. State your initial assumptions for each situation.

a) Coffee is made by boiling water then pouring it through a filter filled with coffee. Milk is boiled and added to the filtered coffee and the coffee is served.

b) Three taxis are to be cleaned by two cleaners. One works on the outside of a taxi and the other works on the inside. The inside must be completely cleaned before any work can start on the outside.

c) A picture is to be framed. Assume that the following activities need to be done.

    A  Obtain the frame materials.
    B  Make the frame.
    C  Obtain the glass.
    D  Fit the glass to the frame.
    E  Mount the picture.
    F  Fit the picture to the assembled frame and glass.
    G  Obtain adhesive tape for the dust excluder.
    H  Fix the adhesive tape.

**4** The tables below give a series of activities, their completion times and immediate preceding activities (**predecessors**). Construct a precedence network.

a)

| Task | A | B | C | D | E | F | G | H | I | J | K | L |
|------|---|---|---|---|---|---|---|---|---|---|---|---|
| Duration (weeks) | 3 | 9 | 3 | 1 | 4 | 3 | 8 | 4 | 5 | 7 | 6 | 10 |
| Preceding activities | | | | | A | B, E | C | C | D | D | H, I | J |

b)

| Task | A | B | C | D | E | F | G | H | I |
|------|---|---|---|---|---|---|---|---|---|
| Duration | 5 | 6 | 10 | 6 | 12 | 4 | 11 | 5 | 5 |
| Preceding activities | | A | A | B | B, C | C | D, E | E, F | G, H |

**5** A company is to manufacture a piece of equipment made from two parts, A and B, which are assembled together. Each part requires the installation of a machine and the assembly then requires the installation of a third machine, C. The completed piece then needs testing for strength and durability on a specially-made testing rig. Assuming all materials are to hand, the activities and their duration times are as listed in this table.

| Activity | Duration (weeks) |
|----------|:----------------:|
| Obtain and install the machine for A. | 5 |
| Obtain and install the machine for B. | 9 |
| Obtain and install machine C. | 16 |
| Make part A. | 4 |
| Make part B. | 7 |
| Assemble the piece of equipment. | 7 |
| Make the test rig. | 25 |
| Test and dispatch. | 8 |

## CRITICAL PATHS

When we are planning a project we usually want to know the minimum time we will need to complete it. We can find this from the precedence network, by finding the longest path through the network from the start vertex to the finish vertex. This path is called the **critical path**. Any delay in the activities along this path would inevitably lead to a delay in completing the whole project.

The algorithm for finding the longest path is in two parts:

■ the forward pass, in which we label each activity with its earliest start and finish times
■ the backward pass, in which we find the latest time that we can risk starting each activity without delaying the whole project.

The **critical activities** can then be identified as being those that must start at a particular time or the whole project will be delayed. The critical path is the path through the network that includes all the critical activities.

We can demonstrate this on the two types of network developed in the previous section.

## Activity on vertex

Consider the problem of decorating Kate's room introduced in Exploration 6.1.

### *Forward pass*

We start by labelling the start vertex 0 (zero). We then label the vertices directly connected to the start with their earliest start time relative to the start time of the project, which is taken as zero. This is shown at the top left-hand corner of the box. On the top right-hand corner we show the earliest finish times which is calculated as:

> *earliest start time + duration of the activity*

and is recorded as the total number of hours from the start of the project.

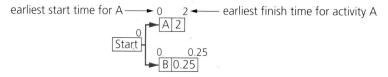

Next we consider activity C, which must be preceded by both A and B; this means that the earliest start time for C is 2 hours after the project start time, since it cannot start until A is completed. The earliest finish time for C is 2 + 1.5 = 3.5 hours.

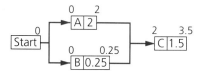

We continue in this way until we have labelled all the vertices with their earliest start and finish times. The network then looks like this.

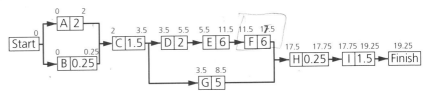

## Backward pass

We now work backwards from the finish time of 19.25 hours, putting the labels below the vertices. The earliest finish time for activity I is 19.25 and the latest start time is 17.75, found by subtracting the duration of activity I from the latest finish time. In the same way the latest finish time and start times for activity H can be calculated and marked on the network.

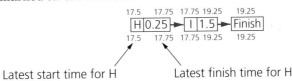

Latest start time for H          Latest finish time for H

Activity H has two activities directly preceding it, F and G. The latest finish time for both of these will be 17.5 since they must both be completed before H begins. Subtracting the duration of these activities from their latest finish times will give us their latest start times.

activity F:    latest start time = 17.5 – 6 = 11.5

activity G:    latest start time = 17.5 – 5 = 12.5

These in turn can be marked on the network, which now looks like this.

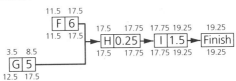

We continue the backward pass until we reach vertex C, which precedes two other activities. Since we are considering latest start and finish times, we must give C a latest finish time of 3.5 since it must be completed before D is begun.

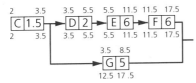

We continue in this way until we have labelled all the vertices with their latest start and finish times.

Our network now looks like this.

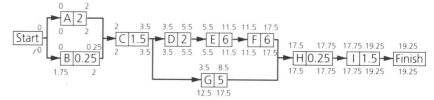

Notice that some of the activities have some flexibility in their start times, for example activity G can be started at any time from 3.5 hours after the start of the project to 12.5 hours after the start of the project without delaying the total time required. The maximum time by which an activity can be delayed without delaying the project is called the **float**. Activity G has a float of 9 hours. The activities with floats are shown below.

| Activity | Latest start time | Earliest start time | Float |
|---|---|---|---|
| B | 1.75 | 0 | 1.75 hours |
| G | 12.5 | 3.5 | 9 hours |

## Activity on arc

### *Forward pass*

*Remember, an event is the finish of one activity and the start of another.*

When the activities are represented on the arcs, the forward pass finds the earliest event times and the backward pass finds the latest event times. We annotate the network using a double box like this.

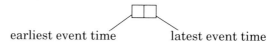

earliest event time          latest event time

The forward pass gives these values for the earliest event times.

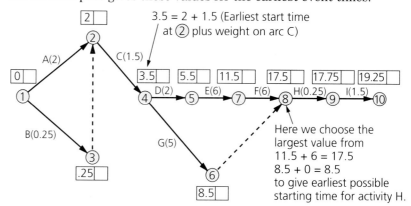

3.5 = 2 + 1.5 (Earliest start time at ② plus weight on arc C)

Here we choose the largest value from
11.5 + 6 = 17.5
8.5 + 0 = 8.5
to give earliest possible starting time for activity H.

### *Backward pass*

The backward pass gives these values for the latest event times.

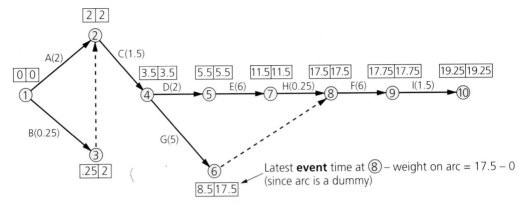

Latest **event** time at ⑧ – weight on arc = 17.5 – 0
(since arc is a dummy)

Once again we can identify the activities with float from the network by considering the event times. For activity $(i, j)$:

$$\text{float} = \frac{\text{latest event}}{\text{time for } j} - \frac{\text{earliest event}}{\text{time for } i} - \frac{\text{duration}}{\text{of activity}}$$

For example, in the case of activities B(1, 3) and G(4, 6):

for B: float = 2 – 0 – 0.25 = 1.75 hours

for G: float = 17.5 – 3.5 – 5 = 9 hours

## Showing the critical path on a network diagram

As we stated earlier, the critical activities *must* start at a particular time or the whole project will be delayed. The activities that have no flexibility in their start times are those with a zero float. In our example these are A, C, D, E, F, H and I, so the critical path must be the path through these activities.

The critical path can be shown on the network diagram like this.

**Activity on vertex**

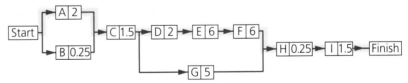

Critical path A, C, D, E, F, H, I

**Activity on arc**

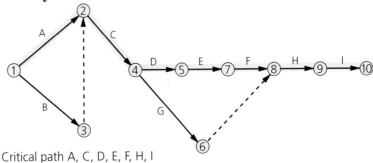

Critical path A, C, D, E, F, H, I

119

In the previous example, the activities which had float were independent of each other. This is called an **independent float**. Now consider this example.

## Example 6.1

*Kieran decides to make a wooden chair for his niece's third birthday. The activities necessary to complete the project are shown in this table.*

| | Activity | Duration (hours) | Preceding activities |
|---|---|---|---|
| **A** | Make the wooden base. | 2 | |
| **B** | Make the back. | 2.5 | |
| **C** | Make the legs. | 5 | |
| **D** | Cut the foam for the cushions. | 1 | |
| **E** | Make covers for the cushions. | 3 | D |
| **F** | Assemble the chair. | 2 | A, B, C |
| **G** | Assemble the cushions. | 1 | E |
| **H** | Paint the chair. | 4 | F |
| **I** | Fix the cushions to the chair. | 0.5 | G, H |

*If we construct the precedence network and find the critical activities and the activities with float, we can see that some of them are linked by precedence, so that if one activity with a float, say D, does not start at its earliest start time, it affects the amount of float available for activities E and G.*

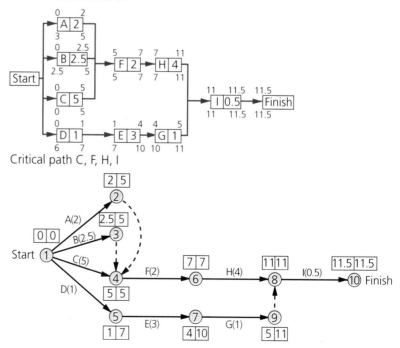

Critical path C, F, H, I

*The type of float in the example above is called an **interfering float**.*

**EXERCISES**

**6.2 A**

Perform a forward and backward pass through the network, for questions 1–5 in Exercises 6.1A. Identify the critical activities. For each network give the critical path.

**EXERCISES**

**6.2 B**

1    Find the critical path and minimum project time for this network.

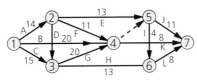

2    A hospital intends to computerise its medical records. A consultancy firm has identified the activities and their duration, in weeks, that are needed to obtain the system and make it ready for operating.

|     | Activity | Duration | Preceding activities |
| --- | --- | --- | --- |
| **A** | Order/delivery of hardware | 6 | – |
| **B** | Order/delivery of software | 4 | – |
| **C** | Preparation of computer site | 5 | – |
| **D** | Installation of hardware | 1 | A, C |
| **E** | Loading  software | 1 | B, D |
| **F** | Preparing data | 10 | K |
| **G** | Loading data | 2 | E, F |
| **H** | Training operators | 3 | D |
| **I** | Initial user training | 7 | K |
| **J** | Hands-on user training | 3 | E, I, H |
| **K** | Agree final specification | 2 | – |
| **L** | Inform managers | 1 | K |
| **M** | System text | 2 | G, H |

The network representing the activities is given below.

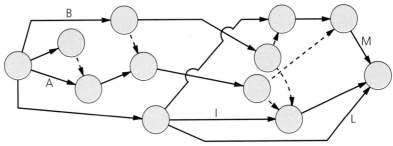

Number the events and complete the labelling of the activities. Determine the critical path and the minimum completion time.

3    Perform a forward and backward pass through the network in each of questions 3–5 in Exercises 6.1B. Identify the critical activities. Give the critical path and minimum project time.

## CASCADE CHARTS

**Exploration 6.2**    *Kate's decorating – 2*

In an unaccustomed fit of generosity, Kate's brother Rowan offers to help her to decorate her room. Rowan wants to know when he will be needed so he can plan his own activities. When should she ask him to be available to help her?

In the previous section, we saw how to find the minimum time needed to complete a project, providing that the non-critical activities could be carried out at the same time as the critical activities. In other words, there was no restriction on the number of workers. Kate's decorating project could be completed in 19 hours 15 minutes, provided someone else can remove the curtains for her while she moves the furniture and then come back and help with the painting or wallpapering. If she has to do the job by herself it will take her up to 24.5 hours, the total time for all the activities (unless she can save a bit of time by starting the wallpapering while the paint is drying).

If a project is to be planned efficiently, we need to take account of the resources (both people and equipment) needed at different stages. **Cascade charts** show the same information as a precedence network but in the form of bars drawn against a time scale to show when the activities will need to take place so as to make best use of resources as well as time. Before we can draw a cascade chart we need to allocate each activity a **cascade activity number** (CAN) to ensure a logical sequence. There is a method for this, and we can relate it to our example.

| Activity | CAN | | |
|---|---|---|---|
| | | | **Step 1:** Set CAN at 1; set time at 0. Begin at the start vertex and number the first activity connected to it 1. |
| A | 1 | These are connected directly to the start vertex. | **Step 2:** Allocate CANs to other activities connected directly to the start vertex. |
| B | 2 | | |
| C | 3 | This is preceded by A and B, but step 4 is satisfied. | **Step 3:** Consider activities that immediately follow those already numbered. allocate CANs to them in turn. |
| D | 4 | | |
| G | 5 | G, like D, is directly connected to C. | |
| E | 6 | | **Step 4:** Where an activity is directly preceded by more than one other, it cannot be allocated a CAN until all directly preceding activities are numbered. |
| F | 7 | | |
| H | 8 | Preceded by F and G – step 4 is satisfied. | |
| I | 9 | Stop. | **Step 5:** Stop when all activities have been numbered. |

Having applied the CAN method to our example we have a chart that looks like this.

| CAN | | Activity | Duration (hours) | Preceding activities |
|---|---|---|---|---|
| 1 | A | Tidy the room and move the furniture. | 2 | |
| 2 | B | Remove the curtains and curtain rail. | 0.25 | |
| 3 | C | Strip the old wallpaper. | 1.5 | A, B |
| 4 | D | Wash and sand the paintwork. | 2 | C |
| 5 | G | Hang the wallpaper. | 5 | C |
| 6 | E | Apply a first coat of paint and let it dry. | 6 | D |
| 7 | F | Apply a second coat of paint and let it dry. | 6 | E |
| 8 | H | Replace the curtain rail and hang the curtains. | 0.25 | F, G |
| 9 | I | Reinstate the furniture. | 1.5 | H |

Now we can display this information on a cascade chart. Activities are represented by bars with length proportional to the duration of the activity. The activities are assumed to start at the earliest start time. The dotted lines show precedence relations between activities, and float times for non-critical activities.

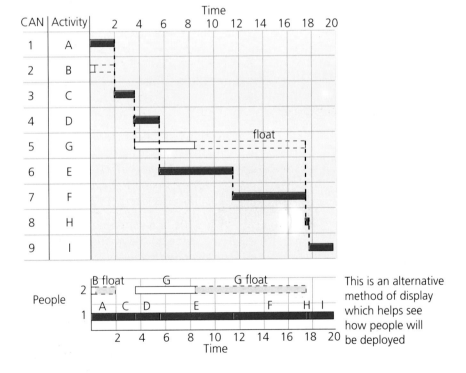

This is an alternative method of display which helps see how people will be deployed

The critical activities have been shaded; we know that these cannot be moved. Other activities could be delayed, but not beyond the dotted lines showing precedence otherwise they will move all subsequent activities along, making the duration of the project greater.

## Resource levelling

Once we have drawn a cascade chart, it is possible to draw a **resource histogram** to show the number of people needed at any stage in the project.

For the relatively simple example of the decorating job, the resource histogram would look like this.

**Resource histogram**

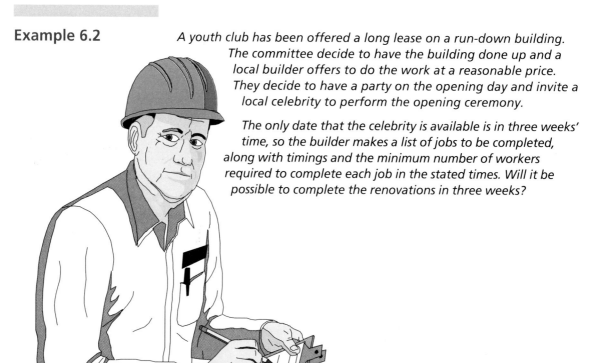

We can see from this that Rowan needs to be available to help Kate for 15 minutes at the start and for 5 hours some time between 3 and 17 hours into the project. If he helps her out she can probably get the job done in a weekend, but it will be hard work.

Suppose Kate starts work at 6 p.m. on Friday. Draw up a timetable for her so that she finishes on Sunday (and manages to eat and sleep).

With a more complicated project a resource histogram can be used as a planning tool, to ensure that there is a balance between the amount of work to be done and the resources available. This process is know as **resource levelling**.

**Example 6.2**

*A youth club has been offered a long lease on a run-down building. The committee decide to have the building done up and a local builder offers to do the work at a reasonable price. They decide to have a party on the opening day and invite a local celebrity to perform the opening ceremony.*

*The only date that the celebrity is available is in three weeks' time, so the builder makes a list of jobs to be completed, along with timings and the minimum number of workers required to complete each job in the stated times. Will it be possible to complete the renovations in three weeks?*

Here is the list the builder prepared.

| | Activity | Duration | Preceding activity | Number of workers required |
|---|---|---|---|---|
| A | Agree the details of the lease. | 4 | | 1 |
| B | Remove fireplaces and fittings. | 2 | A | 2 |
| C | Rewire. | 2 | A | 2 |
| D | Strip the wallpaper. | 1 | A | 1 |
| E | Burn off the old paint. | 2 | A | 2 |
| F | Install central heating. | 3 | B, C | 2 |
| G | Replaster. | 3 | D, F | 2 |
| H | Sand the floor and woodwork. | 3 | G, E | 1 |
| I | Varnish the floor. | 2 | H | 1 |
| J | Paint the ceiling. | 1 | I | 1 |
| K | Paint the woodwork. | 2 | I | 1 |
| L | Repaint the walls. | 2 | J | 1 |
| M | Replace fittings. | 1 | L, K | 1 |
| N | Clean up. | 1 | M | 3 |

### Solution

We start by drawing a precedence network and finding the critical path.

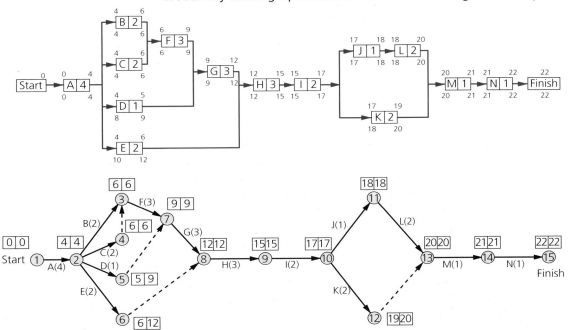

The critical activities are A, B, C, F, G, H, I, J, L, M, N. So there are two critical paths:

$$\left.\begin{matrix} AB \\ AC \end{matrix}\right\} FGHIJLMN$$

and the activities which have float are:

| Activity | Float |
|---|---|
| D | 4 |
| E | 6 |
| K | 1 |

125

*We next allocate the CAN numbers, using the method in the previous section.*

| CAN | | Activity | Duration | Preceding activity | Number of workers required |
|---|---|---|---|---|---|
| 1 | A | Agree the details of the lease. | 4 | | 1 |
| 2 | B | Remove fireplaces and fittings. | 2 | A | 2 |
| 3 | C | Rewire. | 2 | A | 2 |
| 4 | D | Strip the wallpaper. | 1 | A | 1 |
| 5 | E | Burn off the old paint. | 2 | A | 2 |
| 6 | F | Install central heating. | 3 | B, C | 2 |
| 7 | G | Replaster. | 3 | D, F | 2 |
| 8 | H | Sand the floor and woodwork. | 3 | G, E | 1 |
| 9 | I | Varnish the floor. | 2 | H | 1 |
| 10 | J | Paint the ceiling. | 1 | I | 1 |
| 11 | K | Paint the woodwork. | 2 | I | 1 |
| 12 | L | Repaint the walls. | 2 | J | 1 |
| 13 | M | Replace fittings. | 1 | L, K | 1 |
| 14 | N | Clean up. | 1 | M | 3 |

*Now we can display the information on a cascade chart.*

**Cascade chart**

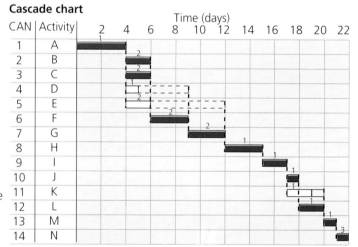

**Note:** The numbers above the bars give the number of workers required for each job

*Then we can use the cascade chart to draw a resource histogram.*

**Resource histogram**

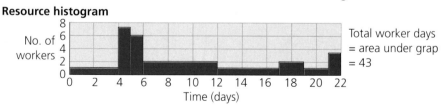

*It would appear from this that the work cannot be done in three weeks, though it is only a day over schedule, so perhaps it is not an insurmountable problem. Looking at the histogram, we might also wonder whether the best use is being made of the workers. For example, is it absolutely necessary to have seven workers on day 5?*

### Applying resource levelling

There are three ways in which resource levelling can be applied to a problem in order to increase the efficiency of a project.

1   Make better use of resources (in this case workers) by using the float to alter the timing of non-critical activities.

**Cascade chart**

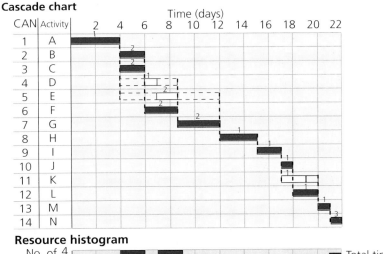

**Resource histogram**

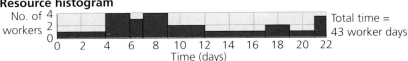

By using the float to change the start timed activities D and E we have *evened out* the number of workers, getting rid of the peak value of 7.

2   We can use the float time on activity E to reduce the number of workers from two to one and increase the duration of the activity from two days to four.

**Cascade chart**

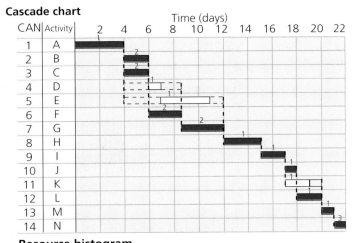

This gives a more even use of workers which is more cost-effective but still does not solve the problem of running over time.

3   If it is important to stay within a time limit, then we can choose to increase the number of workers on some activities, thus reducing the time. If the builder were to put two workers on activities H and I, we could reduce the time for the two from five days to two and a half days.

**Cascade chart**

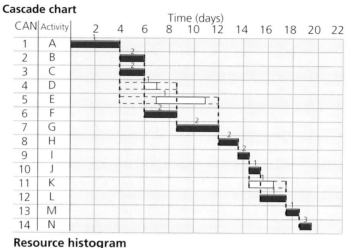

**Resource histogram**

This enables the work to be completed in less than 20 days, leaving time to prepare for the opening ceremony. It also has the advantage of using only the same number of worker-days as the other plans, so should not, in theory, cost more!

## EXERCISES

**6.3 A**

1   Draw a cascade chart for the information given in question 2 of Exercises 6.1A. Use resource levelling to obtain a solution which makes the most efficient use of the workers available. How many workers are needed, if only one person is needed for each activity?

2   Consider question 4 of Exercises 6.1A. If Marc's girlfriend is helping him to cook, how much time could be saved in preparing the meal?

3   Use the information in Example 6.1 to draw a cascade chart for Kieran's chair-building project. Be careful with the interfering floats.

    How many people would need to help in order to finish the project in the minimum time?

**4** Study this chart.

| Activity | Duration | Preceding activity | Number of workers required |
|----------|----------|--------------------|----------------------------|
| A | 4 | | 1 |
| B | 7 | | 1 |
| C | 5 | | 1 |
| D | 2 | B, C | 2 |
| E | 3 | A, B | 1 |
| F | 5 | E | 1 |
| G | 2 | E, H | 1 |
| H | 6 | D | 2 |
| I | 2 | G, F | 1 |
| J | 3 | F | 1 |
| K | 3 | H | 1 |

**a)** Draw a precedence network and find the critical path.

**b)** Use a cascade chart and resource histogram to schedule the activities as efficiently as you can.

**5** The table below shows a simplified schedule for building a house.

| | Activity | Duration | Preceding activity |
|---|----------|----------|--------------------|
| A | Dig the foundations. | 12 | |
| B | Pour the foundations. | 8 | A |
| C | Build the walls. | 10 | B |
| D | Erect the roof. | 6 | C |
| E | Make the windows and doors. | 7 | |
| F | Install the windows and doors. | 2 | C, E |
| G | Install the plumbing. | 4 | C |
| H | Install the electrics. | 3 | D, F |
| I | Plaster the internal walls. | 4 | G, H |
| J | Decorate. | 5 | I |
| K | Landscape the gardens. | 7 | G |

**a)** Produce a network for the project and identify the critical activities.

**b)** Display the information on a cascade chart.

**c)** For activity E find:
   **i)** the independent float   **ii)** the interfering float.

**d)** The cost associated with each task, when completed in the normal duration, is shown in the table overleaf, together with the extra cost per day which would be incurred if the task is to be completed in less time (sometimes called the **crash** cost).

| Activity | Cost normal duration (£) | Extra cost for early completion per day(£) |
|---|---|---|
| A | 12 000 | 2000 |
| B | 9000 | 2000 |
| C | 18 000 | 4000 |
| D | 15 000 | 3000 |
| E | 3000 | 850 |
| F | 2500 | 800 |
| G | 5000 | 1250 |
| H | 4500 | 1000 |
| I | 6000 | 1500 |
| J | 6000 | 1600 |
| K | 8000 | 2000 |

If the project had to be completed in one less day than the completion time found in **a)**, which activity (or activities) could the builder choose to do in one day less. Why?

**e)** What is the percentage increase in cost?

## EXERCISES

6.3 B

1   Determine the critical path for the network below. Draw a cascade chart and use resource levelling to obtain a solution that makes the most efficient use of the workers available. How many workers are needed, if only one person is needed for each activity?

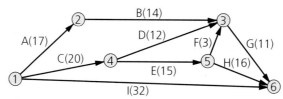

2   Analyse the following network and draw a cascade chart. Use resource levelling to obtain a solution which makes the most efficient use of the workers available.

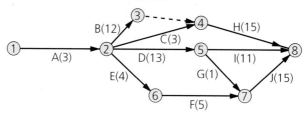

3   Use a cascade chart for the information given in each of the questions in Exercises 6.2B. Use resource levelling to obtain a solution that makes the most efficient use of the workers available. How many works are needed if only one person is needed for each activity?

**4** The table below gives the activities to prepare, cook and serve a plate of fish and chips.

| | Activity | Duration (minutes) | Preceding activity |
|---|---|---|---|
| A | Peel the potatoes. | 6 | – |
| B | Prepare the oil. | 6 | – |
| C | Fillet the fish. | 10 | – |
| D | Prepare the batter. | 4 | – |
| E | Mix the batter. | 4 | D |
| F | Chip the potatoes. | 4 | A |
| G | Fry the chips. | 14 | B, F |
| H | Coat the fish with batter. | 5 | C, E |
| I | Fry the fish. | 12 | B, H |
| J | Serve the fish with the chips. | 3 | G, I |

**a)** Produce a network for the above and identify the critical activities. State the minimum time for completion.

**b)** Display the information on a cascade chart. Use resource levelling to obtain a solution which makes most efficient use of manpower.

## MATHEMATICAL MODELLING ACTIVITY

### Going to your school or college

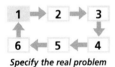

*Specify the real problem*

### Problem statement

Identify the activities you perform from the moment of rising from your bed to leaving for school or college on a typical morning. Identify the minimum time this takes and the critical activities.

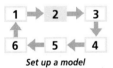

*Set up a model*

### Set up a model

Initially you should take seven or eight activities you perform before leaving home in the morning. Make a record of the time it takes to carry out each activity. You may need to do this on several occasions and take the average time.

Assume initially that you are the only person at home, so you will need to prepare your own breakfast.

### Mathematical model

*Formulate the mathematical problem*

Certain activities should run concurrently, for instance you may complete your assignments or homework whilst eating your breakfast. State clearly all assumptions you make about the activities you selected.

### Mathematical solution

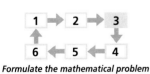

*Solve the mathematical problem*

Construct a precedence network using the activity on vertex method. Determine the critical path and the latest time that you need to get up, to leave by a set time. Draw up a cascade chart and a resource histogram, so

131

that if, at any stage, there is an added resource (for instance someone to prepare breakfast for you), then your activities may be planned more efficiently.

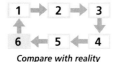

*Compare with reality*

## Refinement to the model

Now re-inspect each activity and consider whether it may be broken down into components, particularly if these components may be performed concurrently. Rework the solution using your revisions.

## CONSOLIDATION EXERCISES FOR CHAPTER 6

1   The two diagrams give alternative but equivalent ways of representing the precedences with which the seven tasks of a project must be performed. In this question you may use whichever one of these two ways you prefer.

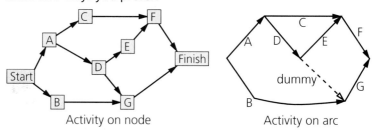

Activity on node          Activity on arc

The times required to complete each task are:

| Task | A | B | C | D | E | F | G |
|------|---|---|---|---|---|---|---|
| Time required (minutes) | 2 | 4 | 4 | 3 | 4 | 6 | 3 |

a) Produce a table showing the immediate predecessor(s) for each activity.

b) Perform a critical path analysis to obtain the critical path and the completion time for the project.

c) What flexibility is there in the scheduling of activity C?

d) Show how to modify your network to incorporate an activity H which cannot start until activities A and B are completed, and which must itself be completed before activity D can start.

e) i)  Produce a modified precedence table incorporating activity H and any consequential changes.

     ii) If the time needed for activity H is two minutes, what effect will your modification have on the completion time and on the critical path?

*(UODLE Question 19, Paper 4, 1995)*

2   The project of assembling a piece of furniture delivered as a 'flat-pack' was broken down into the activities shown in the following table. The table also shows each activity with its preceding activities and the duration of each activity (in minutes).

| | Activity | Duration (minutes) | Preceding activity |
|---|---|---|---|
| A | Open the box. | 10 | |
| B | Check the sections against the parts list. | 30 | A |
| C | Check the fixings against the parts list. | 20 | A |
| D | Assemble the shell. | 40 | B |
| E | Cut back to size. | 40 | B |
| F | Screw the shell together. | 20 | C, D |
| G | Assemble the drawers. | 20 | B |
| H | Glue the drawers together. | 10 | G |
| I | Put the whole together. | 30 | E, F, H |

**a)** Draw the weighted network which models this project, numbering each event.

**b)** Calculate the earliest time after the start of the project at which each activity may start.

**c)** Hence determine the length of the critical path.

**d)** Write down the activities which lie on the critical path.

*(ULEAC Question 8, Specimen Paper D1, 1996)*

**3** **a)** The activity network shows the precedences between seven activities involved in a project. The activities are represented by arcs and are labelled by letters. Their durations are also shown.

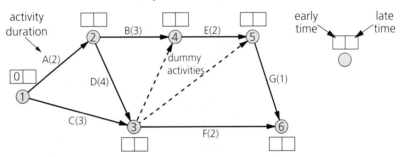

**i)** Explain why the dummy activity shown from event number 3 to event number 5 is not needed.

**ii)** Complete this table showing the immediate predecessor(s) for each activity.

| Activity | A | B | C | D | E | F | G |
|---|---|---|---|---|---|---|---|
| Immediate predecessor(s) | | none | | none | | | |

**iii)** Perform a forward pass and a backward pass on the activity network to determine the earliest and latest event times.

**iv)** Complete the table (overleaf) showing the float for each activity.

| Activity | A | B | C | D | E | F | G |
|----------|---|---|---|---|---|---|---|
| Float    |   |   |   |   |   |   |   |

**v)** State the minimum time for completion and the activities forming the critical path.

**b)** A second activity network with six activities is as follows.

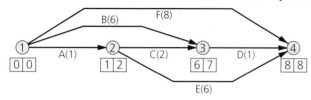

Activity C has interfering float of 4 (because 7 − 1 − 2 = late time of 3 − early time of 2 − duration of C = 4). It has independent float of 2 (because 6 − 2 − 2 = early time of 3 − late time of 2 − duration of C = 2). Explain the relevance of these measures when scheduling activity C.

**c)** The activity network below is the same as that in part **b)**, except that activities B and E have reduced durations, and activity D has increased duration. This has affected some early and late times.

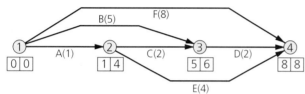

Give the interfering float and the independent float for activity C in this network.

*(O & C MEI Question 1,*
*Decision & Discrete Mathematics*
*Paper 19, January 1996)*

**4**   At 4.30 p.m. one day the BBC news team hear of a Government Minister resigning. They wish to prepare an item on the event for that evening's 6 o'clock news. The table below lists the jobs needed to prepare this new item, the time each job takes and the constraints on when the work can commence.

| | Job | Time needed (in minutes) | Constraints |
|---|-----|------|-------------|
| A | Interview the resigning Minister. | 15 | Starts at 4.30 p.m. |
| B | Film Downing Street. | 20 | None |
| C | Get reactions from regions. | 25 | Cannot start until A and B are completed. |
| D | Review possible replacements. | 40 | Cannot start until B is completed. |
| E | Review the Minister's career. | 25 | Cannot start until A is completed. |
| F | Prepare film from archives. | 20 | Cannot start until C and E are completed. |
| G | Edit. | 20 | Cannot start until A, B, C, D, E and F are completed. |

**a)** Construct an activity network for this problem and, by finding the critical path in your network, show that the news item can be ready before 6.00 p.m. that day.

**b)** If each of the jobs A, B, C, D, E and F needs a reporter, and once a reporter has started a job that same reporter alone must complete it, explain how three reporters can have the news item ready before 6.00 p.m. but that two reporters cannot.

*(AEB Question 8, Specimen Paper P4, 1996)*

**5** An industrial boiler is to be cleaned. The following activities have to be performed and it is your task to schedule them so that the boiler may be brought back into service as quickly as possible.

| | Activity | Preceding activities | Duration (hours) |
|---|---|---|---|
| **A** | Dismantle the pipe connections. | | 4 |
| **B** | Dismantle the pressure seals. | | 6 |
| **C** | Remove the heat exchanger. | A, B | 7 |
| **D** | Clean the inside of the boiler. | C | 9 |
| **E** | Clean the heat exchanger. | C | 10 |
| **F** | Lubricate the pressure seals. | B | 3 |
| **G** | Replace the heat exchanger. | D, E | 6 |
| **H** | Reseal the boiler. | F, G | 5 |
| **I** | Reconnect the pipes. | G | 3 |
| **J** | Test the boiler. | H, I | 2 |

a) Draw a network to represent the problem.

b) Calculate the earliest and latest times that each activity can start if the boiler is to be brought back into service as quickly as possible. You should explain your calculations carefully for activity G. For all other activities you may write your answers onto the network without showing your working.

c) Which are the critical activities? Explain the way in which you recognise them.

d) What is the earliest time that the boiler can be brought back into service?

e) Suppose that you cannot reconnect the pipes (I) until 3 hours of work have been done resealing the boiler (H). Show how you would indicate this on your network. How does this modify your answers to parts **b)**, **c)** and **d)**?

*(UODLE Question 8, Paper 4, 1989)*

**6** The table gives information about a construction project.

| Activity | Duration (days) | Immediate predecessors |
|---|---|---|
| A | 10 | |
| B | 3 | |
| C | 5 | B |
| D | 3 | A, C |
| E | 4 | B |
| F | 7 | D |
| G | 2 | C, E |

a) Complete the activity-on-arc precedence diagram for the project.

135

**b)** Perform a forward and backward pass on your precedence network to determine the earliest and latest event time.

**c)** State the minimum time for completion and the critical path.

**d)** Use an appropriate method to produce an ordering of the activities, and hence draw a cascade chart for the project on the axes given.

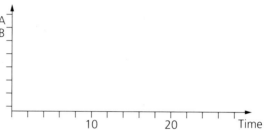

*(O & C MEI Question 1, Specimen Paper, 1994)*

**7** The diagram shows an activity-on-arc precedence diagram. Activity durations are in days, and are shown in brackets against the arc representing the activity. Activity X connects event 6 to event 8.

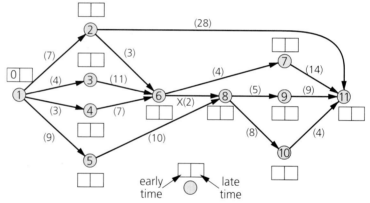

**a)** Perform a forward pass on this network. Give the early time for event number:

**i)** 6      **ii)** 8.

**b)** Perform a backward pass on the network. Give the late time for event number:

**i)** 8      **ii)** 6.

**c)** The independent float for activity X is defined as:

early time for event 8 – late time for event 6 – duration of activity X (provided that it is non-negative).

Give the independent float for activity X.

The total float for activity X is defined as:

late time for event 8 – early time for event 6 – duration of activity X

Give the total float for activity X.

Explain the significance of each of independent float and total float scheduling activity X.

**d)** The network below is part of a larger network. It shows all of the activities which are connected to activity X, *and only those activities*, together with some early and late times.

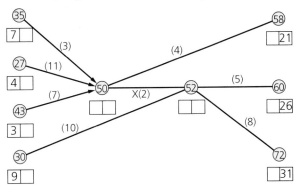

Say which of the following it is possible to compute from the information given on this network. In each case do the computation if it is possible, and say why it is not possible if not.

the early time for event 52
the early time for event 58
the late time for event 50
the late time for event 35

*(O & C MEI Question 2, Decision & Discrete Mathematics Paper 19, June 1995)*

**8**  A boat purchased in kit form is to be assembled and launched as quickly as possible. The following activities have to be performed, according to the given precedences.

| | Activity | Preceding activities | Duration (days) |
|---|---|---|---|
| **A** | Sand the hull. | | 2 |
| **B** | Fit the doors. | | 0.5 |
| **C** | Cut out and fit the windows. | | 1 |
| **D** | Fit the sea cocks and install the toilet. | A, B | 1 |
| **E** | Fit the galley. | B | 0.25 |
| **F** | Paint the hull and deck. | A, D | 3 |
| **G** | Apply anti-fouling. | F | 1 |
| **H** | Fit the winches. | C | 0.5 |
| **I** | Step the mast. | H | 0.25 |
| **J** | Fit internal linings. | C | 2 |
| **K** | Fit the upholstery. | D, E, J | 0.5 |
| **L** | Launch. | G, I, K | 0.5 |

**a)** Draw a network to represent the problem.
**b)** Calculate the earliest and latest times that each activity can start if the boat is to be launched as quickly as possible. You should explain your calculations carefully for activity D. For all other activities you may write your answers onto the network without showing your working.

c) Which are the critical activities? How may they be recognised?

d) What is the shortest time to complete the launch of the boat?

e) The owner does not wish to start fitting the upholstery until the deck paint is dry, that is until four days after starting activity F. Show how to modify the network to account for this. Find the new critical path and the shortest time to complete launching.

*(UODLE Question 19, Paper 4, 1991)*

9    The table shows the activities involved in a project, their durations, and their immediate predecessors.

| Activity | A | B | C | D | E | F | G | H |
|---|---|---|---|---|---|---|---|---|
| Duration (days) | 1 | 2 | 1 | 1 | 4 | 2 | 3 | 2 |
| Immediate predecessors | | | A | B | B | C, D | E, F | E |

a) Complete the activity-on-arc precedence diagram for the project.

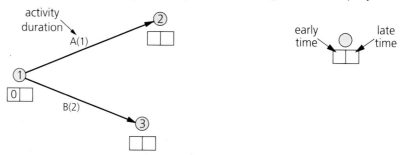

b) Perform a forward pass and a backward pass on your precedence diagram to determine the earliest and latest event times.
State the minimum time for completion and the activities forming the critical path.

c) Complete a cascade chart for the project on the given axes, with the activities ordered as shown.

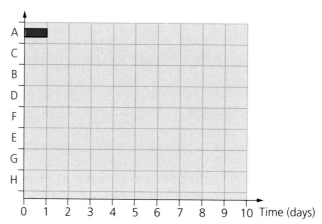

**d)** The numbers of people needed for each activity are as follows.

| Activity | A | B | C | D | E | F | G | H |
|----------|---|---|---|---|---|---|---|---|
| People | 1 | 4 | 2 | 3 | 1 | 2 | 3 | 2 |

Activities C and F are to be scheduled to start later than their earliest start times so that only five people are needed at any one time, whilst the project is still completed in the minimum time. Specify the scheduled start times for activities C and F.

*(O & C MEI, Question 1, Decision & Discrete Mathematics Paper 19, January 1995)*

**10** The following table gives details of a set of eight tasks which have to be completed to finish a project. The 'immediate predecessors' are those tasks which must be completed before a task may be started.

| Task | Duration (days) | Immediate predecessor(s) |
|------|-----------------|--------------------------|
| A | 3 | |
| B | 8 | |
| C | 7 | |
| D | 4 | A |
| E | 2 | D, B |
| F | 6 | C |
| G | 3 | E |
| H | 2 | E |

**a)** Produce either an activity-on-arc or an activity-on-node network for the project.

**b)** Find the minimum time to completion and the critical activities. The cost associated with each task, when completed in the normal duration, is given in the following table, together with the extra cost that would be incurred in using extra resources to complete the task in one day less than the normal duration.

| Task | Cost (£) Normal duration | Extra cost (£) Completion 1 day sooner |
|------|--------------------------|----------------------------------------|
| A | 3000 | 1000 |
| B | 6000 | 800 |
| C | 5000 | 700 |
| D | 5000 | 1200 |
| E | 1200 | 600 |
| F | 5500 | 1000 |
| G | 2800 | 1100 |
| H | 1700 | 500 |

**c)** Find the minimum cost that would be entailed in completing the project in one day less time than the minimum time to completion which you found in **b)**.

Give the percentage increase in cost that this involves.

*(UODLE Question 7, Paper 4, 1994)*

## Summary

*After working through this chapter you should be able to:*

■ draw precedence networks to show the necessary steps in a project, using activity-on-vertices and activity-on-arcs procedures

■ find the critical path, which is the minimum time needed to complete a project, following the longest path through the precedence network and including all the critical activities which are activities that must be started on time to avoid delaying the whole project

■ draw cascade charts showing precedence relations, to make best use of all resources

■ use resource levelling to balance the amount of work to be done and the resources available for a project.

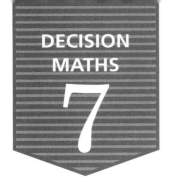

# DECISION MATHS

# 7

# *Matchings*

*By the end of this chapter you should be able to:*

- *understand how a bipartite graph, which is one that can be divided into two subsets in such a way that each edge of the graph joins a vertex from one subset to a vertex in the other, can be applied in modelling*

- *subdivide a bipartite graph by inserting vertices of degree 2*

- *apply a matching to a bipartite graph by linking some of the vertices in one subset to some of the vertices in the other*

- *use the algorithm for maximum matching of the two subsets in a bipartite graph.*

**Exploration 7.1**

## *Building a timetable*

Mrs Frawt, deputy head of Everbrite School, is trying to complete the timetable. She must fit Environmental Studies, Geography, Graphics, Mathematics and Science into the last timetable slot and fortunately she has five suitable teachers available. The table shows which subjects each teacher can cover.

| Teacher | Subjects |
|---------|----------|
| Miss Abraham | Environmental Studies, Geography |
| Mr Bowen | Environmental Studies, Mathematics, Science |
| Mrs Carter | Geography, Graphics |
| Ms Delgardo | Geography, Mathematics, Science |
| Mr Eggham | Graphics, Mathematics, Science |

How should Mrs Frawt allocate her staff? Try solving the problem using your own methods, before reading on. Then you can compare the method you use with the techniques developed in this chapter.

## GRAPH THEORY – BIPARTITE GRAPHS

A **bipartite graph** is a graph in which the vertices can be divided into two subsets, A and B, in such a way that each edge of the graph joins a vertex in A to a vertex in B. We distinguish the vertices in subset A from those in subset B by using a small, solid circle for those in A and an open circle for those in B.

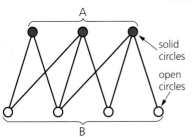

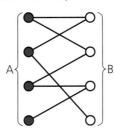

141

A **complete bipartite graph** is one in which each vertex in A is joined to every vertex in B by an edge. Complete bipartite graphs are denoted by $\mathbf{K}_{r,s}$ where $r$ is the number of vertices in A and $s$ is the number of vertices in B.

*Note: $K_{1,5}$ is called a* **star graph**, *for obvious reasons.*

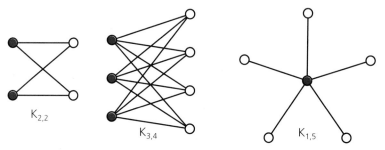

Unlike most of the graphs we have met so far, not all bipartite graphs are planar. For example, is $K_{3,3}$ planar?

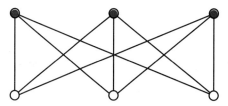

If $K_{3,3}$ is planar we should be able to draw it with no edges crossing. If you try it, you will soon see that it is not possible.

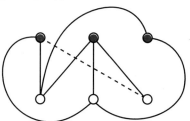

A **subdivision** of a graph is formed by inserting vertices of degree 2.

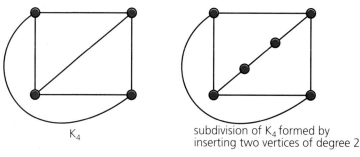

subdivision of $K_4$ formed by inserting two vertices of degree 2

If a graph is non-planar, every subdivision of that graph must also be non-planar. The Polish mathematician Kuratowski proved that all non-planar graphs contain subgraphs which are subdivisions of either $K_{3,3}$ or $K_5$ (we proved that $K_5$ is non-planar in Chapter 4, *Networks 3: Route inspection problems*), thus we can show that a graph is non-planar by finding a subdivision of $K_{3,3}$ or $K_5$ in it.

**Example 7.1**

*Show which of these graphs are non-planar because they contain subgraphs which are subdivisions of either $K_{3,3}$ or $K_5$.*

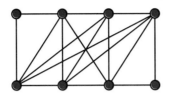  

**Solution**

**a)** *This graph contains $K_{3,3}$ and is non-planar.*

**b)** *This graph contains a subdivision of $K_5$ and is non-planar.*

**c)** *This graph does not contain subdivisions of either $K_{3,3}$ or $K_5$ and thus is planar.*
$v - e + f = 6 - 11 + 7 = 2$
*Euler's formula is satisfied.*

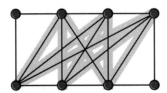  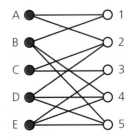

## Modelling with bipartite graphs

A situation such as the one in Exploration 7.1 can be modelled using a bipartite graph. We represent the teachers by the vertices in subset A and the subjects by the vertices in subset B. To simplify the graph we will abbreviate the teachers' names to A, B, C, D and E and denote the subjects by numbers.

| Teacher | Subjects |
|---|---|
| Miss Abraham (A) | Environmental Studies (1), Geography (2) |
| Mr Bowen (B) | Environmental Studies (1), Mathematics (4), Science (5) |
| Mrs Carter (C) | Geography (2), Graphics (3) |
| Ms Delgardo (D) | Geography (2), Mathematics (4), Science (5) |
| Mr Eggham (E) | Graphics (3), Mathematics (4), Science (5) |

Now the edges of the graph represent 'can teach' and they join each teacher to the subjects which they are able to teach.

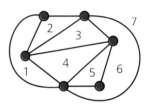

## EXERCISES

7.1A

**1**   **a)**   Draw the following complete bipartite graphs.
      **i)** $K_{2,3}$    **ii)** $K_{4,4}$    **iii)** $K_{1,6}$

     **b)**   How many vertices and edges does the complete bipartite graph $K_{r,s}$ have?
       What are the degrees of the vertices?

**2**   Show which of the following graphs are non-planar because they contain subdivisions of either $K_{3,3}$ or $K_5$.

*The graph in part **a)** is called a **Petersen graph** and is named after Danish mathematician Julius Petersen (1839–1910) who published a paper on its properties in 1898.*

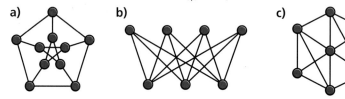

**3**   This bipartite graph shows five people (A, B, C, D and E) and five tasks which they are qualified to do (1–5). Use the graph to state which tasks each person is qualified to do.

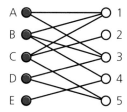

**4**   Everbrite School has entered a team in the local schools' swimming gala. They make a list of their best female and male swimmers and the events in which they can compete.

| | Student | Event |
|---|---|---|
| **Girls** | Meena | breaststroke, backstroke |
| | Lauren | freestyle, breaststroke |
| | Tara | backstroke, breaststroke, freestyle |
| | Kerry | butterfly, freestyle |
| **Boys** | Gary | butterfly, freestyle, breaststroke |
| | Paul | backstroke, freestyle |
| | Eddie | butterfly, backstroke |
| | James | freestyle, backstroke, breaststroke |

Show the two sets of information on bipartite graphs.

**5**   The manager on a building site has to assign five tasks to five members of the workforce. The workers who are qualified to do the jobs are shown in this table.

| Worker | Jobs for which they are qualified |
|---|---|
| Anne | plastering, joinery, glazing |
| Brian | bricklaying, roofing |
| Carl | bricklaying, plastering, joinery |
| Daljit | joinery, roofing, glazing |
| Errol | bricklaying, plastering, roofing |

Show the information on a bipartite graph.

## EXERCISES

1   Determine which of the following are non-planar graphs by trying to find subdivisions of $K_{3,3}$ or $K_5$.

a)                    b)                    c)

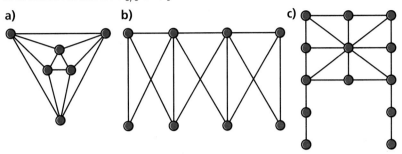

2   a)  Draw the following complete bipartite graphs.
       i) $K_{4,2}$    ii) $K_{2,4}$
       Comment on the resulting graphs.
    b)  If one edge is removed from $K_5$ show that the resulting graph is planar.
    c)  If one edge is removed from $K_{3,3}$ show that the resulting graph is planar.

3   a)  How many vertices and edges does the graph $K_{2n}$ have? For what values of $n$ is $K_{2n}$ planar?
    b)  Show that $K_{2,4}$ is a planar graph by giving a planar drawing of it and verify that it satisfies Euler's theorem.

4   In an oral examination, five students (A, B, C, D and E) are to be examined by four lecturers.
       Lecturer 1 examines A, D, and E. Lecturer 2 examines B, C and E.
       Lecturer 3 examines A, C and D. Lecturer 4 examines B, C and D.
    Show the information on a bipartite graph.

5   Give examples of:
    a)  a subgraph of a non-planar graph which is planar
    b)  a non-planar graph for which Euler's formula is true.

## AN ALGORITHM FOR MAXIMUM MATCHING

Let us return to Mrs Frawt, in Exploration 7.1. Having represented her timetabling problem using a bipartite graph, Mrs Frawt can use it to help find a solution. Although this example can easily be solved by inspection, we shall use it to demonstrate an algorithm that will always provide an optimal solution. A graph that links some of the vertices in A with some of the vertices in B is called a **matching**. The optimal solution is called the **maximum matching**.

The first part of this algorithm involves using a labelling procedure. We start with any matching, **M**. An example follows.

## Step 1:

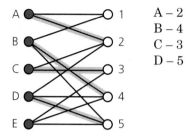

A – 2
B – 4
C – 3
D – 5

This is not a maximum matching, since Mr Eggham has not been allocated a subject and there is no one to teach Environmental Studies, so we proceed as follows.

## Step 2:

Label all the vertices on the left (subset A) which are not in the initial matching with a star (*). In this example the only vertex in subset A *not* in the matching is E.

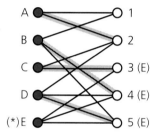

Then we find all the vertices on the right (subset B) that are directly connected to E by edges *not* in the initial matching. We label these (E) since they are connected to vertex E.

If there were more than one right-vertex not in our initial matching, we should repeat this process until they were all labelled in this way.

## Step 3:

Next we choose one of the newly-labelled right-vertices, say 3, and label any left-vertices which are connected directly to it by an edge which is in the initial matching. So we should label vertex C with 3.

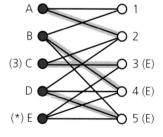

We then repeat this process with other newly-labelled right-vertices until all the labelling at this stage is complete.

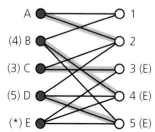

## Step 4:

We now look in turn at the newly-labelled left-vertices, B, C and D, and label the right-vertices connected to them but not in the initial matching which are not already labelled, giving this situation.

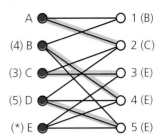

## Step 5:

We repeat steps 3 and 4 until all possible vertices have been labelled.

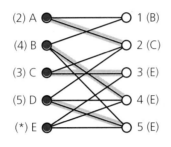

When no further labelling is possible, we move on to the next stage of the algorithm.

### *Finding an alternating path*

An **alternating path** in a bipartite graph is one which satisfies the following conditions:

- ■ it joins vertices from subset A to those in subset B
- ■ alternate edges of the path are in **M**, and the other edges are not in **M**
- ■ the initial and final vertices are not incident with an edge of **M**.

## Step 6:

We now look for a labelled vertex in subset B which is not in our initial matching, in this case vertex 1.

## Step 7:

Starting here, we go to the vertex indicated by its label (B), and from B we go to the vertex indicated by its label and so on until we reach the vertex labelled (*):

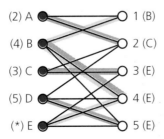

The alternating path is 1–B–4–E.

Our new matching consists of:

■ the edges of the initial matching that are not in the alternating path:
A – 2, C – 3, D – 5

■ the edges in the alternating path that are not in the initial matching:
B – 1, E – 4
giving this solution.

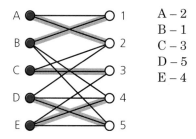

A – 2
B – 1
C – 3
D – 5
E – 4

This is a maximum matching, since all five vertices in A have been matched to vertices in B.

**Note:** *Don't forget to interpret your solution in the context of the original problem.*

So Mrs Frawt should timetable Miss Abraham to teach Geography, Mr Bowen to teach Environmental studies, Mrs Carter to teach Graphics, Ms Delgardo to teach Science and Mr Eggham to teach Mathematics.

It is important to note that this solution is not unique. If we had started with a different initial matching we may well have obtained a different solution.

The algorithm for finding a maximum matching is listed below.

**Step 1:** Find an initial matching (**M**).

**Step 2:** Label with (*) all the vertices in A which are not in **M**.

**Step 3:** Choose a newly-labelled vertex in A, say $a$, and label with ($a$) all the unlabelled vertices in B joined to $a$ by an edge not in **M**. Repeat for all newly-labelled vertices in A before going on to step 4.

**Step 4:** Choose a newly-labelled vertex in B, say $b$, and label with ($b$) all the unlabelled vertices in A joined to $b$ by an edge in **M**. Repeat for all newly-labelled vertices in B before going on to steps 5.

**Step 5:** Repeat steps 3 and 4 until no further labelling is possible.

**Step 6:** Look for a labelled vertex in B which is not in **M**. If such a vertex is found, go to step 7. If there is no such vertex, the matching **M** is optimal. STOP!

**Step 7:** Find an alternating path, starting at the labelled vertex in B and ending at (*).

**Step 8:** The new matching consists of:
■ the edges of **M** which are not in the alternating path
■ the edges in the alternating path but not in **M**.

The previous section was set out with a diagram at each step so that it would be easy to see what was happening. In setting out questions of this type it is not necessary to draw each step, though coloured pens are useful to indicate the initial matching and the alternating path, as shown below.

**Example 7.2**

*A youth club wishes to enter a team at the regional athletics meeting. They have six good runners, A, B, C, D, E and F, who can take part and there are six track events. Runner A can run the 100 m and 200 m, runner B can run the 800 m and 1500 m, C the 200 m and 100 m hurdles, D the 400 m, 800 m and 1500 m, E the 400 m and 800 m and F the 100 m and 100 m hurdles. How should the captain pick the team?*

**Solution**
Initial matching:

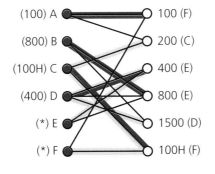

Initial matching **M**
A – 100
B – 800
C – 100H
D – 400

**Note:** This leaves out E *and* F. Thus we get two alternating paths – one starting at 200 (pale highlighter) and one starting at 1500 (dark highlighter).

The alternating paths are:
■ 200 – C – 100H – F
■ 1500 – D – 400 – E

Edges of **M** that are not in either alternating path are:
■ A – 100
■ B – 800

Edges in alternating paths but not in **M** are:
■ C – 200
■ F – 100H
■ D – 1500
■ E – 400

So a solution is:
A – 100 m
B – 800 m
C – 200 m
D – 1500 m
E – 400 m
F – 100 m hurdles

## EXERCISES

**7.2 A**

**1**   Repeat Mrs Frawt's timetabling problem using the initial matching:
A – 1, C – 2, D – 4, E – 5.

**2**   Use the maximum matching algorithm to solve question 3 of
Exercises 7.1A.

**3**   Use the maximum matching algorithm to choose the teams which
Everbrite School will enter for the swimming gala in question 4 of
Exercises 7.1A.

**4**   Using the information in question 5 of Exercises 7.1A, allocate the
jobs on the building site.

**5**   The managing director of a store is seeking to appoint five supervisors
for five of his departments: Electrical goods, Furnishings, Hardware,
Men's wear and Ladies' fashions. The short-listed applicants have a
variety of skills.

Annabel has experience of retailing in both ladies' and men's wear.
Bettina has never worked in clothing, but could work in any of the
other departments.
Chris has spent his working life in electrical and hardware shops.
Doreen's last post was as supervisor in a small hardware shop and
before that she spent several years in a ladies' fashion store.
Imran has worked in men's tailoring and in furnishings.
Who should be appointed to supervise each department? Is your
solution unique?

## EXERCISES

**7.2 B**

**1**   In this bipartite graph, the thick
lines denote an initial matching.
Using this initial matching, find a
maximum matching.

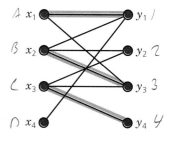

**2**   In this bipartite graph, the thick lines
denote an initial matching. Write down
an alternating path with respect to this
matching and hence find a maximum
matching in the graph.

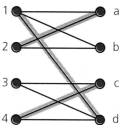

**3**   A university has seven vacancies (a, b, c, d, e, f and g) left on their
courses. There are seven applicants (A, B, C, D, E, F and G) left in
clearing, and their qualifications for the different vacancies are shown
in the following table.

| Course | Suitable applicant |
|--------|--------------------|
| a | A, D, E |
| b | B, E, G |
| c | D |
| d | B, C, E, G |
| e | D, F |
| f | B, D, F |
| g | A, B |

Draw a bipartite graph to represent the problem of allocating courses to applicants. Write down an initial matching and hence find a maximum matching. Is this solution unique?

4   Angela, Beth, Caroline, Denise and Erica are going clubbing on St Valentine's Day and the table below gives the males they would like to accompany them. Determine a suitable scheme for partners for the night out (if one exists).

| | |
|---------|-------------------|
| Angela | Dan, Mike, Gavin |
| Beth | Simon, John |
| Caroline | Simon, Mike |
| Denise | Dan, Gavin, John |
| Erica | Mike, John, Dan |

5   There are five clues left, to complete a crossword.
1 across begins with D.
16 across ends in the letter G.
20 across needs a word with letter A in the middle.
8 down requires an R in the second position and a T in the last position.
12 down is the clue 'student monetary award'.
The five words that come to mind are DOING, DRAFT, FRAME, GOING and GRANT. Can these words be uniquely fitted into the crossword? If so, in which positions should they be placed?

## MATHEMATICAL MODELLING ACTIVITY
## Supply and demand

## Problem statement

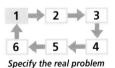
*Specify the real problem*

Five builders' merchants, A, B, C, D and E, supply building sites P, Q, R, S and T. The cost of supplying bricks from each builders' merchant to each building site is shown in the table (in £s per tonne).

| | P | Q | R | S | T |
|---|----|----|----|----|----|
| A | 15 | 14 | 16 | 15 | 10 |
| B | 12 | 10 | 13 | 14 | 10 |
| C | 13 | 10 | 12 | 13 | 12 |
| D | 10 | 16 | 11 | 13 | 13 |
| E | 14 | 12 | 14 | 13 | 15 |

The sites all require the same amount of bricks. Allocate a builders' merchant to supply each site so that the total cost is minimised.

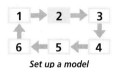

*Set up a model*

# Set up a model

A good starting point is to reduce the table to a 'relative cost' matrix by subtracting the minimum cost from all entries. List any assumptions which have been made.

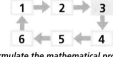

*Formulate the mathematical problem*

# Mathematical model

We will try to allocate as many as possible using the routes with zero relative cost, so we can use the relative cost matrix to produce a bipartite graph showing zero cost edges.

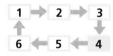

*Solve the mathematical problem*

# Mathematical solution

Obtain an optimal matching using the bipartite graph and consider how you would deal with any builders' merchants or building sites not in this solution.

Inspect your possible solutions to consider which gives a minimum delivery cost.

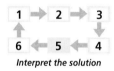

*Interpret the solution*

# Refinements to the model

Instead of simply subtracting the minimum cost from all entries, try firstly subtracting the minimum value in each row from all entries in the row, then subtracting the minimum value from each column of your new matrix from all the values in that column. Use this new matrix to draw a bipartite graph using maximum matching algorithm. Check your solution against your solution from the first attempt.

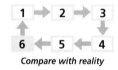

*Compare with reality*

# Interpret the solution

Interpret your final solution in the context of the original problem.

## CONSOLIDATION EXERCISES FOR CHAPTER 7

1  Draw the complete bipartite graphs for **a)** $K_{4, 3}$  **b)** $K_{5, 2}$.

2  Which of the following graphs are planar?

**a)** 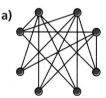  **b)**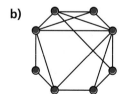

3  Find a maximum matching for the problem modelled by this bipartite graph.

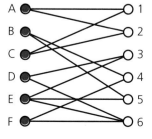

4  Draw a flow chart to show the algorithm for maximum matching.

**5** When Mr Brown moves to a new house, he decides to devote part of the garden to growing his own vegetables. He would like to grow carrots, peas, tomatoes, courgettes and herbs. The area is divided into five plots and, due to the sunlight and drainage,

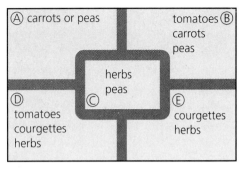

he decides he could plant the crops shown in the diagram. Represent the information on a bipartite graph and decide which crop he should plant in each area.

**6** A small firm gets a rush order. To finish the job on time, the manager asks each of the five employees to work overtime to watch the machines on one evening during the week. Due to other commitments, the employees are only available on the following evenings.

A can work late on Monday or Wednesday.
B can work late on Monday or Thursday.
C can work late on Tuesday, Wednesday or Friday.
D can only work late on Thursday.
E can work late on Tuesday or Friday.

In how many ways can the manager organise the overtime rota?

**7** A manager wishes to appoint supervisors to four of his departments: Electrical, Food, Stationery and Watches. He has four candidates for these posts. The table shows which departments the candidates are qualified to supervise.

| Ahmed | Stationery, Watches |
|---|---|
| Bridget | Electrical, Food, Stationery |
| Chris | Food, Stationery, Watches |
| Diane | Electrical, Stationery |

**a)** Model this situation with a bipartite graph.

The manager allocates Ahmed, Chris and Diane to the first department in their individual lists.

**b)** Starting from this matching show clearly how a matching in which each candidate gets a department can be obtained.

*(ULEAC Question 4, Specimen Paper D1, 1996)*

**8** Everbrite School is mounting its annual play and several staff have volunteered to take charge of various aspects of the production. Mrs Frawt and Miss Delgardo have both said they will do either publicity or ticket sales, Miss Perry can do make-up and costumes, Mr Quince offers help with lighting or scenery, Ms Rouse would like to be the producer but wouldn't mind doing make-up, Mrs Standish enjoys dressmaking but can also paint scenery, Mr Taylor has produced shows before and can also do sound or lighting and Mr Unwin says he'll have a go at anything except sewing! Who should do which task?

**9**    One Saturday six potential customers visit the local travel agent looking for 'last minute' bargain holidays. The agent has holidays available with the following accommodation.

    1 hotel with a twin room     1 hotel with a family room (for four)
    1 apartment sleeping four     1 flat sleeping up to four
    1 apartment sleeping six     1 flat sleeping up to eight

The customers are: one married couple, a group of three friends, a family of four, a family of five, two families (seven people in all) who wish to stay together and a group of five friends. Can the agent arrange holidays for all the customers? If not, who will be disappointed?

**10**    A medley relay swimming team consists of Mike, Neal, Oliver and Pete. The legs which they can swim are:

    Mike: backstroke or butterfly
    Neal: breaststroke, backstroke or freestyle
    Oliver: breaststroke, freestyle or butterfly
    Pete: freestyle or backstroke.

**a)**    In how many ways can the swimming order be decided?
**b)**    The table below gives each swimmer's times for the different legs. Which solution gives the minimum time?

|  | Breaststroke | Backstroke | Butterfly | Freestyle |
|---|---|---|---|---|
| Mike |  | 66 | 64 |  |
| Neal | 66 | 64 |  | 60 |
| Oliver | 63 |  | 65 | 61 |
| Pete |  | 65 |  | 63 |

## Summary

*After working through this chapter you should:*

■ understand that a bipartite graph is a graph in which the vertices can be divided into two subsets such that each edge of the graph joins a vertex in one subset to a vertex in the other subset

■ understand the term 'subdivision' to mean the insertion into a bipartite graph of vertices of degree 2

■ be able to show when a graph is non-planar by finding a subdivision of $K_{3,3}$

■ know how to use bipartite graphs to model real problems such as matching workers with particular skills to a job that needs to be done

■ use the maximum matching algorithm to find the best solution to the matching problem.

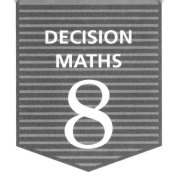

# Linear programming 1: Modelling and graphical solution

*By the end of this chapter you should be able to:*

■ *recognise problems that involve linear programming and know how to formulate them as mathematical problems*

■ *define the variables in a problem and formulate the objective function which relates these variables*

■ *establish the constraints which are limits on the values and ranges of the variables*

■ *use a graphical method for solving linear programming problems in two variables by identifying the feasible region which includes all the possible combinations of values of the variables.*

## LINEAR PROGRAMMING PROBLEMS

**Exploration 8.1**

### A balanced diet

The parents of a two-year-old child want to ensure that they include at least the minimum satisfactory amount of protein, energy and vitamin C. The daily amounts recommended by good health guides are 32 g of protein, 5500 kilojoules (kJ) of energy and 30 mg of vitamin C. Suppose that the parents decide to achieve the daily amounts through milk and orange juice. The following table shows the amount of protein, energy and vitamin C available in these drinks.

| Item | Measure of drink | Protein (g) | Energy (kJ) | Vitamin C (mg) |
|------|------------------|-------------|-------------|----------------|
| milk | 100 ml | 3.0 | 300 | 1 |
| juice | 100 ml | 0.2 | 40 | 10 |

Milk costs 38p per litre and orange juice costs 48p per litre. The parents want to find the minimum costs to achieve the recommended daily intake.

If $x$ and $y$ are the number of measures of milk and juice, write down:
■ the total costs of the drinks
■ three inequalities for the amount of protein, energy and vitamin C so that the minimum satisfactory amounts are achieved.

## Cost function

The **cost function** is the relationship between the cost and the variable components included in the cost. One measure of milk or orange juice is 0.1 litre so the cost of a measure of milk is 3.8p and the cost of a measure of juice is 4.8p.

$x$ measures of milk will cost $3.8x$ and $y$ measures of orange juice will cost $4.8y$.

The total cost function is given by $C = 3.8x + 4.8y$.

## Constraints

For protein, the number of grams in $x$ measures of milk is $3x$ and in $y$ measures of orange juice it is $0.2y$. The minimum amount of protein required is 32 g. So:
   $3x + 0.2y \geq 32$

Similarly, the amount of energy in $x$ measures of milk and $y$ measures of orange juice is $300x + 40y$. Since the minimum requirement is 5500 kJ:
   $300x + 40y \geq 5500$

For vitamin C:
   $x + 10y \geq 30$

Finally, since it is not possible for the child to drink negative amounts of milk and orange juice:
   $x \geq 0$ and $y \geq 0$

The mathematical problem that we need to solve can be summarised like this.

| | | |
|---|---|---|
| Find a minimum value for | $C = 3.8x + 4.8y$ | **1** |
| where | $3x + 0.2y \geq 32$ | **2** |
| | $300x + 40y \geq 5500$ | **3** |
| | $x + 10y \geq 30$ | **4** |
| | $x \geq 0, y \geq 0$ | **5** |

This is an example of a linear programming problem. It is made up of two parts:
■ line 1: the **objective function**
■ lines 2–5: the **constraints**.

The objective function is a **linear function** that has to be minimised or maximised depending on the nature of the problem.

The constraints are linear equalities or linear inequalities that restrict the values of the variables $x$ and $y$. The constraints given by line 5 are called the **trivial constraints**.

The above mathematical problem will be solved graphically and algebraically later in this chapter.

## Formulating linear programming problems

**Example 8.1**

*Ben has two evening and weekend jobs, mowing lawns and cleaning cars. He can spend a maximum of 15 hours per week on these jobs. He has several regular clients for whom he spends six hours per week mowing lawns and three hours per week cleaning cars. Ben is paid £2.50 per hour for mowing lawns and £3.50 per hour for cleaning cars.*

*Formulate a linear programming problem for Ben to maximise his earnings.*

### Solution
*Start by defining the variables.*
- *Let $x$ be the number of hours per week spent mowing lawns.*
- *Let $y$ be the number of hours per week spent on cleaning cars.*
- *Let $I$ be the income per week.*

*Now set up the objective function.*
*The amount Ben earns by mowing lawns is 2.5x and from cleaning cars it is 3.5y. The objective function is:*
$$I = 2.5x + 3.5y$$

*Establish the constraints.*
*Since the maximum number of hours available is 15:*
$$x + y \leq 15$$

*Ben is committed to 6 hours of mowing lawns and 3 hours of cleaning cars:*
$$x \geq 6, y \geq 3$$

*The linear programming problem is:*

| | | |
|---|---|---|
| *Find a maximum value of* | $I = 2.5x + 3.5y$ | **1** |
| *where* | $x + y \leq 15$ | **2** |
| | $x \geq 6, y \geq 3$ | **3** |

Notice that in this example we do not need the trivial constraints because they are satisfied by line 3.

## EXERCISES

**8.1 A**

1   Chris is making coffee cakes and chocolate cakes for a cake stall at the village fete. Each chocolate cake needs 2 eggs and 250 g of margarine. Each coffee cake needs 3 eggs and 150 g of margarine. Chris buys three dozen eggs and a 3 kg tub of margarine. Formulate a linear programming problem so that Chris can make as many cakes as possible.

2   A factory produces two types of toys: trucks and bicycles. In the manufacturing process of these toys three machines are used. These are a moulder, a lathe and an assembler. The table shows the length of time needed for each toy.

| | Moulder | Lathe | Assembler |
|---|---|---|---|
| bicycle | 1 hour | 3 hours | 1 hour |
| truck | 0 hours | 1 hour | 1 hour |

The moulder can be operated for 3 hours per day, the lathe for 12 hours per day and the assembler for 7 hours per day. Each bicycle made gives a profit of £15 and each truck made gives a profit of £11. Formulate a linear programming problem so that the factory maximises its profit.

3    A physical fitness enthusiast, Karen, decided to exercise by a combination of jogging and cycling. She jogs at 6 mph and cycles at 18 mph. A measure of the benefit of these activities is given through aerobic points. An hour of jogging gains 4 aerobic points and an hour of cycling gains 3 aerobic points. Each week she would like to earn at least 12 aerobic points, cover at least 54 miles and cycle at least twice as much as she jogs.

Formulate a linear programming problem so that Karen minimises the total time spent exercising while achieving her goals.

4    A house builder builds two types of home.

The first type requires one plot of land, £40 000 capital, 150 worker days of labour and is sold for £8000 profit.
The second type requires two plots of land, £105 000 capital, 200 worker days of labour to build and is sold for £11 500 profit.
The house builder owns 150 plots of land, has available £9 600 000 and 24 000 worker days of labour.
Formulate a linear programming problem to advise the builder on a strategy to maximise the profit.

5    A food manufacturer makes Alma margarine from vegetable and non-vegetable oils. The raw oils are refined by the manufacturer and blended to form the margarine. One objective of the manufacturer is to make as large a profit as possible; however, there are various constraints. It is important that the raw oils are refined separately to avoid contamination and the final blend of oils must be soft enough to spread, but not too runny. To achieve this quality each ingredient and the final product has a hardness factor. The following table gives the cost and hardness of the raw materials used to make Alma margarine.

|  | Oil | Cost (£ per kg) | Hardness factor |
|---|---|---|---|
| Vegetable | Groundnut | 0.7 | 1.2 |
|  | Soya bean | 1.05 | 3.4 |
|  | Palm | 1.05 | 8.0 |
| Non-vegetable | Lard | 1.30 | 10.8 |
|  | Fish | 1.10 | 8.3 |

Alma margarine sells for £1.55 per kg and to ensure good spreading, its hardness must be between 5.6 and 7.4. The manufacturers can refine up to 40 000 kg of vegetable oil and up to 32 000 kg of non-vegetable oil per day. Formulate the problem of maximising the profit as a linear programming problem.

## EXERCISES

1   As part of a charity stunt a student wishes to travel around a circular route in 80 minutes or less. The first five miles are to be travelled on top of a horse-drawn tram and the final three miles in a pink hearse. If the speeds of tram and hearse are $x$ mph and $y$ mph respectively, write down the constraints for the above.

2   For a dog to remain healthy its daily diet must provide it with at least 2 megajoules (MJ) of energy and at least 0.3 kg of protein, but not more than 0.04 kg of salt. There are two items in the diet of most dogs, biscuits and meat.
1 kg of meat costs 90p and provides 5 MJ of energy, 0.5 kg of protein and 0.04 kg of salt.
1 kg of biscuits costs 25p and provides 20 MJ of energy and 0.2 kg of protein and 0.08 kg of salt.
Formulate the linear programming problem to enable you to find the cheapest combination of biscuits and meat in the dog's daily diet, that will keep it healthy.

3   A soap company manufactures two powders: XTRA for extra cleaning power and YTER for a whiter wash. These powders contain ingredients A and B in the following proportions.

|       | A   | B   |
|-------|-----|-----|
| XTRA  | 90% | 10% |
| YTER  | 50% | 50% |

The company can store up to 20 000 kg of ingredient A and up to 7000 kg of B. To obtain favourable price reductions their supplier requires a purchase of at least 5000 kg of A and at least 2000 kg of B per week. The maximum amounts of XTRA and YTER in total that can be produced by the plant is 25 000 kg per week. The profit on XTRA is £1.50 per kg and on YTER it is £2.10 per kg.
Devise a set of inequalities to represent all of the above constraints in terms of the number of thousand kg of each of XTRA and YTER produced each week. Write down the algebraic expression representing the weekly profit.

4   A fruit farmer has two orchards A and B, both of which produce apples of class 1, class 2 and class 3.

In orchard A, the labour and equipment costs for picking are £20 per hour and picking for one hour produces six bushels of class 1 apples, two bushels of class 2 apples and four bushels of class 3 apples.
In orchard B, the labour and equipment costs for picking are £16 per hour and picking for one hour produces two bushels of class 1 apples, two bushels of class 2 apples and twelve bushels of class 3 apples.

The farmer is under contract to supply twelve bushels of class 1 apples, eight bushels of class 2 apples and fourteen bushels of class 3 apples to a market each day.

Formulate a linear programming problem to determine how many hours each day the labour force should be consigned to each orchard if the total costs are to be minimised, given that the labour force will not work more than eight hours per day.

The farmer may offset his costs by selling picked apples, which are surplus to those under contract, to a cider manufacturer at £1 per bushel (regardless of class). Determine the new cost function in this case.

5　New World Wines Ltd produce a wine by blending 'Chateau Australia House' (CAH) with an inferior 'Bondi Beach Muddie' (BBM). The blend becomes unstable if the proportion of BBM exceeds 80 per cent or falls below 25 per cent of the mixture, so this is to be avoided. The maximum wine-handling capacity per day is 5000 litres and the plant is so constrained that certain pipes and mixing vessels must always be kept full, so that the total amount of BBM processed and half the amount of CAH processed is at least 2000 units. The cost to the company of BBM is 22 cents per litre and that of CAH is 30 cents per litre and the blended wine is sold at 200 cents per litre.

Formulate a linear programming problem that will:
- minimise the costs for the company,
- maximise the profit for the company.

## SOLVING LINEAR PROGRAMMING PROBLEMS

We can find an optimal solution for linear programming problems with two variables by drawing straight-line graphs for the constraints. This defines an acceptable region called the **feasible region**. Then an optimal solution can be found at one of the points of the feasible region.

**Exploration 8.2**

## *A graphical solution*

A linear programming problem is given by the following summary.

| Maximise the function | $f = 80x + 70y$ | 1 |
| where | $2x + y \leq 32$ | 2 |
| | $x + y \leq 18$ | 3 |
| | $x \geq 0, y \geq 0$ | 4 |

- Draw the straight-line graphs $2x + y = 32$ and $x + y = 18$.
- Shade in the region of the $x$–$y$ plane in which inequalities 2–4 hold.
- Investigate at which point the objective function $f$ takes its largest value.

## Solving graphically

Referring to Exploration 8.2, consider the constraint 2 in the form:

$$y \le 32 - 2x$$

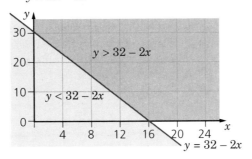

For all points on the line, $y = 32 - 2x$. Therefore the points satisfying $y < 32 - 2x$ must lie below the line i.e. for $y$ values that are smaller than $32 - 2x$.

Repeating the same argument for inequalities 3 and 4, the allowable region that satisfies the constraints is shown in the diagram. This is the feasible region and in this case it includes the edges of the region.

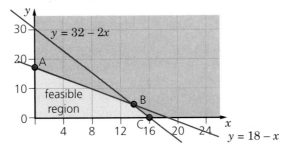

The objective function is $f = 80x + 70y$. By trial and error you may have found that this function takes its largest value at point B(14, 4). At this point:

$$f = 80 \times 14 + 70 \times 4 = 1400$$

The result that the optimal value occurred at one of the vertices of the feasible region is no mere chance, as the following theorem shows.

## The fundamental theorem of linear programming

The maximum (or minimum) value of the objective function occurs at one of the vertices of the feasible region.

We can verify this theorem using the feasible region and objective function for Exploration 8.2. Consider the objective function:

$$f = 80x + 70y$$

Rearranging this gives:

$$y = \tfrac{1}{70} f - \tfrac{8}{7} x$$

For given values of $f$ this is an equation of a straight line with slope $-\tfrac{8}{7}$.    161

The diagram shows the feasible region and the graph of $y = \frac{1}{10}f - \frac{8}{7}x$ for $f = 700, f = 1400$ and $f = 2100$.

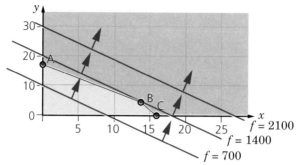

The three lines are all parallel. As $f$ increases in value the line $y = \frac{1}{10}f - \frac{8}{7}x$ moves through the feasible region until it reaches B when it is about to leave the feasible region. Beyond B the constraints are no longer satisfied. So the maximum value of $f$ for points inside the feasible region occurs at vertex B.

## Exploration 8.3

### *Investigating the fundamental theorem*

A linear programming problem is given by the following summary.

| | | |
|---|---|---|
| Maximise the function | $f = x + y$ | 1 |
| where | $y - 2x \leq 0$ | 2 |
| | $x + 2y \leq 100$ | 3 |
| | $4x + 3y \leq 300$ | 4 |
| | $x \geq 0, y \geq 0$ | 5 |

■  Draw the feasible region.
■  Verify the fundamental theorem of linear programming for $f = x + y$.

### Investigating the theorem

The diagram shows the feasible region and the line $y = f - x$ for $f = 100$.

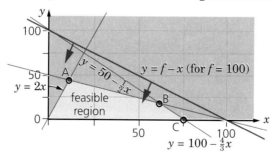

As the line $y = f - x$ moves towards the feasible region the value of $f$ decreases until the line reaches the vertex B. At this point $x = 60$ and $y = 20$ and $f = x + y = 80$. If the line moves into the feasible region then $f$ decreases further.

Hence the maximum value of $f$ occurs at point B and has value 80.

**Example 8.2**

*Solve this linear programming problem.*

*Minimise the function* $f = 2x + 3y$
*where*
$$x + y \geq 4$$
$$3x + 5y \geq 16$$
$$x \geq 0, y \geq 0$$

**Solution**
*The feasible region is the region bounded by the straight lines*
$x + y = 4$, $3x + 5y = 16$ *and* $y = 0$.

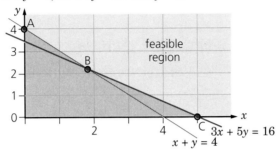

*This table shows the coordinates of the vertices and the value of the objective function.*

| Vertex | $x$ | $y$ | $f = 2x + 3y$ |
|:---:|:---:|:---:|:---:|
| A | 0 | 4 | 12 |
| B | 2 | 2 | 10 |
| C | $\frac{16}{3}$ | 0 | $\frac{32}{3}$ |

*The minimum value of the objective function is $f = 10$ and occurs at $x = 2$, $y = 2$.*

**Example 8.3**

*Solve this linear programming problem.*

*Minimise the function* $f = 7x + 4y$
*where*
$$2x + y \geq 11$$
$$x + y \leq 10$$
$$x + 3y \leq 18$$
$$x + 4y \geq 16$$

**Solution**
*The feasible region is the region bounded by the straight lines*
$2x + y = 11$, $x + y = 10$, $x + 3y = 18$ *and* $x + 4y = 16$.

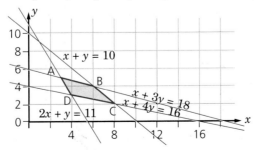

*The following table shows the coordinates of the vertices and the value of the objective function f.*

| Vertex | $x$ | $y$ | $f = 7x + 4y$ |
|:---:|:---:|:---:|:---:|
| A | 3 | 5 | 41 |
| B | 6 | 4 | 58 |
| C | 8 | 2 | 64 |
| D | 4 | 3 | 40 |

*The minimum value of f is 40 and occurs at x = 4, y = 3.*

**Exploration 8.4**

## *A non-unique solution*

■ Investigate the fundamental theorem applied to this problem.

Maximise the function   $f = x + y$
where
$$y - 4x \leq 0$$
$$x + y \leq 10$$
$$4x + y \leq 20$$
$$x \geq 0, y \geq 0$$

## More than one solution

In this problem the slope of the objective function straight line $y = f - x$ is equal to the slope of the boundary of the feasible region. The maximum value of $f$ is 10 and occurs at any point of the boundary between A and B.

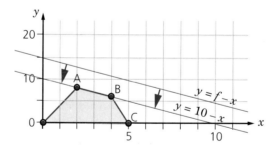

## EXERCISES

**8.2 A**

1   Solve the following linear programming problems using the graphical method of solution.

   **a)** Maximise the function   $f = 2x + 3y$
      where
$$x + 3y \leq 120$$
$$7x + 2y \leq 156$$
$$x \leq 20$$
$$x \geq 0, y \geq 0$$

**b)** Maximise the function    $f = x + 2y$
where

$x + y \leq 100$
$y - 3x \geq 60$
$x - y \geq 0$

**c)** Minimise the function    $f = x + y$
where

$2x + y \leq 6$
$2x + 3y \leq 12$
$4x + y \leq 10$
$x \geq 0, y \geq 0$

**d)** Minimise the function    $f = 5x + 2y$
where

$x + y \leq 20$
$2x + y \geq 20$
$4x - 3y \geq 10$
$y \geq 4$

**e)** Maximise the function    $f = \frac{2}{3}x + y$
where

$x + y \geq 15$
$x + y \leq 45$
$x \leq 30$
$y \leq 25$
$x \geq 0, y \geq 0$

**2**  Maximise the function    $c = 10x + 17y - 1500$
where

$3x + 2y \leq 3600$
$x + 4y \leq 1800$
$x \geq 0, y \geq 0$

**3**  Minimise the function    $f = 2x + y$
where

$x + y \leq 27$
$x + y \geq 15$
$y \leq 2x$
$x \geq 3, y \geq 3$

**4**  Maximise the function    $f = x + \frac{1}{2}y$
where

$2x + y \leq 36$
$x + y \geq 15$
$y \leq 2x$
$x \geq 3, y \geq 3$

**5**  The constraints of a linear programming problem are given by:

$2y - x < 40$
$3x + 4y \geq 180$
$x < 90$
$y \geq 0$

**a)** Find integer values of $x$ and $y$ which maximise the objective function
$f = 10x + 10y$.

**b)** Find integer values of $x$ and $y$ which minimise the objective
function $c = x + 5y$.

**6**  In a developing nation the government wants to encourage everyone
to make rice and soya beans part of the staple diet. The object is to
design a lowest-cost diet that provides certain minimum levels of

protein, calories and vitamin B12 (riboflavin). Suppose that one cup of uncooked rice costs 27 pence and contains 15 grams of protein, 810 calories and $\frac{1}{9}$ milligram of riboflavin. On the other hand, one cup of uncooked soya beans costs 21 pence and contains 22.5 grams of protein, 270 calories and $\frac{1}{3}$ milligram of riboflavin. Suppose that the minimum daily requirements are 90 grams of protein, 1620 calories and 1 milligram of riboflavin. Formulate and solve a linear programming problem to minimise the cost of a diet to satisfy the requirements.

7   An oil company owns two refineries. Refinery 1 produces each day 100 barrels of high-grade oil, 200 barrels of medium-grade oil and 300 barrels of low-grade oil and costs £10 000 to operate. Refinery 2 produces each day 200 barrels of high-grade oil, 100 barrels of medium-grade oil and 200 of low-grade oil and costs £9000 to operate. An order is received for 1000 barrels of high-grade oil, 1000 barrels of medium-grade oil and 1800 barrels of low-grade oil. How many days should each refinery be operated in order to fill the order at the least cast?

8   A publisher, MathGem, publishes 40 new books each year which are classified as sixth-form level, undergraduate level and graduate level. The policy of the company is to publish at least three times as many sixth-form books as undergraduate books, and at least twice as many undergraduate books as graduate books. To raise its profit, MathGem has decided to publish at least four graduate books per year.

On average the annual profits are £9000 for each sixth-form book, £5000 for each undergraduate book and £1000 for each graduate book. How many books of each type should be published for MathGem to maximise its annual profit?

9   CTM foods Ltd is designing a new breakfast cereal using wheatgerm and enriched oat flour as the basic ingredients. Each ounce of wheatgerm contains 2 milligrams of niacin, 3 milligrams of iron and 9.5 milligrams of thiamine, and costs 5 pence. Each ounce of enriched oat flour contains 3 milligrams of niacin, 3 milligrams of iron and 0.25 milligrams of thiamine, and costs 8 pence. The nutritionist wants the cereal to have at least 7 milligrams of niacin, 8 milligrams of iron and 1 milligram of thiamine per ounce. How many ounces of wheatgerm and how many ounces of enriched oat flour should be used in each serving in order to meet the nutritional requirements at the least cost?

10   Solve questions 1–4 from Exercises 8.1A. Why can you not solve question 5?

## EXERCISES

### 8.2B

1   Shade in the feasible region determined by the simultaneous inequalities:
$3y + 2x \le 12,\ x - 5y \le 5,\ y - 3x \le 3$.
Which of the points (0, 0), (–2, 3), (2, –3), (0, –1), (2, 2), (3, 5) lie in the feasible region?

**2**  A feasible region is defined by the area inside and on the square with the four vertices:
A(1, 1), B(–1, 1), C(–1, –1), D(1, –1).

At which corner does the maximum of the objective function
$z = -2x - 3y$ occur? State the maximum value of $z$.

**3**  Draw the line segments joining the following points.
A(0, 0) to B(5, 9)   B(5, 9) to C(10, 7)   C(10, 7) to D(15, 0)
D(15, 0) to A(0, 0)
A feasible region is defined as that region on or inside these line segments. Where is the minimum of the objective function $f = y - 2x$ attained? What is the minimum value of $f$?

**4**  Maximise $P = 4x + 2y$ subject to the constraints:
$x - 2y \le 1, x + y \le 6, 2x + y \le 6, x \ge 0, y \ge 0.$
At which point does the maximum value of $P$ occur?

**5**  The feasible region M is defined by $y \ge 2, x \le 1, y \le 2x + 4$. Where does the objective function $f = -6x + 2y$ attain:
**a)** a minimum       **b)** a maximum value in M?

**6**  Sketch the feasible region defined by:
$x \ge 0, y \ge 0, 2y + x \le 0, x + y \le 6, y + 2x \le 10.$
Determine the vertices and label them O(0, 0), A, B, C and D in a clockwise sense from O (so A is (0, 5)). If $m$ and $n$ are fixed positive real numbers, maximise the objective function $f = mx + ny$. Show that the maximum value of $f$ will occur at A if $\dfrac{m}{n} < \dfrac{1}{2}$. Determine inequalities satisfied by $\dfrac{m}{n}$ if $f$ is maximised at:

**a)** B     **b)** C     **c)** D.

What occurs if $\dfrac{m}{n} = \dfrac{1}{2}$?

**7–10**  Find solutions to questions 2–5 of Exercises 8.1B.

# MATHEMATICAL MODELLING ACTIVITY

## Curve fitting

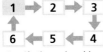

**1 → 2 → 3**
**6 ← 5 ← 4**
*Specify the real problem*

## Problem statement

A pump's performance may be represented by three curves:
- production head* against flow rate
- absorbed power against flow rate
- efficiency against flow rate.

*Head refers to the height to which the pump can lift a column of water and is the most common way of describing the device's pressure-generating capability.

At a fixed speed of shaft rotation, these parameters are linked by:

$$\eta = 0.001 \frac{\rho g h q}{P}$$

where $\eta$ is the efficiency (%), $\rho$ is the density (assumed as 1000 kg/m³), $g$ is the gravitational acceleration (taken as 9.81 m/s²), $h$ is the head generated in metres (m), $q$ is the flow rate in litres per second (l/s) and $P$ is the power in kilowatts (kW).

When a pump is being mathematically tested during the design process, the 'best efficiency point' (BEP) is first fixed in order to specify the optimum flow, head, efficiency and consequent power absorption.

For a particular pump, the BEP head = 20 m, BEP flow = 50 l/s, BEP efficiency = 80% and so BEP power = 12.26 kW.

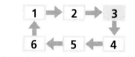

*Formulate the mathematical problem*

# Mathematical problem

■ In addition to the BEP, the head may be specified at two additional points, one at zero flow and another at, for example, 140% of BEP flow. A curve can be defined to join these three specified points of head against flow. Assume that at zero flow $h = 25$ m and at 140% of BEP flow $h = 15$ m. Based on these figures, deduce a method of joining the three points with a smooth curve and determine its equation.

■ The efficiency will also be represented by a curve that is fixed by three conditions:
   **a)** its value will be zero at zero flow
   **b)** its value at BEP is known
   **c)** it must peak at BEP.
   Deduce a method of meeting these three efficiency conditions with another smooth curve and determine the equation relating to $q$.

■ Given that head–flow and efficiency–flow curves are now defined in equation form, can the power be represented as another curve for which the equation can be found? If so, calculate this equation. Hence or otherwise calculate the power which is absorbed at intervals of 5 l/s between zero flow and 140% of BEP flow. Draw the graph of these results and interpret your results.

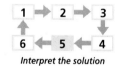

*Interpret the solution*

# Refinement of the model

■ Additional efficiency points may be specified through which the efficiency curve must pass (for instance $n = 15$, $q = 15$). Investigate the effect this has on the modelling process.

■ The head–flow curve of a pump is often best represented by a straight line at low flow, say linear up to flow 20 l/s and head 24 m, becoming curved at higher flow. How would this affect your modelling process?

## CONSOLIDATION EXERCISES FOR CHAPTER 8

1   A farmer grows two crops of wheat and beet. The number of hectares
    of wheat, $x$, and the number of hectares of beet, $y$, grown must
    satisfy the constraints:
    $$10x + 3y \leq 52, 2x + 3y \leq 18, y \leq 4, x \geq 0, y \geq 0.$$
    a) Determine by graphical means the values of $x$ and $y$ for which the
       profit function $P = 7x + 8y$ is a maximum. State the maximum
       value of $P$.
       To ease administration and processing the farmer decides that
       $x$ and $y$ must be integers.
    b) Determine the values of $x$ and $y$ which make $P$ a maximum in this
       case. State this maximum value.
       *(ULEAC Question 6, Decision Mathematics Specimen Paper D1, 1996)*

2   In order to supplement his daily diet Harry wishes to take some
    Xtravit and some Yeastalife tablets. Their contents of iron, calcium
    and vitamins (in milligrams per tablet) are shown in the table.

| Tablet | Iron content | Calcium content | Vitamin content |
|--------|--------------|-----------------|-----------------|
| Xtravit | 6 | 3 | 2 |
| Yeastalife | 2 | 3 | 4 |

    a) By taking $x$ tablets of Xtravit and $y$ tablets of Yeastalife Harry
       needs to receive at least 18 milligrams of iron, 21 milligrams of
       calcium and 16 milligrams of vitamins. Write these conditions
       down as three inequalities in $x$ and $y$.
    b) In a coordinate plane illustrate the region of those points $(x, y)$ which
       simultaneously satisfy $x \geq 0$, $y \geq 0$ and the three inequalities in a).
    c) If the Xtravit tablets cost 10p each and the Yeastalife tablets cost
       5p each, how many tablets of each should Harry take in order to
       satisfy the above requirements at the least cost?
       *(AEB Question 9, Discrete Mathematics, Specimen Paper, 1996)*

3   A firm of biscuit bakers make two varieties of custard creams:
    ■ type A, in packets of 12
    ■ type B, in packets of 20.
    Type-A biscuits are for the upper end of the market, having twice as
    much filling as type-B biscuits. If only type-A biscuits were made,
    there would be enough filling to make 2000 packets per day.

    Each type-A biscuit requires 25 per cent more biscuit crumbs for its
    manufacture than does a type-B biscuit. If only type-A biscuits were
    made, there would be enough crumbs for 2400 packets per day.

    Customer demand requires that for every two packets of type-A
    biscuits there must be at least one packet of type-B biscuits. The profit
    on each packet of biscuits sold is 10p regardless of type, and the firm
    wishes to maximise its profits. Assuming that every packet produced is
    sold, formulate and solve the linear programming problem that
    would help to decide how to do this.
    *(Open University)*

4    A craftworker makes cushions and rag dolls to sell to a shop. Each
     cushion uses £15 worth of materials and takes 40 minutes to make.
     Each rag doll uses £4.50 worth of materials and takes one hour to
     make. The craftworker has only enough material to make at most
     four rag dolls.
   **a)** Given that the craftworker has £78 to buy materials and 6 hours in
        which to make his products, show that this situation can be
        modelled by the following constraints:
        $10x + 3y \leq 52$
        $2x + 3y \leq 18$
        $y \leq 4, x \geq 0, y \geq 0$
        where $x$ is the number of cushions made and $y$ the number of rag
        dolls made.
        The craftworker makes £7 profit for each cushion he sells and £8
        profit for each rag doll he sells.
   **b)** Write down an expression for the profit $P$.
   **c)** Given that the craftworker can only sell completed work, use a
        graphical method to obtain his most profitable strategy.
                    *(ULEAC Question 5, Decision Mathematics Specimen Paper D1, 1996)*

5    A baker has 8.5 kg of flour, 5.5 kg of butter and 5 kg of sugar
     available at the end of a working day. With these ingredients, biscuits
     and/or buns can be made. The recipes are as follows.

| 30 biscuits | 200 g flour | 40 buns | 200 g flour |
|---|---|---|---|
|  | 120 g butter |  | 200 g butter |
|  | 100 g sugar |  | 200 g sugar |

   **a)** Biscuits are sold for 5p each and buns for 7p each. Let $x$ be the
        number of biscuits baked and $y$ be the number of buns baked.
        Assuming that any number of biscuits and/or buns can be baked,
        show that the amount of flour used, in g, is given by $6\frac{2}{3}x + 5y$.
        Hence write down and simplify an inequality constraining the
        values of $x$ and $y$.
   **b)** Produce two further inequalities relating to the availability of
        butter and of sugar.
   **c)** Assuming that all biscuits and buns can be sold, produce a function
        which gives the income from the sales.
   **d)** Formulate and solve graphically the linear program: 'Maximise
        income subject to the availability of flour, butter and sugar.'
        State the maximum income expected, and the numbers of biscuits
        and buns baked.
                                              *(UODLE Paper 4, 1991)*

6    A paper manufacturer has a roll of paper to cut up. It is 40 cm wide
     and 200 metres long, and is to be cut along its length to produce
     widths of 11 cm for toilet rolls, and widths of 24 cm for kitchen rolls.
     The diagram shows two possible cutting plans. Some of the roll may
     be cut to plan A, some to plan B, and some may be left uncut.

**Plan A**                    **Plan B**

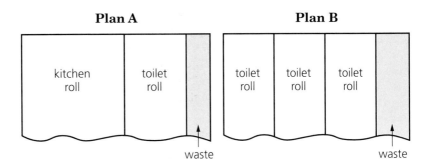

Kitchen roll paper sells for 7p a metre. Toilet roll paper sells for 4p a metre. Any of the roll that remains uncut can be sold for 8p a metre. The manufacturer is keen to ensure that no more than 15 per cent of the roll is wasted.

**a)** Let $a$ metres be the length that is cut to plan A, and $b$ metres be the length cut to plan B, so that $(200 - a - b)$ metres remains uncut. Formulate a linear programme to find the length of roll which should be cut to each plan so as to produce the maximum income within the given constraints.

**b)** Solve the problem graphically.

*(UODLE Paper 4, 1992)*

**7**    James Bond is very particular about his cocktails. He has them mixed from gin and martini and he insists that they satisfy the following constraints:

**Dryness**
Gin has a dryness rating of 1. Martini has a dryness rating of 3. Dryness blends linearly, i.e. a mixture of $x$ ml of gin and $y$ ml of martini has a dryness of $\dfrac{x + 3y}{x + y}$.

James insists that the dryness rating of his cocktail is $\leq 2$.

**Alcohol**
Gin is 45 per cent alcohol by volume. Martini is 15 per cent. Alcohol also blends linearly. James insists that his cocktail is between 18 per cent and 36 per cent alcohol.

James has ordered a 200 ml cocktail. Let the amount of gin in it be $x$ ml, and let the amount of martini be $y$ ml, so that $x + y = 200$.

**a)** Explain why the dryness constraint may be expressed as $x + 3y \leq 400$.

**b)** In a similar way produce and simplify two constraints relating to alcohol content.

**c)** Graph all three constraints.

**d)** Given that gin is more expensive than martini, and remembering $x + y$ must equal 200, give the cheapest and most expensive cocktails that will satisfy James's requirements.

**e)** An alternative approach to this problem is to let $p$ be the proportion of gin in the cocktail, so the $1 - p$ is the proportion of martini. Using this approach the first constraint becomes: $p + 3(1 - p) \leq 2$; that is, $p \geq \frac{1}{2}$.

Express the other constraints in this way and compare the acceptable values for $x$ with those which you obtained for $x$ in **d)**.

*(UODLE Paper 4, 1994)*

8    James T Squirt and Co., toy manufacturers, produce two different model spaceships, the Alpha and the Beta. There are 160 units of materials, 120 robot hours and 90 hours for hand-finishing are available. The Alpha uses 15 units of materials, 20 robot hours and takes 12 hours to finish. The Beta uses 20 units of materials, 5 robot hours and takes 9 hours to finish. The profits on the Alpha and Beta are £8.50 and £11.75 respectively. How many of each model should Squirt and Co. produce?

## Summary

*After working through this chapter you should know that:*

■ in a linear programming problem we find the maximum or minimum of a linear function (called the objective function) which is subject to a set of constraints which are linear inequalities or equalities

■ the steps in solving a linear programming type problem are:
Step 1: Formulate the linear programming problem:
   a) define variables
   b) form objective function
   c) form the constraints.
Step 2: Write down the equations of the boundary of the feasible region.
Step 3: Determine the vertices of the feasible region.
Step 4: Evaluate the objective function at each vertex and determine the optimal value.

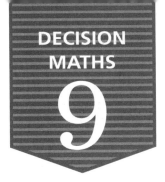

# 9

# *Linear programming 2: the simplex method*

*In this chapter we continue the investigation of linear programming problems and introduce:*

■ *the method of Gaussian elimination for solving simultaneous linear equations*

■ *the simplex method for solving linear programming problems which uses the elimination ideas.*

The fundamental theorem of linear programming tells us that an important step in the solution process is the need to find the coordinates of the vertices of the feasible region. The optimal solution occurs at one of these vertices. The graphical approach adopted in Chapter 8, *Linear programming 1: Modelling and graphical solution*, works well for problems with two variables but cannot easily be extended to problems with three or more variables. In these cases we need an algebraic method of solution that can be used in a computer algorithm.

## THE GAUSSIAN ELIMINATION METHOD

### An algebraic solution

**Example 9.1**

*Find the solution of this pair of linear equations.*
$x + 2y = 7$
$2x + 7y = 23$

**Solution**
*Suppose we label the two equations like this.*
$x + 2y = 7$        $E_1$
$2x + 7y = 23$      $E_2$

*Now subtract $2E_1$ from $E_2$.*
$E_2 - 2E_1 \Rightarrow 7y - 4y = 23 - 14$
$3y = 9$

*Solving for $y \Rightarrow y = 3$*
*Substitute for $y$ into $E_1$ then solve for $x$ to find $x = 1$.*

*The solution is $x = 1$ and $y = 3$.*

The number 2 that multiplies $E_1$ in $E_2 - 2E_1$ is called the **multiplier**.

**Exploration 9.1**

## *The method of elimination*

Consider this pair of linear equations.

$$3x + 4y = 5 \qquad \mathbf{E_1}$$
$$2x - 3y = 9 \qquad \mathbf{E_2}$$

- Find the multiplier $r$ so that $E_2 - rE_1$ is an equation without $x$.
- Solve for $y$.
- Solve for $x$.

## Elimination

Choosing $r = \frac{2}{3}$ then $E_2 - \frac{2}{3}E_1$

$$\Rightarrow \quad -3y - \frac{2}{3} \times 4y = 9 - \frac{2}{3} \times 5 \qquad \mathbf{E_{2a}}$$

$$\Rightarrow \quad -\frac{17}{3}y = \frac{17}{3} \qquad \mathbf{E_{2a}}$$

Solving for $y$:
$$y = -1$$

Substituting for $y$ into $E_1$ and solving for $x$:
$$x = 3$$

This method of solving simultaneous linear equations is called the **Gaussian elimination method**, after the mathematician Carl Friedrich Gauss.

The method is a two-stage process.

- The first stage is the process of subtracting a multiple of $E_1$ from $E_2$ to obtain a new equation $E_{2a}$ that does not contain $x$. This is the **elimination stage**.
- The second stage of solving for $y$ and substituting back into $E_1$ to find $x$ is the **back substitution stage**.

For a pair of linear equations written in more general form:

$$a_1x + b_1y = c_1 \ \mathbf{E_1}$$
$$a_2x + b_2y = c_2 \ \mathbf{E_2}$$

in which $x$ and $y$ are the variables and $a_1$ is not zero, the multiplier is $\dfrac{a_2}{a_1}$ so that for the elimination stage we calculate:

$$E_2 - \frac{a_2}{a_1}E_1$$

to give an equation without a term in $x$.

***Carl Friedrich Gauss
(1777–1855)***

The Gaussian elimination method can be applied to sets of equations containing more than two variables.

This forms an important part of the method of solving linear programming problems introduced in the next two sections.

**Example 9.2**

*Solve these equations by the Gaussian elimination method.*

$$3x + y - z = 1 \qquad \mathbf{E_1}$$
$$5x + y + 2z = 6 \qquad \mathbf{E_2}$$
$$4x - 2y - 3z = 3 \qquad \mathbf{E_3}$$

### Solution

*The elimination stage:*

$$\mathbf{E_2} - \tfrac{5}{3}\mathbf{E_1} \Rightarrow 0x - \tfrac{2}{3}y + \tfrac{11}{3}z = \tfrac{13}{3} \quad \mathbf{E_{2a}}$$

$$\mathbf{E_3} - \tfrac{4}{3}\mathbf{E_2} \Rightarrow 0x - \tfrac{10}{3}y - \tfrac{5}{3}z = \tfrac{5}{3} \quad \mathbf{E_{3a}}$$

*We now have three equations but two do not contain $x$.*

$$3x + y - z = 1 \qquad\qquad \mathbf{E_1}$$
$$-\tfrac{2}{3}y + \tfrac{11}{3}z = \tfrac{13}{3} \qquad\qquad \mathbf{E_{2a}}$$
$$-\tfrac{10}{3}y - \tfrac{5}{3}z = \tfrac{5}{3} \qquad\qquad \mathbf{E_{3a}}$$

*Now we eliminate $y$ using equations $\mathbf{E_{3a}}$ and $\mathbf{E_{2a}}$.*

$$\mathbf{E_{3a}} - 5\mathbf{E_{2a}} \Rightarrow 0y - \frac{60}{3}z = -\frac{60}{3}$$

$$\Rightarrow 20z = 20 \qquad \mathbf{E_{3b}}$$

*Now we apply the back substitution stage.*
*Solve equation $\mathbf{E_{3b}}$ for $z$:*
$$z = 1$$
*Substitute into $\mathbf{E_{2a}}$ for $z$:*
$$-\tfrac{2}{3}y + \tfrac{11}{3} = \tfrac{13}{3} \Rightarrow -\tfrac{2}{3}y = \tfrac{2}{3}$$

*Solve for $y$:*
$$y = -1$$
*Substitute into $\mathbf{E_1}$ for $z$ and $y$:*
$$3x - 1 - 1 = 1 \Rightarrow 3x = 3$$
*Solve for $x$:*
$$x = 1$$

*The solution for the equations is $x = 1$, $y = -1$, $z = 1$.*

In this example note the systematic method of approach. Equation $\mathbf{E_1}$ remains unchanged; equation $\mathbf{E_{2a}}$ does not contain $x$ and equation $\mathbf{E_{3b}}$ does not contain $x$ or $y$. We then work backwards through the equations to find the variables: $\mathbf{E_{3b}}$ then $\mathbf{E_{2a}}$ then $\mathbf{E_1}$.

## EXERCISES

**9.1A**

**1** Use the method of Gaussian elimination to solve the following equations. In each case write down the multiplier.

**a)** $x + 3y = 11$
   $3x + y = 9$

**b)** $x - y = 8$
   $4x + y = 23$

**c)** $2x - y = 15$
   $x + 5y = 19$

**d)** $3x - 5y = 8$
   $4x + 7y = 11$

**e)** $2x + y = 3$
   $3x + 2y = 5$

**f)** $0.5x - 0.3y = -0.5$
   $0.1x + 0.4y = 2.2$

**2**   Use Gaussian elimination to solve the following equations. In each case write down the multiplier for each of the elimination processes.

**a)** $3x - 6y + 9z = 0$
$4x - 6y + 8z = -4$
$-2x - y + z = 7$

**b)** $3x - y + z = 6$
$6x - z = 3$
$y + z = 0$

**c)** $x - 3y + 4z = 1$
$4x - 10y + 10z = 4$
$-3x + 9y - 5z = -6$

**d)** $x + 2y + 2z = 11$
$2x + 5y + 9z = 39$
$x - y - z = -4$

**3**   Solve the following using Gaussian elimination.

**a)** $x_1 - 2x_2 + 5x_3 = 6$
$x_1 + 3x_2 - 4x_3 = 7$
$2x_1 + 7x_2 - 12x_3 = 12$

**b)** $2x_1 - 3x_2 + x_3 + x_4 = 10$
$x_1 + x_2 + 4x_3 - x_4 = 5$
$3x_1 - x_2 - x_3 + 2x_4 = 8$
$4x_1 + 2x_2 - 3x_3 + 5x_4 = 11$

**4**   Solve the pair of simultaneous linear equations:
$x + 4y = 9$
$-2x - \mu y = 7$
for the two cases:   **a)** $\mu = 1$   **b)** $\mu = 4$.
Interpret your results in terms of the graphs of appropriate linear functions.

**5**   It is required to find the real coefficients $a$, $b$, $c$ and $d$ of the following quartic equation.
$$y = ax^4 + bx^3 + cx^2 + dx + e$$
If it is known that the graph of $y$ against $x$ cuts the $y$-axis at 4, evaluate one of the coefficients immediately.
If we further know that the graph passes through the points (–2, 0), (1, 0), (2, 16) and (–1, 4) obtain and solve four linear simultaneous equations in the other four coefficients.

**6**   A stone is thrown vertically upwards and its height $h$ metres above the ground is given by the equation:
$$h = a + bt - ct^2$$
where $a$, $b$ and $c$ are constants and $t$ is the time, in seconds, the stone has been in the air.
Given that $h$ = 27, 42, 47 after 1, 2, 3, seconds respectively, determine the values of $a$, $b$ and $c$. From what height was the stone launched?

## EXERCISES

**9.1B**

**1**   Use the method of Gaussian elimination to solve the following equations. Sketch the graphs of each pair of equations and confirm geometrically the validity of your solution.

**a)** $x + y = 3$
$2x + y = 4$

**b)** $3x + 4y = 6$
$x - y = 2$

**c)** $3x + y = x + 2y - 1 = 6$

**d)** $-4x + y = -9$
$x - 2y = 4$

**e)** $-2x - 3y = 1$
$5x - y = 6$

**f)** $4x - y = 7a + 4b$
$5x - 4y = 6a + 5b$
where $a$ and $b$ are constants

**2**   Sketch the graphs of each pair of equations.

    **a)** $4x - 6y = 5$                       **b)** $4x - 6y = 5$

         $6x - 9y = 7.5$                        $6x - 9y = 10$

    How many solutions do **a)** and **b)** have?

    Use the method of Gaussian elimination in the algebraic solution of these equations, explicitly stating equation $E_{2a}$. Use these results to confirm your conclusions from the sketches.

**3**   Solve the pair of simultaneous linear equations:

$$2x - 3y = 4$$
$$-4x + \lambda y = -7$$

    for the two cases:  **a)** $\lambda = 1$   **b)** $\lambda = 6$.

    Interpret your results in terms of the graphs of appropriate linear functions.

**4**   Solve the following using Gaussian elimination.

    **a)** $x + y - z = 3$                   **a)** $-2x_1 + x_2 - 3x_3 = 7$

         $2x - 2y - 3z = 3$                  $3x_1 + 2x_2 - x_3 = -5$

         $4x - 3y - 2z = 3$                  $2x_1 + x_2 = -4$

**5**   Use the process of Gaussian elimination to obtain equation $E_{2b}$ for each case below. Hence, suggest the number of solutions each set of equations has.

    **a)** $x + 2y + 3z = 6$              **b)** $2x - y + z = 1$

         $2x - y - 3z = 3$               $3x + y + 2z = 6$

         $3x + y + 4z = 8$               $x + 2y + z = 4$

**6**   It is required to find the real coefficients $a$, $b$, $c$ and $d$ of the cubic equation:

$$y = ax^3 + bx^2 + cx + d$$

    If it is known that the graph of $y$ against $x$ cuts the $y$-axis at 8, evaluate one of the coefficients immediately.

    If we further know that the graph passes through the points $(-1, 0)$, $(1, 6)$ and $(2, 6)$ obtain and solve three linear simultaneous equations in the other three coefficients.

## ORGANISING THE WORK IN A TABLE

The important part of the Gaussian elimination method is the elimination stage of the process. We can lay out the information and equation manipulations in the form of a tabular block called a **matrix**. The coefficients of the linear equations are the elements of the matrix.

Consider the pair of equations in Example 9.1.

$$x + 2y = 7$$
$$2x + 7y = 23$$

We repeat the solution of these equations showing on the left-hand side the manipulation of the equations and on the right-hand the equivalent operation on the rows of the matrix.

| Equation manipulations | | Row operations |
|---|---|---|

**Equation manipulations**

$x + 2y = 7$     $\mathbf{E}_1$

$2x + 7y = 23$     $\mathbf{E}_2$

**Row operations**

$$\begin{bmatrix} 1 & 2 & | & 7 \\ 2 & 7 & | & 23 \end{bmatrix} \begin{matrix} R_1 \\ R_2 \end{matrix}$$

Coefficients of $x$ and $y$    Right-hand side of linear equations

$\mathbf{E}_1 \Rightarrow x + 2y = 7$     $\mathbf{E}_1$

$\mathbf{E}_2 - 2\mathbf{E}_1 \Rightarrow 3y = 9$     $\mathbf{E}_{2a}$

$$\begin{matrix} R_1 \\ R_2 - 2R \end{matrix} \Rightarrow \begin{bmatrix} 1 & 2 & | & 7 \\ 0 & 3 & | & 9 \end{bmatrix} \begin{matrix} R_1 \\ R_{2a} \end{matrix}$$

The back substitution is carried out in the same way as before. Operating on the rows of the matrix turns out to be less tedious for the linear programming problems than writing out the full equations.

## Exploration 9.2

### Row operations

Consider the equations in Example 9.2.

$3x + y - z = 1$
$5x + y + 2z = 6$
$4x - 2y - 3z = 3$

- Write the equations in tabular form.
- Write the equation manipulations as row operations on the table.

## Using the matrix approach

**Equation manipulations**

$3x + y - z = 1$     $\mathbf{E}_1$

$5x + y + 2z = 6$     $\mathbf{E}_2$

$4x - 2y - 3z = 3$     $\mathbf{E}_3$

**Row operations**

$$\begin{bmatrix} 3 & 1 & -1 & | & 1 \\ 5 & 1 & 2 & | & 6 \\ 4 & -2 & -3 & | & 3 \end{bmatrix} \begin{matrix} R_1 \\ R_2 \\ R_3 \end{matrix}$$

$\mathbf{E}_2 - \frac{5}{3}\mathbf{E}_1 - \frac{2}{3}y + \frac{11}{3}z = \frac{13}{3}$     $\mathbf{E}_{2a}$

$\mathbf{E}_3 - \frac{4}{3}\mathbf{E}_1 - \frac{10}{3}y + \frac{5}{3}z = \frac{5}{3}$     $\mathbf{E}_{3a}$

$$\begin{matrix} R_1 \\ R_2 - \frac{5}{3}R_1 \\ R_3 - \frac{4}{3}R_1 \end{matrix} \begin{bmatrix} 3 & 1 & -1 & | & 1 \\ 0 & -\frac{2}{3} & \frac{11}{3} & | & \frac{13}{3} \\ 0 & -\frac{10}{3} & -\frac{5}{3} & | & \frac{5}{3} \end{bmatrix} \begin{matrix} R_1 \\ R_{2a} \\ R_{3a} \end{matrix}$$

$\mathbf{E}_{3a} - 5\,\mathbf{E}_{2a} - \frac{60}{3}z = -\frac{60}{3}$     $\mathbf{E}_{3b}$

$$R_{3a} - 5R_{2a} \begin{bmatrix} 3 & 1 & -1 & | & 1 \\ 0 & -\frac{2}{3} & \frac{11}{3} & | & \frac{13}{3} \\ 0 & 0 & -\frac{60}{3} & | & -\frac{60}{3} \end{bmatrix} \begin{matrix} R_1 \\ R_{2a} \\ R_{3b} \end{matrix}$$

Note that in the example and in Exploration 9.2 the final table has 0 in each position below the main diagonal.

## EXERCISES

**9.2A**

For each question in Exercises 9.1A:
a) write the equations in tabular form
b) carry out the row operations on the table to give zeros in all the positions below the diagonal.

## EXERCISES

1 Write question 2 of Exercises 9.1B in tabular form and carry out the row operations on the table to obtain zeros in all positions below the diagonal.

Note the form of the bottom line in the final table. In which way will this line indicate whether there is a unique solution, an infinity of solutions or no solutions?

2 For question 1 of Exercises 9.1B:
a) write the equations in tabular form
b) carry out the row operations on the table to give zeros in all the positions below the diagonal.

3 In each of three sets of simultaneous equations in $x$, $y$ and $z$, Gaussian elimination has been performed in tabular form. After reduction to diagonal form we have these arrays.

a) $\begin{bmatrix} 4 & -1 & -2 & 0 \\ 0 & -\frac{3}{4} & \frac{1}{2} & -2 \\ 0 & 0 & \frac{1}{3} & \frac{2}{3} \end{bmatrix}$
b) $\begin{bmatrix} 2 & 1 & 4 & 2 \\ 0 & 3 & 2 & 5 \\ 0 & 0 & 0 & 1 \end{bmatrix}$
c) $\begin{bmatrix} 1 & 2 & 4 & 5 \\ 0 & 1 & 3 & 3 \\ 0 & 0 & 0 & 1 \end{bmatrix}$

For cases a) and b), state the number of solutions and determine the solution where it exists.

For case c) state why there are infinitely many solutions. To obtain a solution in general form, let $z = t$ say, where $t$ can take any real value. Determine $y$ and $x$ in terms of $t$.

4 Solve the pairs of simultaneous equations using Gaussian elimination in tabular form. In each case, determine the value of which will give:
i) a unique solution
ii) an infinity of solutions
iii) no feasible solutions.

a) $x + y = 1$
   $x + \lambda^2 y = \lambda$

b) $x + y + z = 3$
   $4x + 5y + 2z = 11$
   $2x + y + (\lambda^2 + 6)z = \lambda + 12$

For those cases where solutions exist determine the solutions.

5 Repeat question 5 of Exercises 9.1B using the tabular method with row operations.

## LINEAR PROGRAMMING AND THE SIMPLEX TABLEAU

### *Maximising profit*

A manufacturer makes three products A, B and C. Each product requires time in the cutting, assembly and finishing departments. The times in each department, total time available and profit-margin per item are shown in the table overleaf.

| | | **Product** | | | |
|---|---|---|---|---|---|
| | | A | B | C | Time available (hours per week) |
| Time | Cutting | 2 | 3 | 5 | 204 |
| (hours) | Assembling | 4 | 3 | 6 | 260 |
| | Finishing | 3 | 1 | 10 | 304 |
| | Profit (£) | 10 | 12 | 16 | |

How many of each product should be made each week in order to maximise profits?

## Slack variables

If we set up the exploration as a linear programming problem it becomes:

Maximise $P = 10A + 12B + 16C$

subject to:

$$2A + 3B + 5C \leq 204 \quad \text{(cutting)}$$
$$4A + 3B + 6C \leq 260 \quad \text{(assembling)}$$
$$3A + B + 10C \leq 304 \quad \text{(finishing)}$$

It would not be possible to solve this problem graphically, as it contains three variables ($A$, $B$ and $C$). So we need to consider an alternative approach.

We can use the row operation process to solve linear programming problems. The first step is to convert the inequalities in the constraints into a system of linear equations. We do this by introducing **slack variables**.

Consider this problem.

Maximise $\quad f = 80x + 70y \qquad$ **1**

where $\qquad 2x + y \leq 32 \qquad$ **2**

$\qquad\qquad x + y \leq 18 \qquad$ **3**

$\qquad\qquad x \geq 0, y \geq 0 \qquad$ **4**

If we were to graph this problem we would obtain a feasible region bounded by four lines. We know that the optimal solution lies at a vertex, so we need only consider the value of $f$ at the four vertices O, A, B and C.

Each vertex is at the intersection of two lines, for example O is where $x = y = 0$.

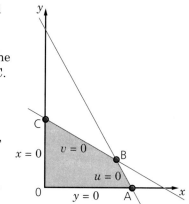

We introduce the slack variables, $u$ and $v$, so that we can define AB as $u = 0$ and BC as $v = 0$. In this way we say that each vertex corresponds to two of the variables being zero:

at O  $x = y = 0$

at A  $y = u = 0$

at B  $u = v = 0$

at C  $v = x = 0$

In effect, $u$ and $v$ represent the slack between the total available and what is actually being used, and they enable us to replace the inequalities with linear equations.

$2x + y + u = 32$　　**2A**
$x + y + v = 18$　　**3A**

Because of the way they are defined $u \geq 0$ and $v \geq 0$ at the optimal solution. The linear programming problem can be redefined as follows.

> Among all the solutions of the system of the linear equations:
> $f = 80x + 70y$
> $2x + y + u = 32$
> $x + y + v = 18$
>
> find the one for which $u \geq 0, v \geq 0, x \geq 0, y \geq 0$ and $f$ is as large as possible.

This is an example of the standard form of linear programming problem in which:
1. the objective function is to be maximised
2. each variable (actual and slack) $\geq 0$.

**Exploration 9.4**

## *Standard form of linear programming problems*

Consider this linear programming problem.

Maximise　$f = 2x + y + 3z$　　　**1**
where　　　$x + y + z \leq 6$　　　**2**
　　　　　　$x + 3y \leq 6$　　　　**3**
　　　　　　$2x - z \geq 9$　　　　**4**
　　　　　　$x \geq 0, y \geq 0, z \geq 0$　**5**

■ Introduce slack variables so that the constraints **2 – 4** become equalities. (Be careful with constraint **4**.)
■ Write the linear programming problem in standard form.

## Using slack variables

Constraints **2** and **3** are easily written in standard form by introducing the slack variables $u$ and $v$ to give:
$x + y + z + u = 6$
$x + 3y + v = 6$

*Remember: Multiplying an equation throughout by a negative number reverses the inequality sign.*

Constraint **4** has $\geq$ as the inequality. It can be rewritten as:
$-2x + z \leq -9$

Now we can introduce slack variable $w$ to give:
$-2x + z + w = -9$

The problem 'Find the solution of the linear equations:
$f = 2x + y + 3z$
$x + y + z + u = 6$
$x + 3y + v = 6$
$-2x + z + w = -9$

181

such that   $x \geq 0, y \geq 0, z \geq 0,$
$$u \geq 0, v \geq 0, w \geq 0$$

and for which $f$ is as large as possible' is now written in standard form.

If the objective function is to be minimised then the standard form can be obtained by maximising $-f$.

## The simplex method

We are now ready to introduce the **simplex method** of solving linear programming problems, which uses the elimination steps of the Gaussian elimination method. In demonstrating the method the graphs show a comparison with the graphical method of Chapter 8, *Linear programming 1: Modelling and graphical solution*.

Consider the problem in standard form:

Maximise  $f = 80x + 70y$
where      $2x + y + u = 32$
$$x + y + v = 18$$
$$x \geq 0, y \geq 0, u \geq 0, v \geq 0$$

*Note: The objective equation is found by rearranging the objective function to give $f - 80x - 70y = 0$.*

We write the equations in a tabular form called the **simplex tableau**.

| $x$ | $y$ | $u$ | $v$ | $f$ | |
|---|---|---|---|---|---|
| 2 | 1 | 1 | 0 | 0 | 32 |
| 1 | 1 | 0 | 1 | 0 | 18 |
| −80 | −70 | 0 | 0 | 1 | 0 |

$\left.\begin{array}{l}\phantom{x}\\\phantom{x}\end{array}\right\}$ Constraint equations

Objective equation (third row)

The heading over each column shows the variable with which the elements are associated. The table has three types of information.

| $x$ | $y$ | $u$ | $v$ | $f$ | |
|---|---|---|---|---|---|
| 2 | 1 | 1 | 0 | 0 | 32 |
| 1 | 1 | 0 | 1 | 0 | 18 |
| −80 | −70 | 0 | 0 | 1 | 0 |

right-hand side of constraints

original variables      slack variables and objective
$x, y$: group I    function $u, v, f$: group II

At the start of the problem-solving process the array for the group II variables is a diagonal array of 1s with 0s on the off-diagonal elements. This property allows us to read off a particular value of the original variables by assigning zero values to $x$ and $y$.

$x = 0, y = 0, u = 32, v = 18, f = 0$

The idea behind the simplex method is that we move round the feasible region, travelling along the edges and visiting the vertices in turn until a maximum value of $f$ is reached. A row operation on the simplex tableau is equivalent to jumping from one vertex to another.

The edges of the feasible region are defined by $x = 0$, $y = 0$, $u = 0$ and $v = 0$. A vertex is a solution of two of these equations.

Consider the bottom row of the tableau:

$f - 80x - 70y = 0$

If we start at $x = 0$, $y = 0$ then $f = 0$.

We can increase $f$ by increasing $x$ or $y$ (or both but we choose one or the other).

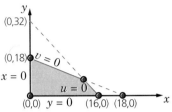

Since increasing $x$ leads to a larger change in $f$ than increasing $y$, we choose to increase $x$ and keep $y = 0$.

The effect is to move along the $x$-axis. From the graph we see that the vertex $(16, 0)$ is visited first. This is the intersection of $y = 0$ and $u = 0$.

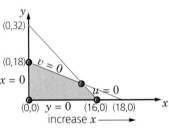

We can find this information from the tableau by dividing the right-hand column by the $x$-column entries.

$$
\begin{array}{ccccc|c}
x & y & u & v & f & \\
2 & 1 & 1 & 0 & 0 & 32 \\
1 & 1 & 0 & 1 & 0 & 18 \\
\hline
-80 & -70 & 0 & 0 & 1 & 0
\end{array}
\qquad
\begin{array}{l}
32 \div 2 = 16 \\
18 \div 1 = 18 \\
\\
\end{array}
$$

$\uparrow$

The smallest value gives the **pivot row** since we arrive at $x = 16$ before $x = 18$.

This is called the **pivot column**.

In this case row 1 is the pivot row. The element in the intersection of the pivot row and pivot column is the **pivot element**. We make this element 1 and then carry out row operations to make each other element in the pivot column equal to zero.

*What happens to the equations*

$$
R_1 \div 2 \quad
\begin{array}{ccccc|c}
x & y & u & v & f & \\
1 & \frac{1}{2} & \frac{1}{2} & 0 & 0 & 16 \\
-\ 1 & 1 & 0 & 1 & 0 & 18 \\
\hline
-80 & -70 & 0 & 0 & 1 & 0
\end{array}
$$

$$
\begin{aligned}
x + \tfrac{1}{2}y + \tfrac{1}{2}u &= 16 \quad &\mathbf{E_1} \\
x + y + v &= 18 \quad &\mathbf{E_2} \\
-80x - 70y + f &= 0 \quad &\mathbf{E_3}
\end{aligned}
$$

$$
\begin{array}{c}
\\
R_1 - R_2 \\
R_3 + 80R_1
\end{array}
\begin{array}{ccccc|c}
x & y & u & v & f & \\
1 & \frac{1}{2} & \frac{1}{2} & 0 & 0 & 16 \\
0 & \frac{1}{2} & -\frac{1}{2} & 1 & 0 & 2 \\
0 & -30 & 40 & 0 & 1 & 1280
\end{array}
$$

$$
\begin{aligned}
x + \tfrac{1}{2}y + \tfrac{1}{2}u &= 16 \\
\tfrac{1}{2}y - \tfrac{1}{2}u + v &= 2 \quad (\mathbf{E_2 - E_1}) \\
-30y + 40u + f &= 1280 \quad (\mathbf{E_3 + 80E_1})
\end{aligned}
$$

$\uparrow$

new value of $f$

The new value of $f$ is 1280 and occurs at the vertex where the lines $y = 0$, $u = 0$ meet. Note that the $x$, $v$ and $f$-columns are now the group II variables. Also, $y$ and $u$ are the group I variables. In the new simplex tableau the value of $f$ has increased.

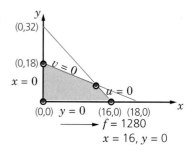

(0,0) $y = 0$   (16,0) (18,0)
$f = 1280$
$x = 16$, $y = 0$

## Exploration 9.5

### Increasing $y$

Consider this simplex tableau.

$$
\begin{array}{ccccc|c}
x & y & u & v & f & \\
\hline
1 & \frac{1}{2} & \frac{1}{2} & 0 & 0 & 16 \\
0 & \frac{1}{2} & -\frac{1}{2} & 1 & 0 & 2 \\
0 & -30 & 40 & 0 & 1 & 1280
\end{array}
$$

■ Explain why $y$ should be increased next.
■ As $y$ is increased what is happening on the graph of the feasible region?
■ Find the pivot element.
■ Construct a new simplex tableau using row operations based on the pivot element.
■ Is this the maximum value of $f$?

### Increasing $y$

You should have found that the maximum value of $f$ is 1400 and it occurs at the vertex $(14, 4)$. The steps in the simplex method are as follows.

From the bottom row:

$$f - 30y + 40u = 1280$$

To increase $f$ further we hold $u = 0$ and increase $y$. Graphically this means that we are moving along the edge $u = 0$.

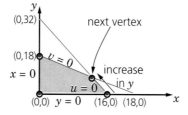

The pivot column is the $y$-column.

$$
\begin{array}{ccccc|c}
x & y & u & v & f & \\
\hline
1 & \frac{1}{2} & \frac{1}{2} & 0 & 0 & 16 \\
0 & \frac{1}{2} & -\frac{1}{2} & 1 & 0 & 2 \\
0 & -30 & 40 & 0 & 1 & 1280
\end{array}
\quad
\begin{array}{l}
16 \div \frac{1}{2} = 32 \\
2 \div \frac{1}{2} = 4 \quad \leftarrow \text{pivot row} \\
\\
\end{array}
$$

↑
pivot column

Arrive at $y = 4$ before $y = 32$.

$$
\begin{array}{r}
\\
\\
R_2 \div \tfrac{1}{2}
\end{array}
\begin{array}{ccccc}
x & y & u & v & f \\
\end{array}
\left[\begin{array}{ccccc|c}
1 & \tfrac{1}{2} & \tfrac{1}{2} & 0 & 0 & 16 \\
0 & \boxed{1} & -1 & 2 & 0 & 4 \\
0 & -30 & 40 & 0 & 1 & 1280
\end{array}\right]
$$

$\bigcirc$ = pivot element

Now pivot about the pivot element using row operations.

$$
\begin{array}{r}
R_1 - \tfrac{1}{2}R_2 \\
\\
R_3 + 30R_2
\end{array}
\begin{array}{ccccc}
x & y & u & v & f \\
\end{array}
\left[\begin{array}{ccccc|c}
1 & 0 & 1 & -1 & 0 & 14 \\
0 & 1 & -1 & 2 & 0 & 4 \\
0 & 0 & 10 & 60 & 1 & 1400
\end{array}\right]
\left.\begin{array}{l}
\\
\\
\\
\end{array}\right\}
\begin{array}{l}
x + u - v = 14 \\
y - u + 2v = 4 \\
10u + 60v + f = 1400
\end{array}
$$

The equation for $f$ is now:

$$f + 10u + 60v = 1400$$

Increasing $u$ and $v$ will decrease $f$, so we have found the maximum value for $f$.

It is $f = 1400$ and occurs when $u = 0$ and $v = 0$.

From the table we can read off the values of $x$ and $y$ for the maximum value of $f$.

It is where $x = 14$ and $y = 4$.

## Example 9.3

*This is a two variable, two constraint problem.*

*Maximise the objective function $f = 10x + 7y - 50$*
*where*
$$3x + 2y \le 42$$
$$2x + 5y \le 50$$
$$x \ge 0, y \ge 0.$$

### Solution
Introduce slack variables $u$ and $v$ to give:

$$3x + 2y + u = 42$$
$$2x + 5y + v = 50$$
$$-10x - 7y + f = -50$$

The simplex tableau is:

$$
\begin{array}{ccccc}
x & y & u & v & f \\
\end{array}
\left[\begin{array}{ccccc|c}
3 & 2 & 1 & 0 & 0 & 42 \\
2 & 5 & 0 & 1 & 0 & 50 \\
-10 & -7 & 0 & 0 & 1 & -50
\end{array}\right]
\begin{array}{l}
\tfrac{42}{3} = 14 \quad \leftarrow \textit{pivot row} \\
\tfrac{50}{2} = 25 \\
\\
\end{array}
$$

$\uparrow$
*pivot column*

Divide the row by the pivot element 3.

$$
R_1 \div 3
\begin{array}{ccccc}
x & y & u & v & f \\
\end{array}
\left[\begin{array}{ccccc|c}
1 & \tfrac{2}{3} & \tfrac{1}{3} & 0 & 0 & 14 \\
2 & 5 & 0 & 1 & 0 & 50 \\
-10 & -7 & 0 & 0 & 1 & -50
\end{array}\right]
$$

$$
\begin{array}{c}
 \\
R_2 - 2R_1 \\
R_3 + 10R_1
\end{array}
\begin{array}{cccccc}
x & y & u & v & f & \\
\end{array}
\left[
\begin{array}{ccccc|c}
1 & \frac{2}{3} & \frac{1}{3} & 0 & 0 & 14 \\
0 & \frac{11}{3} & -\frac{2}{3} & 1 & 0 & 22 \\
-10 & -\frac{1}{3} & \frac{10}{3} & 0 & 1 & 90
\end{array}
\right]
\begin{array}{l}
14 \div \frac{2}{3} = 21 \\
22 \div \frac{11}{3} = 6 \quad \leftarrow pivot\ row \\
\phantom{x}
\end{array}
$$

$\uparrow$
*pivot column*

*Divide the pivot row by* $\frac{11}{3}$.

$$
\begin{array}{c}
 \\
R_2 \div \frac{11}{3} \\
 \\
\end{array}
\begin{array}{ccccc}
x & y & u & v & f \\
\end{array}
\left[
\begin{array}{ccccc|c}
1 & \frac{2}{3} & \frac{1}{3} & 0 & 0 & 14 \\
0 & 1 & -\frac{2}{11} & \frac{3}{11} & 0 & 6 \\
0 & -\frac{1}{3} & \frac{10}{3} & 0 & 1 & 90
\end{array}
\right]
$$

$$
\begin{array}{c}
R_1 - \frac{2}{3}R_2 \\
 \\
R_3 + \frac{1}{2}R_2
\end{array}
\begin{array}{ccccc}
x & y & u & v & f \\
\end{array}
\left[
\begin{array}{ccccc|c}
1 & 0 & \frac{5}{11} & -\frac{2}{11} & 0 & 10 \\
0 & 1 & -\frac{2}{11} & \frac{3}{11} & 0 & 6 \\
0 & 0 & \frac{36}{11} & \frac{1}{11} & 1 & 92
\end{array}
\right]
$$

*The elements in the bottom row are all positive so:*

$f = 92 - \frac{36}{11}u - \frac{1}{11}v$

*The maximum value of f is 92 and occurs when $u = 0$, $v = 0$ and from the table when $x = 10$ and $y = 6$.*

## Example 9.4

*This is a two variable, three constraint problem.*

*Use the simplex tableau to solve this linear programming problem.*

*Maximise* $f = 9x + 4y$
*where* $\quad 3x + 4y \le 48$
$\qquad\quad 2x + y \le 17$
$\qquad\quad 3x + y \le 24$
$\qquad\quad x \ge 0, y \ge 0$

### Solution
*Introducing slack variables $u$, $v$ and $w$ the system of linear equations is:*

$3x + 4y + u = 48$
$2x + y + v = 17$
$3x + y + w = 24$
$-9x - 4y + f = 0$

*The simplex tableau is:*

*Pivot about one of the x-values in order to increase f.*

$$
\begin{array}{cccccc}
x & y & u & v & w & f \\
\end{array}
\left[
\begin{array}{cccccc|c}
3 & 4 & 1 & 0 & 0 & 0 & 48 \\
2 & 1 & 0 & 1 & 0 & 0 & 17 \\
3 & 1 & 0 & 0 & 1 & 0 & 24 \\
-9 & -4 & 0 & 0 & 0 & 1 & 0
\end{array}
\right]
\begin{array}{l}
48 \div 3 = 16 \\
17 \div 2 = 8.5 \\
24 \div 3 = 8 \quad \text{Pivot about row 3.} \\
\phantom{x}
\end{array}
$$

$$\begin{array}{cccccc} & x & y & u & v & w & f \\ & \begin{bmatrix} 3 & 4 & 1 & 0 & 0 & 0 \\ 2 & 1 & 0 & 1 & 0 & 0 \\ 1 & \frac{1}{3} & 0 & 0 & \frac{1}{3} & 0 \\ -9 & -4 & 0 & 0 & 0 & 1 \end{bmatrix} & \begin{matrix} 48 \\ 17 \\ 8 \\ 0 \end{matrix} \end{array}$$

$R_1 \div 3$ on row 3.

*Pivot about y.*

$$\begin{array}{l} \\ R_1 - 3R_3 \\ R_2 - 2R_3 \\ \\ R_4 - 9R_3 \end{array} \begin{array}{cccccc} x & y & u & v & w & f \\ \begin{bmatrix} 0 & 3 & 1 & 0 & -1 & 0 \\ 0 & \frac{1}{3} & 0 & 1 & -\frac{2}{3} & 0 \\ 1 & \frac{1}{3} & 0 & 0 & \frac{1}{3} & 0 \\ 0 & -1 & 0 & 0 & 3 & 1 \end{bmatrix} & \begin{matrix} 24 \\ 1 \\ 8 \\ 72 \end{matrix} \end{array}$$

$24 \div 3 = 8$

$1 \div \frac{1}{3} = 3$  *Pivot about row 2.*

$8 \div \frac{1}{3} = 24$

$$\begin{array}{l} \\ R_2 \div \frac{1}{3} \\ \\ \\ \end{array} \begin{array}{cccccc} x & y & u & v & w & f \\ \begin{bmatrix} 0 & 3 & 1 & 0 & -1 & 0 \\ 0 & 1 & 0 & 3 & -2 & 0 \\ 1 & \frac{1}{3} & 0 & 0 & \frac{1}{3} & 0 \\ 0 & -1 & 0 & 0 & 3 & 1 \end{bmatrix} & \begin{matrix} 24 \\ 3 \\ 8 \\ 72 \end{matrix} \end{array}$$

$$\begin{array}{l} R_1 - 3R_2 \\ \\ R_3 - \frac{1}{3}R_2 \\ R_4 + R_2 \end{array} \begin{array}{cccccc} x & y & u & v & w & f \\ \begin{bmatrix} 0 & 0 & 1 & -9 & 5 & 0 \\ 0 & 1 & 0 & 3 & -2 & 0 \\ 1 & 0 & 0 & -1 & 1 & 0 \\ 0 & 0 & 0 & 3 & 1 & 1 \end{bmatrix} & \begin{matrix} 15 \\ 3 \\ 7 \\ 75 \end{matrix} \end{array}$$

Since each entry in the last row is positive we have achieved the maximum value for $f$. If we increase $v$ and $w$ then we decrease $f$, since the bottom row gives:

$f = 75 - 3v - w$

The maximum value of $f$ is 75 and occurs when $x = 7$ and $y = 3$.

## Example 9.5

*This is a three variable, two constraint problem.*

Maximise the objective function $f = -x + 8y + z$
where
$$x + 2y + 9z \le 10$$
$$y + 4z \le 12$$
$$x \ge 0, y \ge 0, z \ge 0$$

### Solution
Introduce two slack variables $u$ and $v$ since we have two constraints.
$$x + 2y + 9z + u = 10$$
$$y + 4z + v = 12$$

The simplex tableau is:

*Pivot about y.*

$$\begin{array}{cccccc} x & y & z & u & v & f \\ \begin{bmatrix} 1 & 2 & 9 & 1 & 0 & 0 \\ 0 & 1 & 4 & 0 & 1 & 0 \\ 1 & -8 & -1 & 0 & 0 & 1 \end{bmatrix} & \begin{matrix} 10 \\ 12 \\ 0 \end{matrix} \end{array}$$

$10 \div 2 = 5$  *Pivot about row 1.*

$12 \div 1 = 12$

*First divide the pivot row by the pivotal element.*

$$R_1 \div 2 \begin{bmatrix} & x & y & z & u & v & f & \\ & \frac{1}{2} & 1 & \frac{9}{2} & \frac{1}{2} & 0 & 0 & 5 \\ & 0 & 1 & 4 & 0 & 1 & 0 & 12 \\ & 1 & -8 & -1 & 0 & 0 & 1 & 0 \end{bmatrix}$$

$$\begin{matrix} & x & y & z & u & v & f & \\ & \frac{1}{2} & 1 & \frac{9}{2} & \frac{1}{2} & 0 & 0 & 5 \\ R_2 - R_1 & -\frac{1}{2} & 0 & -\frac{1}{2} & -\frac{1}{2} & 1 & 0 & 7 \\ R_3 + 8R_1 & 5 & 0 & 35 & 4 & 0 & 1 & 40 \end{matrix}$$

*The bottom row contains positive elements and so the maximum value for f has been reached. The bottom row gives:*

$$f = 40 - 5x - 35z - 4u$$

*Hence the maximum value is f = 40 and occurs when x = 0, z = 0 and u = 0 and from the tableau y = 5.*

You can now use the simplex tableau to solve the problem which was posed at the beginning of this section on page 179.

The answer is   Item A – 14
                  Item B – 12    P = £724
                  Item C – 25

## EXERCISES

**9.3 A**

1 Use the simplex method to solve the following linear programming problems.

a) Maximise $f = 2x + 3y$
   subject to  $x + 3y \le 120$
               $7x + 2y \le 154$
               $x \ge 0, y \ge 0$

b) Maximise $c = x + \frac{1}{2}y$
   subject to  $2x + y \le 36$
               $x + y \le 15$
               $x \ge 0, y \ge 0$

c) Maximise $f = 8x + 13y$
   subject to  $2x + 3y \le 350$
               $5x + y \le 500$
               $x \ge 0, y \ge 0$

d) Maximise $g = 3x + 4y$
   subject to  $x + y \le 20$
               $x + 2y \le 25$

e) Maximise $M = x + 3y$
   subject to  $-x + y \le 160$
               $3x - y \le 3$
               $x \ge 0, y \ge 0$

2 Use the simplex method to solve the following linear programming problems.

a) Maximise $f = x + 3y$
   subject to  $5x + y \le 30$
               $3x + 2y \le 60$
               $x + y \le 50$
               $x \ge 0, y \ge 0$

b) Maximise $f = x + y$
   subject to  $2x + y \le 6$
               $2x + 3y \le 12$
               $4x + y \le 10$
               $x \ge 0, y \ge 0$

c) Maximise $c = x + \frac{1}{2}y$
subject to $2x + y \le 24$
$x + y \le 15$
$y \le 2x$
$x \ge 3, y \ge 3$

d) Maximise $c = 10x + 7y - 1500$
subject to $3x + 2y \le 3600$
$x + 4y \le 1800$
$y \le 4x$
$x \ge 0, y \ge 0$

3  Use the simplex method to solve the following linear programming problems.

a) Maximise $f = x + 3y + 5z$
subject to $x + y + 2z \le 10$
$3y + z \le 24$
$x \ge 0, y \ge 0, z \ge 0$

b) Maximise $c = 5x + 6y + 5z$
subject to $x - 2y \le 6$
$3x + y + z \le 9$
$x \ge 0, y \ge 0, z \ge 0$

c) Maximise $M = 2x + 3y + 4z + 15$
subject to $x - y + 2z \le 10$
$2x + y + 3z \le 12$
$x \ge 0, y \ge 0, z \ge 0$

d) Maximise $f = 4x - y + 5z + 50$
subject to $x + 2z \le 5$
$3x + z \le 12$
$x \ge 0, y \ge 0, z \ge 0$

You may have seen this question in Chapter 8. You should now use the simplex method to tackle it.

4  A factory produces two types of toys: trucks and bicycles. In the manufacturing process of these toys three machines are used. These are a moulder, a lathe and an assembler. The table shows the length of time needed for each toy.

|         | Moulder | Lathe   | Assembler |
|---------|---------|---------|-----------|
| bicycle | 1 hour  | 3 hours | 1 hour    |
| truck   | 0 hours | 1 hour  | 1 hour    |

The moulder can be operated for 3 hours per day, the lathe for 12 hours per day and the assembler for 7 hours per day. Each bicycle made gives a profit of £15 and each truck made gives a profit of £11. Formulate a linear programming problem so that the factory maximises its profit.

You may have seen this question in Chapter 8. You should now use the simplex method to tackle it.

5  Chris is making coffee cakes and chocolate cakes for a cake stall at the village fete. Each chocolate cake needs 2 eggs and 250 g of margarine. Each coffee cake needs 3 eggs and 150 g of margarine. Chris buys three dozen eggs and a 3 kg tub of margarine. Formulate and solve a linear programming problem so that Chris can make as many cakes as possible.

You may have seen this question in Chapter 8. You should now use the simplex method to tackle it.

6  A publisher, MathGem, publishes 40 new books each year which are classified as sixth-form level, undergraduate level and graduate level. The policy of the company is to publish at least three times as many sixth-form books as undergraduate books, and at least twice as many undergraduate books as graduate books. To raise its profit, MathGem has decided to publish at least four graduate books per year.

On average the annual profits are £9000 for each sixth-form book, £5000 for each undergraduate book and £1000 for each graduate book. How many books of each type should be published for MathGem to maximise its annual profit?

7 A food manufacturer makes Alma margarine from vegetable and non-vegetable oils. The raw oils are refined by the manufacturer and blended to form the margarine. One objective of the manufacturer is to make as large a profit as possible; however, there are various constraints. It is important that the raw oils are refined separately to avoid contamination and the final blend of oils must be soft enough to spread, but not too runny. To achieve this quality each ingredient and the final product has a hardness factor. The following table gives the cost and hardness of the raw materials used to make Alma margarine.

*You may have seen this question in Chapter 8. You should now use the simplex method to tackle it.*

| | Oil | Cost £ per kg | Hardness factor |
|---|---|---|---|
| Vegetable | Groundnut | 0.7 | 1.2 |
| | Soya bean | 1.05 | 3.4 |
| | Palm | 1.05 | 8.0 |
| Non-vegetable | Lard | 1.30 | 10.8 |
| | Fish | 1.10 | 8.3 |

Alma margarine sells for £1.55 per kg and to ensure good spreading, its hardness must be between 5.6 and 7.4. The manufacturers can refine up to 40 000 kg of vegetable oil and up to 32 000 kg of non-vegetable oil per day. Formulate the problem of maximising the profit as a linear programming problem.

## EXERCISES

**9.3B**

1 Use the simplex method to solve the following linear programming problems.

a) Maximise $f = 20x + 30y$
subject to $7x + 10y \le 80$
$x + 2y \le 12$
$x \ge 0, y \ge 0$

b) Maximise $g = x + \frac{1}{10}y$
subject to $2x + 3y \le 6$
$5x + 2y \le 10$
$x \ge 0, y \ge 0$

c) Maximise $h = 2x + 7y$
subject to $x + 2y \le 30$
$2x + 3y \le 35$
$x \ge 0, y \ge 0$

d) Maximise $M = 13x + 25y$
subject to $-8x + 5y \le 35$
$3x \le y$
$x \ge 0, y \ge 0$

e) Maximise $N = 11x + 9y$
subject to $5x - 4y \le 10$
$-x + 2y \le 10$
$x \ge 0, y \ge 0$

2 Use the simplex method to solve the following linear programming problems.

a) Maximise $f = 17x + 3y$
subject to $5y + 8x \le 100$
$3y + 5x \le 61$
$2y + 7x \le 70$
$x \ge 0, y \ge 0$

b) Maximise $g = 8x + 5y$
subject to $x + y \le 100$
$5x + 4y \le 420$
$3x + 2y \le 240$
$x \ge 0, y \ge 0$

**c)** Maximise $h = 9x + 2y + 100$
subject to $5x \leq 5y$
$6x + y \leq 44$
$3x + y \leq 38$
$x \geq 0, y \geq 0$

**d)** Maximise $M = 4x + y$
subject to $2x + 7y \leq 114$
$5x + y \leq 54$
$2x \leq 5y$
$x \geq 1, y \geq 2$

**3** Use the simplex method to solve the following linear programming problems.

**a)** Maximise $f = 7x + 2y + z$
subject to $x + y + z \leq 12$
$7x + y + z \leq 8$
$x \geq 0, y \geq 0, z \geq 0$

**b)** Maximise $g = 3x + 5y + 4z$
subject to $3x + 2y + z \leq 12$
$2x + y + 2x \leq 10$
$x \geq 0, y \geq 0, z \geq 0$

**c)** Maximise $h = 5x - y + 2z - 10$
subject to $2x + y \leq 10$
$-x + 4y \leq 3$
$x \geq 0, y \geq 0, z \geq 0$

**d)** Maximise $M = 10x + y - 2z + 20$
subject to $5x + y + 8z \leq 15$
$x + 2y + z \leq 6$
$x \geq 0, y \geq 0, z \geq 0$

**4** A shop sells a mixture of two types of coffee beans designated A and B. Bean A costs £1.50 per kilogram, Bean B costs £2 per kg and £1000 has been set aside for purchasing the beans. For a palatable mixture, the weight of bean B plus twice the weight of bean A must not exceed 550 kg. If the shop sells the mixture at £6 per kg, how many kilograms of each type of coffee bean should be purchased to maximise the income based on the assumption that the entire mixture can be sold?

**5** A haulage firm is negotiating a contract with a company which uses two sizes of containers: large 0.4 m³ containers weighing 15 kg and small 0.2 m³ containers weighing 4 kg. The haulage firm will use a vehicle that can handle a maximum load of 1640 kg and a cargo size of up to 100 m³. For stability of load the vehicle should carry no more than 300 small containers. if the companies have agreed on a transport rate of £1 for each large container and 75p for each small container, how many containers of each type should the haulage company place on a vehicle to maximise the income?

**6** A company manufactures two types of PCs: a 16-bit and 32-bit machine. In the manufacturing cycle the company can produce no more than 200 16-bit machines and 300 32-bit machines. The company production cycle means the total number cannot exceed 400. Chip requirements entail that twice the number of 16-bit plus three times the number of 32-bit machines made has a maximum of 1050 machines. The profit is £300 on each 16-bit machine and £400 on each 32-bit machine. Determine the number of each machine that should be manufactured to maximise the profit.

**7** A bakery makes two types of cake, A and B. Each kilogram of cake A requires $\frac{1}{4}$ kilogram of flour, $\frac{1}{4}$ kilogram of sugar and 2 eggs; each kilogram of cake B requires 0.3 kg of flour, 0.35 kg of sugar and 3 eggs. The baker has 90 kg of flour, 90 kg of sugar and 900 eggs. If cake A sells at £6 per kg and cake B at £7 per kg, how many kilograms of each cake should the bakery produce to maximise the income? State the amount of eggs, flour and sugar which should be used.

## SOLVING NON-STANDARD PROBLEMS

The simplex method needs to be amended for non-standard problems. These are problems in which the objective function is to be minimised and/or one or more of the constraints is in the form:

linear polynomial ≥ constant

i.e., there is a ≥ sign instead of a ≤ sign. On page 181 we discussed the technique of transforming such a problem to standard form. Now we amend the simplex method to handle the negative constraints that appear.

Consider the transformed problem of Exploration 9.4 on page 181.

$x + y + z + u = 6$
$x + 3y + v = 6$
$-2x + z + w = -9$
$f = 2x + y + 3z$
$x \geq 0, y \geq 0, z \geq 0, u \geq 0, v \geq 0, w \geq 0.$

The simplex tableau is:

$$
\begin{array}{ccccccc}
x & y & z & u & v & w & f \\
\end{array}
$$
$$
\left[
\begin{array}{ccccccc|c}
1 & 1 & 1 & 1 & 0 & 0 & 0 & 6 \\
1 & 3 & 0 & 0 & 1 & 0 & 0 & 6 \\
-2 & 0 & 1 & 0 & 0 & 1 & 0 & -9 \\
\hline
-2 & -1 & -3 & 0 & 0 & 0 & 1 & 0 \\
\end{array}
\right]
$$

The method would proceed as before except for the –9 in the end column. The problem is that when $x = 0$, $y = 0$ and $z = 0$ the value of $w$ is –9. This violates the trivial constraint $w \geq 0$.

Before we proceed we pivot about an appropriate element to remove the negative entry in the end column.

This is done in the following way. Look along a row with a negative entry in the end column (the third row in this case) and choose any negative element to provide the pivot column. In this case –2 is the only negative element so the first column is the pivot column.

Now compare ratios as in the previous examples.

$$
\begin{array}{ccccccc}
x & y & z & u & v & w & f \\
\end{array}
$$
$$
\left[
\begin{array}{ccccccc|c}
1 & 1 & 1 & 1 & 0 & 0 & 0 & 6 \\
1 & 3 & 0 & 0 & 1 & 0 & 0 & 6 \\
-2 & 0 & 1 & 0 & 0 & 1 & 0 & -9 \\
\hline
-2 & -1 & -3 & 0 & 0 & 0 & 1 & 0 \\
\end{array}
\right]
\begin{array}{l}
6 \div 1 = 6 \\
6 \div 1 = 6 \\
-9 \div -2 = 4\frac{1}{2} \quad \leftarrow \text{pivot row} \\
\\
\end{array}
$$

pivot column        pivot element

This gives $4\frac{1}{2}$ as the smallest positive ratio, so the third row is the pivot. This makes $-2$ the pivot element.

|  | $x$ | $y$ | $z$ | $u$ | $v$ | $w$ | $f$ |  |
|---|---|---|---|---|---|---|---|---|
|  | 1 | 1 | 1 | 1 | 0 | 0 | 0 | 6 |
|  | 1 | 3 | 0 | 0 | 1 | 0 | 0 | 6 |
| $R_3 \div -2$ | 1 | 0 | $-\frac{1}{2}$ | 0 | 0 | $-\frac{1}{2}$ | 0 | $\frac{9}{2}$ |
|  | $-2$ | $-1$ | $-3$ | 0 | 0 | 0 | 1 | 0 |

|  | $x$ | $y$ | $z$ | $u$ | $v$ | $w$ | $f$ |  |
|---|---|---|---|---|---|---|---|---|
| $R_1 - R_3$ | 0 | 1 | $\frac{3}{2}$ | 1 | 0 | $\frac{1}{2}$ | 0 | $\frac{3}{2}$ |
| $R_2 - R_3$ | 0 | 3 | $\frac{1}{2}$ | 0 | 1 | $\frac{1}{2}$ | 0 | $\frac{3}{2}$ |
|  | 1 | 0 | $-\frac{1}{2}$ | 0 | 0 | $-\frac{1}{2}$ | 0 | $\frac{9}{2}$ |
| $R_4 + 2R_3$ | 0 | $-1$ | $-4$ | 0 | 0 | $-1$ | 1 | 9 |

Now all the elements in the end column are positive and so we can proceed exactly as before. The $z$-column is the pivot column and the first row is the pivot row.

|  | $x$ | $y$ | $z$ | $u$ | $v$ | $w$ | $f$ |  |
|---|---|---|---|---|---|---|---|---|
| $R_1 \div \frac{3}{2}$ | 0 | $\frac{2}{3}$ | 1 | $\frac{2}{3}$ | 0 | $\frac{1}{3}$ | 0 | 1 |
|  | 0 | 3 | $\frac{1}{2}$ | 0 | 1 | $\frac{1}{2}$ | 0 | $\frac{3}{2}$ |
|  | 1 | 0 | $-\frac{1}{2}$ | 0 | 0 | $-\frac{1}{2}$ | 0 | $\frac{9}{2}$ |
|  | 0 | $-1$ | $-4$ | 0 | 0 | $-1$ | 1 | 9 |

|  | $x$ | $y$ | $z$ | $u$ | $v$ | $w$ | $f$ |  |
|---|---|---|---|---|---|---|---|---|
|  | 0 | $\frac{2}{3}$ | 1 | $\frac{2}{3}$ | 0 | $\frac{1}{3}$ | 0 | 1 |
| $R_2 - \frac{1}{2}R_1$ | 0 | $\frac{8}{3}$ | 0 | $-\frac{1}{3}$ | 1 | $\frac{1}{3}$ | 0 | 1 |
| $R_3 + \frac{1}{2}R_1$ | 1 | $\frac{1}{3}$ | 0 | $\frac{1}{3}$ | 0 | $-\frac{1}{3}$ | 0 | 5 |
| $R_4 + 4R_1$ | 0 | $\frac{5}{3}$ | 0 | $\frac{8}{3}$ | 0 | $\frac{1}{3}$ | 1 | 13 |

Since all the entries in the main part of the bottom row are positive, the optimal solution has been found. The maximum value of $f$ is 13 and occurs when $x = 5$, $y = 0$, $z = 1$ (and $u = 0$, $v = 1$, $w = 0$).

## Example 9.6

*This is a minimisation problem.*

*Find a minimum of the objective function $f = 5x + 4y$ subject to the constraints:*

$x + y \geq 10$
$x - y \leq 15$
$x \geq 0, y \geq 0$

### Solution
*The first step is to rewrite the problem in standard form.*
*Maximise   $c = -5x - 4y$*
*subject to   $-x - y \leq -10$   (Multiply the inequality by –1.)*
*$x - y \leq 15$*
*$x \geq 0, y \geq 0$*

*Now introduce slack variables u and v.*

$-x - y + u = -10$

$x - y + v = 15$

*The simplex tableau is:*

$$\begin{array}{ccccc} x & y & u & v & c \\ \left[\begin{array}{ccccc|c} -1 & -1 & 1 & 0 & 0 & -10 \\ 1 & -1 & 0 & 1 & 0 & 15 \\ \hline 5 & 4 & 0 & 0 & 1 & 0 \end{array}\right] \end{array}$$ Eliminate $-10$

*We have chosen the y-element as the pivot. However we could have chosen the x-element as pivot just as well.*

$$\begin{array}{c} \\ R_1 \div -1 \\ \\ \\ \end{array} \begin{array}{ccccc} x & y & u & v & c \\ \left[\begin{array}{ccccc|c} 1 & 1 & -1 & 0 & 0 & 10 \\ 1 & -1 & 0 & 1 & 0 & 15 \\ \hline 5 & 4 & 0 & 0 & 1 & 0 \end{array}\right] \end{array}$$

$$\begin{array}{c} \\ R_1 + R_1 \\ R_3 - 4R_1 \end{array} \begin{array}{ccccc} x & y & u & v & c \\ \left[\begin{array}{ccccc|c} 1 & 1 & -1 & 0 & 0 & 10 \\ 2 & 0 & -1 & 1 & 0 & 25 \\ \hline 1 & 0 & 4 & 0 & 1 & -40 \end{array}\right] \end{array}$$

*The bottom row gives:*

$c = -40 - x - 4u$

*The maximum value of c is –40 and occurs when $x = 0$, $u = 0$ and, from the tableau, when $y = 10$. Hence the minimum value of f will be +40.*

## EXERCISES

1    Use the simplex method to solve the following linear programming problems.

    a) Maximise   $f = 4x + 3y$          b) Maximise   $c = 3x - 2y$

       subject to   $x + y \leq 5$              subject to   $2x + 5y \leq 100$

                $-2x + 3y \geq 12$                    $x \geq 0, y \geq 0$

                $x \geq 0, y \geq 0$

    c) Maximise   $M = 6x + 5y$

       subject to   $x + 2y \leq 4$

                $3x + 2y \geq 4$

                $x \geq 0, y \geq 0$

2    Use the simplex method to solve the following linear programming problems.

    a) Minimise   $f = 3x + 4y$         b) Minimise   $M = 3x + y$

       subject to   $-x + y \leq 0$              subject to   $x + y \geq 3$

                $-3x + 4y \leq 0$                   $2x \geq 5$

                $x \geq 0, y \geq 2$                    $x \geq 0, y \geq 0$

    c) Minimise   $f = x + y$             d) Minimise   $f = 2x + y$

       subject to   $2x + y \geq 6$            subject to   $x + y \leq 27$

                $2x + 3y \leq 12$                  $x + y \geq 15$

                $4x + y \leq 10$                     $y \leq 2x$

                                                  $x \geq 3, y \geq 3$

*You may have seen this question in Chapter 8. You should now use the simplex method to tackle it.*

**3** CTM foods Ltd is designing a new breakfast cereal using wheatgerm and enriched oat flour as the basic ingredients. Each ounce of wheatgerm contains 2 milligrams of niacin, 3 milligrams of iron and 9.5 milligrams of thiamine, and costs 5 pence. Each ounce of enriched oat flour contains 3 milligrams of niacin, 3 milligrams of iron and 0.25 milligrams of thiamine, and costs 8 pence. The nutritionist wants the cereal to have at least 7 milligrams of niacin, 8 milligrams of iron and 1 milligram of thiamine per ounce. How many ounces of wheatgerm and how many ounces of enriched oat flour should be used in each serving in order to meet the nutritional requirements at the least cost?

*You may have seen this question in Chapter 8. You should now use the simplex method to tackle it.*

**4** A physical fitness enthusiast, Karen, decided to exercise by a combination of jogging and cycling. She jogs at 6 mph and cycles at 18 mph. A measure of the benefit of these activities is given through aerobic points. An hour of jogging gains 4 aerobic points and an hour of cycling gains 3 aerobic points. Each week she would like to earn at least 12 aerobic points, cover at least 54 miles and cycle at least twice as much as she jogs.

Formulate and solve a linear programming problem so that Karen minimises the total time spent exercising while achieving her goals.

**5** A firm requires coal with a phosphorus content no more than 0.03%, and no more than 3.25% ash impurity. Three grades of coal A, B and C are available at the prices shown below.

| Grade | Phosphorus (%) | Ash (%) | Cost (£/tonne) |
|-------|----------------|---------|----------------|
| A | 0.06 | 2.0 | 30 |
| B | 0.04 | 4.0 | 30 |
| C | 0.02 | 3.0 | 42 |

How should these be blended to meet the impurity restrictions at minimum cost?

*You may have seen this question in Chapter 8. You should now use the simplex method to tackle it.*

**6** In a developing nation the government wants to encourage everyone to make rice and soya beans part of the staple diet. The object is to design a lowest-cost diet that provides certain minimum levels of protein, calories and vitamin B12 (riboflavin). Suppose that one cup of uncooked rice costs 27 pence and contains 15 grams of protein, 810 calories and $\frac{1}{9}$ milligram of riboflavin. On the other hand, one cup of uncooked soya beans costs 21 pence and contains 22.5 grams of protein, 270 calories and $\frac{1}{3}$ milligram of riboflavin. Suppose that the minimum daily requirements are 90 grams of protein, 1620 calories and 1 milligram of riboflavin. Formulate and solve a linear programming problem to minimise the cost of a diet to satisfy the requirements.

## EXERCISES

**1** Use the simplex method to solve the following linear programming problems.

**a)** Maximise $f = 10x + 5y$
subject to $2x + y \leq 8$
$3x - 2y \geq 5$
$x \geq 0, y \geq 0$

**b)** Maximise $g = 20x - 2y$
subject to $5x + 4y \geq 20$
$2x + 5y \leq 50$
$x \geq 0, y \geq 0$

**c)** Maximise $h = 20x + 25y$
subject to $2z - 3y \geq 12$
$3x + 4y \leq 60$
$x \geq 0, y \geq 0$

**d)** Maximise $M = 6x + y$
subject to $y \geq 3x$
$4x + 5y \leq 38$
$x \geq 0, y \leq 0$

**2** Use the simplex method to solve the following linear programming problems.

**a)** Minimise $f = x + y$
subject to $x \geq 3, y \leq 0$
$y \geq 1.5x$
$3x + 2y \leq 24$

**a)** Minimise $g = x + 4y$
subject to $4y \geq 9$
$x + y \geq 6$
$x \geq 0, y \geq 0$

**c)** Minimise $h = 2x - y$
subject to $2y \geq x + 2$
$x + y \leq 7x$
$2x + y \leq 10$

**d)** Minimise $M = 7x + 7y$
subject to $x \geq 2, y \geq 4$
$y \leq 4x$
$3y + x \geq 10$
$2x + 3y \leq 22$

**3** Two warehouses, A and B, currently hold stocks of 30 and 15 washing machines. They are to supply machines to three distributors, I, II and III, that have placed orders for 20, 15 and 10 machines respectively. The costs in £s of transporting a machine from a warehouse to a distributor is given in the table below.

|   | I | II | III |
|---|---|----|-----|
| **A** | 6 | 8 | 2 |
| **B** | 4 | 2 | 10 |

If $x$ machines are shipped from A to I and $y$ machines from A to II, formulate the above as a linear programming problem in order to minimise the transport cost.

**4** The coffee shop in question 4 of Exercises 9.3B is now to relax its taste mixture constraint and replace it with the restrictions that the weight of bean A will be at least twice and no more than four times the weight of bean B. Reformulate and solve this revised problem.

**5** A budding impresario decides to make a 30-minute television programme for a cable TV channel. The licensing authorities require all programmes to have a cultural element. The impresario plans to employ a musical trio and to intercut this with a variety act. The impresario wants at least three minutes for commercials, the variety act insists on performing for at least half as long as the orchestra and the Musicians' Union demands that the musical trio occupies at least

four times as many minutes as the commercials. The respective costs per minute of the commercials, the musical trio and variety act are £175, £200 and £250. What is the lowest possible outlay for the 30 minutes and how is this achieved?

6  A brother and sister enter into a partnership to produce wooden bowls, coffee tables and desks. With five days' effort from the brother and three days' effort from the sister 500 bowls can be produced; with two days' effort from the brother and six days' effort from the sister 100 tables can be produced and with three days' efforts from the brother and five days' effort from the sister 20 desks can be produced. Each bowl is sold at £1 profit, each table at £10 profit and each desk at £50 profit. If the numbers of bowls, tables and desks which the partnership makes per week, on average, are $x$, $y$ and $z$ respectively, and given that the brother works at most five days a week and the sister at most four, show that the constraints may be written as:
$$x + 2y + 15z \leq 500, \ 3x + 30y + 125z \leq 2000, \ x \geq 0, \ y \geq 0, \ z \geq 0.$$

Determine the maximum profit and the numbers of bowls, tables and desks that should be produced on average per week to achieve this.

If the production of desks is discontinued, find the number of bowls and tables which the partners should produce per week, on average, to maximise their profits now. What is the decrease in maximum profits per week by using this strategy?

## MATHEMATICAL MODELLING ACTIVITY

## Linear programming

### *Introduction*
MJH Extensions Ltd. is a company employing in excess of 400 people and it has a large turnover. One of its major activities is the production of aluminium extrusions for customers who subsequently convert them into aluminium products such as windows, doors, telephone booths etc. Part of the services it offers its customers is to supply lengths of aluminium extrusion pre-finished, either with a powder paint coating in various colours or with an anodic film in different film thicknesses.

The management of the company wish to investigate the effect of various volumes of different film thicknesses being processed in its anodising plant; in particular it wishes to optimise the earnings of the plant. The three main film thicknesses to be considered are designated AA25, AA10, AA5 with AA25 being the thickest film and hence taking the longest time to process.

There are three stages to consider:

### *Pre-treatment*
1  Cleaning to remove soils and lubricants from the aluminium components.
2  Etching to remove the outer surface of the aluminium components and give a matt or roughened appearance to the surface.
3  De-smutting to remove any smut or loose films from previous stage.
4  Rinsing to avoid chemical carry over and to rinse the components clean.

### Anodising

This is an electrolytic process. At the correct voltage potential the oxygen combines with aluminium to form aluminium oxide. The process is time-based and longer immersions yield thicker anodic films.

### Post-treatment

Thorough rinsing is required prior to the final operation of sealing, which causes the aluminium oxide film previously produced to swell. This final stage ensures complete protection and prevention of environmental corrosion and the product is then finished.

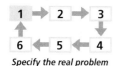

*Specify the real problem*

## Problem statement

There is a plentiful supply of historical production data relating to the various activities within the anodising plant. However, it is believed that the main areas which will have an impact on the throughput of the plant are as follow.

- Loading lengths of extrusion onto jigs to enable an electrical contact to be made within each extrusion: it takes two men 20 minutes to load a jig for AA25 film thickness so that the surface area of the length of extrusions on the jig is 45 m², for AA10 film thickness two men take 25 minutes to load the jig with 50 m², for AA5 film thickness two men take 30 minutes to load with 55 m². The company employs six men carrying out this activity working a 37-hour week.
- In the etching tank the immersion time per jig is ten minutes for AA25, five minutes for AA10 and eight minutes for AA5. This tank is available for $39\frac{1}{2}$ hours per week.
- The plant has three processing vats, two drawing 8000 amps and one drawing 6000 amps. These vats are available for 39.5 hours per week. The processing time in a vat for a jig of film thickness AA25 is 66 minutes, for AA10 35 minutes and for AA5 20 minutes.

The company can earn £4/m² for film thickness AA25, £3/m² for AA10 and £1.50/m² for AA5. However, there is a maximum of 5500 m² available in the marketplace to MJH of AA5 film thickness in any one week.

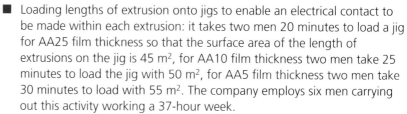

*Set up a model*

## Set up a model

Define clearly the problem under investigation, stating all your simplifying assumptions and choose the appropriate data available and discuss its purpose (particularly the units of measure).

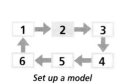

*Formulate the mathematical problem*

## Mathematical problem

Formulate the mathematical problem.

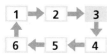

*Solve the mathematical problem*

## Mathematical solution

Solve the problem by an appropriate method.

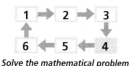

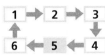

*Interpret the solution*

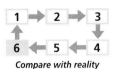

*Compare with reality*

# Interpretation

Interpret your solution. Do the results seem sensible when compared with reality?

# Refinement of the model

Revise the problem if necessary, redefining assumptions made in the initial approach.

The results should be presented in report form to the management in non-mathematical terms, stating clearly the conclusion of your investigation.

## CONSOLIDATION EXERCISES FOR CHAPTER 9

1  In the course of an investigation the following linear programming problem arose.
   Maximise    $P = x + 4y + 10z$
   subject to   $x + 4y + 2z \leq 100$
               $2x + 8z \leq 40$
               $x \geq 0, y \geq 0, z \geq 0$
   Use the simplex algorithm to determine the maximum value of $P$ and the values of $x$, $y$ and $z$ for which this occurs.
   *(ULEAC Question 7, Decision Mathematics, Specimen Paper D1, 1996)*

2  Use the simplex method to solve the following linear programming problem.
   Minimise    $f = 500 - 10x - 3y$
   subject to   $x + y \leq 30$
               $3x + 2y \geq 50$
               $x \geq 0, y \geq 0$

3  Lose Weight Foods Ltd produce a slimming drink consisting of vitamin-enriched milk. To satisfy dietary requirements the drink should contain 300 units of vitamin A, 150 units of vitamin B and 200 units of vitamin C. Three additives are available to provide the vitamins.

   Each millilitre of Additive I contains 1 unit of vitamin A, 2 units of vitamin B, 2 units of vitamin C and costs 1p.
   Each millilitre of Additive II contains 3 units of vitamin A, 1 unit of vitamin B, 2 units of vitamin C and costs 2p.
   Each millilitre of Additive III contains 2 units of vitamin A, 3 units of vitamin B, 5 units of vitamin C and costs 3p.
   Formulate and solve by the simplex method the problem as a linear programming problem in which the manufacturer wishes to minimise the cost of the additives.

*You met this family in Exploration 8.1. Now you are required to solve the problem using the simplex method.*

4  The parents of a two-year-old child want to ensure that they include at least the minimum satisfactory amount of protein, energy and vitamin C. The daily amounts recommended by good health guides are 32 g of protein, 5500 kilojoules (kJ) of energy and 30 mg of vitamin C. Suppose that the parents decide to achieve the daily

amounts through milk and orange juice. The following table shows the amount of protein, energy and vitamin C available in these drinks.

| Item | Measure of drink | Protein (g) | Energy (kJ) | Vitamin C (mg) |
|------|------------------|-------------|-------------|----------------|
| milk | 100 ml | 3.0 | 300 | 1 |
| juice | 100 ml | 0.2 | 40 | 10 |

Milk costs 38p per litre and orange juice costs 48p per litre. The parents want to find the minimum costs to achieve the recommended daily intake.

If $x$ and $y$ are the number of measures of milk and juice, write down:

a) the total costs of the drinks
b) three inequalities for the amount of protein, energy and vitamin C so that the minimum satisfactory amounts are achieved.

Solve the linear programming problem using the simplex method.

*You met Ben in Example 8.1. Now you are required to solve the problem using the simplex method.*

5   Ben has two evening and weekend jobs, mowing lawns and cleaning cars. He can spend a maximum of 15 hours per week on these jobs. He has several regular clients for whom he spends six hours per week mowing lawns and three hours per week cleaning cars. Ben is paid £2.50 per hour for mowing lawns and £3.50 per hour for cleaning cars. Formulate and solve a linear programming problem for Ben to maximise his earnings.

*You may have seen this question in Chapter 8. You should now use the simplex method to tackle it.*

6   A house builder builds two type of home.

The first type requires one plot of land, £40 000 capital, 150 worker days of labour and is sold for £8000 profit.
The second type requires two plots of land, £105 000 capital, 200 worker days of labour to build and is sold for £11 500 profit.
The house builder owns 150 plots of land, has available £9 600 000 and 24 000 worker days of labour.
Formulate and solve a linear programming problem to advise the builder on a strategy to maximise the profit.

7   In the course of an investigation of the following problem arose.

Maximise    $f = 3x + 2y + 7z$
subject to    $x + y + 2z \le 50$
$x + 4z \le 20$
$x \ge 0, y \ge 0, z \ge 0$

Use the simples algorithm to determine the maximum value of $P$ and the values of $x$, $y$ and $z$ for which this occurs.

8   A production unit makes two types of product, X and Y. Production levels are measured in tonnes and are constrained by availability of

finance, staff and storage space. Requirements for, and daily availability of, each of these resources are summarised in the table.

| | Finance required (£/tonne) | Staff time required (hours/tonne) | Storage space requirements (m³/tonne) |
|---|---|---|---|
| Product X | 200 | 8 | 1 |
| Product Y | 100 | 8 | 3 |
| Resource availability each day | £1000 | 48 hours | 15 m³ |

Profits on these products are £160 per tonne for X and £120 per tonne for Y.

a) Express the three resource constraints as inequalities, and write down two further inequalities indicating that production levels are non-negative.
b) Given that the objective is to maximise profit, write down the objective function.
c) Illustrate your inequalities graphically, and use your graph to find the best daily production plan.
d) Set up an initial tableau for the problem and use the simplex algorithm to solve the problem. Relate each stage of the tableau to its corresponding point on your graph.

*(UODLE Question 22, Paper 4, 1993)*

9   In an executive initiative course, participants are asked to travel as far as possible in three hours using a combination of moped, car and lorry. The moped can be carried in the car and the car can be carried on the lorry.

The moped travels at 20 miles per hour (mph) with a petrol consumption of 60 miles per gallon (mpg). The car travels at 40 mph with a petrol consumption of 40 mpg. The lorry travels at 30 mph with a petrol consumption of 20 mpg. $2\frac{1}{2}$ gallons of petrol are available.

The moped must not be used for more than 55 miles, and a total of no more than 55 miles must be covered using the car and/or lorry.

a) Formulate the above specifications as a linear programming problem, stating clearly your choice of variables and your objective function.
b) Set up a simplex tableau to solve the problem.
c) Perform two iterations of the simplex algorithm.
d) Is the tableau resulting from part c) optimal?
   What strategy is indicated by the tableau, and how much time and petrol would be used?

*(UODLE Question 19, Paper 4, 1990)*

**10** A diet planning problem involves mixing two foodstuffs, X and Y, to satisfy three dietary constraints at minimum cost, $C$. The amounts of X and Y used are $x$ and $y$ kg respectively. The costs are £1 per kg for each. The problem is formulated as a linear programme.

Minimise $\qquad\qquad\qquad C = x + y$
subject to the constraints $2x + y \geq 8$
$$3x + 2y \geq 14$$
$$x + 2y \geq 6$$
$$x \geq 0$$
$$y \geq 0$$

a) Draw a graph to illustrate the inequalities and the feasible region.

b) The first three inequalities are converted into equalities by introducing slack variables, $s_1$, $s_2$ and $s_3$, and are laid out in a tableau.

| $C$ | $x$ | $y$ | $s_1$ | $s_2$ | $s_3$ | RHS |
|---|---|---|---|---|---|---|
| 1 | −1 | −1 | 0 | 0 | 0 | 0 |
| 0 | 2 | 1 | −1 | 0 | 0 | 8 |
| 0 | 3 | 2 | 0 | −1 | 0 | 14 |
| 0 | 1 | 2 | 0 | 0 | −1 | 6 |

Explain why it is not possible to apply the simplex algorithm to this tableau.

c) A modification to the simplex algorithm is applied to produce a feasible solution to the above problem. Expressed in tableau form the solution is:

| $C$ | $x$ | $y$ | $s_1$ | $s_2$ | $s_3$ | RHS |
|---|---|---|---|---|---|---|
| 1 | 1 | 0 | −1 | 0 | 0 | 8 |
| 0 | 2 | 1 | −1 | 0 | 0 | 8 |
| 0 | 1 | 0 | −2 | 1 | 0 | 2 |
| 0 | 3 | 0 | −2 | 0 | 1 | 10 |

i) Interpret this tableau by stating the values given for the variables $x$ and $y$ and the cost $C$, and show that the constraints are satisfied.

ii) What are the values taken by $s_1$, $s_2$ and $s_3$ in this solution? How do these values relate to the constraints?

d) Starting from the tableau in c) two iterations of the simplex algorithm are needed to achieve the optimal solution. With this formulation you will choose to pivot on a column (not the $C$-column) that has a positive value in the first row of the table. Thus the first iteration will involve increasing the value of $x$. Perform the two iterations, interpret the result, and mark the point to which it relates on your graph.

*(UODLE Question 20, Paper 4, 1995)*

## Summary

*After working through this chapter you should know the following.*

■ The simplex method is used for linear programming problems in standard form.

**Step 1:** Introduce slack variables so that the constraints become linear equations and form the simplex tableau.

**Step 2:** If the left part of the bottom row contains negative values then the optimal value has not been found. Perform a new iteration in the following way.
  a) Choose the pivot row by calculating the ratio of the end column with the positive entries in the pivot column. The element that gives the smallest positive ratio is the pivot element.
  b) Create a new simplex tableau by pivoting about the pivot element and using row operations.

**Step 3:** Repeat step 2 until the elements in the left part of the bottom row are all positive. The maximum value has then been found.

■ The simplex method can be adapted for linear programming problems in non-standard form.

**Step 1:** Convert all the constraints (except the trivial constraints) into the form
  linear polynomial ≤ constraint
and define an objective function to be maximised.

**Step 2:** If the upper part of the end column contains a negative value then remove it by pivoting in the following way.
  a) Choose a negative element in main part of the row, this element is the pivot.
  b) Create a new simplex tableau by pivoting about the pivot element and using row operations.

**Step 3:** Repeat step 2 until there are no negative entries in the upper part of the end column.

**Step 4:** Carry out steps 2 and 3 of the standard form method to obtain the optimal value of the objective function.

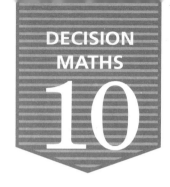

# *Recurrence relations*

*By the end of this chapter you should:*

■ know that a discrete system with a recognised pattern of growth can be represented using a recurrence relation

■ be able to use the recurrence relation as a model for discrete systems

■ be able to carry out iterations, by substituting into the recurrence relation to find successive terms

■ be able to classify a recurrence relationship by its order

■ know how to solve first-order and second-order recurrence relations.

## DEFINING A RECURRENCE RELATION

**Exploration 10.1**

### *Modelling a seal population*

Each year a survey is made of the number of seals in the Orkney Islands. During the year the number of seals born is approximately 40 per cent of the population at the beginning of the year and the number of seals that die is approximately 15 per cent of the population at the beginning of the year.

At the beginning of the 1997 survey the population of seals was 10 000.

■ What is the expected size of the population at the beginning of 1998?
■ Estimate the population in January 2005.
■ In what year will the population of seals reach 30 000?

### *Notation*

Before analysing the solution of this exploration we need to introduce some notation used in modelling problems of this type. Suppose that we represent the population in year $n$ as $P_n$. Then $P_1$ is the initial population at 1 January 1997, given by $P_1 = 10\,000$. Then $P_2$ is the population on 1 January 1998 and so on. The population of seals forms a sequence:

$$P_1, P_2, P_3, \dots, P_n, \dots$$

Now we can set out the problem in Exploration 10.1 as an equation in terms of this sequence.

# Seal population

During each year the population increases by 40 per cent of the population at the beginning of the year. So if $P_{n-1}$ is the population at the beginning of year $n - 1$, the number of seals born is $0.4P_{n-1}$. Similarly the number of seals that die is $0.15P_{n-1}$. The number of seals at the beginning of year $n$ is then:

$$P_n = P_{n-1} + 0.4P_{n-1} - 0.15P_{n-1}$$

population at beginning of year      number of births      number of deaths

This can be simplified to give:

$$P_n = 1.25P_{n-1}$$

*Data that can only take on whole-number values, such as population sizes, are called discrete.*

This is an example of a discrete model called a **recurrence relation**. The number of seals at the beginning of each year can now be found using the recurrence relation to generate a sequence showing the increasing population.

| Year | 1997 | 1998 | 1999 | 2000 | 2001 | |
|---|---|---|---|---|---|---|
| Population | 10 000 | 12 500 | 15 625 | 19 531.25 | 24 414.0625 | ... |

It is important to interpret the answer carefully. We cannot have a fraction of a seal. On 1 January 2001 the number of seals is predicted to be 24 414.

The process of obtaining the solution of a problem from a recurrence relation is called **iteration**.

**Example 10.1**

*On her 18th birthday, Wendy won £12 000 on the National Lottery. She invested it in a building society at a fixed rate of interest of six per cent compounded annually. How much would Wendy have on her twenty-fifth birthday?*

**Solution**
*Let $x_n$ be the total money, in pounds, in year $n$, so that $x_1$ is the amount that Wendy won. At the beginning of the second year:*
*$x_2$ = initial amount + interest*

$$x_2 = 12\ 000 + \left(\frac{6}{100} \times 12\ 000\right) = £12\ 720$$

*At the beginning of the third year she has:*

$$x_3 = 12\ 720 + \left(\frac{6}{100} \times 12\ 720\right) = £13\ 483.20$$
*and so on.*

This sequence can be written more generally as the recurrence relation:

$$x_n = x_{n-1} + \frac{6}{100} x_{n-1} = 1.06 x_{n-1}$$

Wendy starts with £12 000 so $x_1 = 12\,000$. Iteration generates the following sequence.

| $n$ | 1 | 2 | 3 | 4 | 5 | 6 | 7 | 8 |
|---|---|---|---|---|---|---|---|---|
| $x_n$ | 12 000 | 12 720 | 12 483.20 | 14 292.19 | 15 149.72 | 16 058.71 | 17 022.23 | 18 043.56 |

Remember, $x_2$ is the amount in the account when Wendy is 19, $x_3$ is the amount when she is 20 and so on, and $x_8$ is the amount when Wendy is 25 years old.

## EXERCISES

**10.1 A**

1 Claire invests £5000 in a building society at a fixed interest rate of $4\frac{1}{2}$ per cent compounded annually.

a) How much will Claire have invested after ten years?
b) If Claire withdraws £1000 after five years how much will she now have after ten years?
c) Claire's brother, Pasi, invests £5000 in an account which offers an annual interest rate of $4\frac{1}{2}$ per cent compounded every three months. How does Pasi's investment compare with Claire's after ten years, assuming that neither of them withdraws any money during the period?

2 A forest of trees is being carefully managed so that felled trees are continually replaced. At the beginning of 1990 the forest had 20 000 trees. In the summer of each year five per cent of the trees are felled and removed to make paper products. In the autumn of each year 800 new trees are planted.

a) Formulate a recurrence relation to model the number of trees in the forest at the beginning of each year.
b) By iteration find the number of trees at the beginning of year 2000.
c) Investigate what happens to the number of trees in the forest after a long period of time.

3 Find the monthly repayment on a credit card balance of £800 over a period of six months if the rate of interest is $1\frac{1}{2}$ per cent per month and the interest on the balance is added at the end of each month.

4 For each of the following recurrence relations use a calculator to generate the first ten terms in the iteration process. What can you deduce about the long-term behaviour of the sequence?

a) $x_n = 0.5x_{n-1} - 1, x_1 = 4$       b) $x_n = 2x_{n-1} + 3, x_1 = 0$

c) $x_{n+1} = -x_n + 1, x_0 = 2$       d) $x_{n+1} = 4 - \dfrac{x_n - 1}{x_n^2}, x_0 = 1$

5  The figure shows three rings, of different sizes, stacked on a rod. The aim of the puzzle is to move the rings from one rod to another in as few moves as possible, never allowing a larger ring to sit on top of a smaller ring.

*This problem models a puzzle called the Tower of Hanoi.*

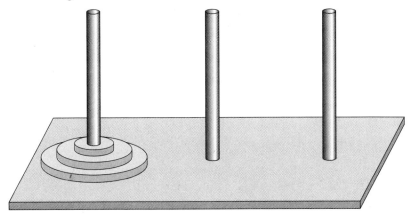

a) How many moves does it take for an initial pile of:
   i) 3 rings     ii) 4 rings     iii) 5 rings?
b) Formulate a recurrence relation to predict the number of moves for $n + 1$ rings in terms of the number of moves for $n$ rings.
c) Using iteration find the number of moves for ten rings.
d) The puzzle originates in the Far East where it is said that if priests move one disc each second from an original pile of 64 discs, then when the pile is reformed the world will end. A priest in the temple started the puzzle 5000 years ago. When should the world end?

## EXERCISES

**10.1 B**

1  The population of a country is currently 5.2 million. The number of births each year is four per cent of the population and the number of deaths each year is three per cent of the population. Also each year 20 000 immigrants arrive in the country.

a) Formulate a recurrence relation model to describe the population size.
b) Use iteration to find the population in 20 years' time.

2  Katrina has £1000 in her bank account which does not give interest. Due to inflation the value of her money is decreasing at a rate of 0.5 per cent each month.

a) Formulate a recurrence relation model to determine the value of the money each month.
b) What would Katrina's investment be worth in ten years' time if she left it in the bank?
c) Suppose instead that Katrina moved the money to a building society giving eight per cent annual interest, compounded monthly. What is the value of her investment after ten years in this case?

3    By iterating an appropriate recurrence relation, find how many years it would take to double your savings if the interest rate was:

   a)  five per cent per year compounded annually
   b)  five per cent per year compounded monthly
   c)  nine per cent per year compounded annually.

4    If $x_{n+1} = ax_n + b$, $n \geq 0$, and $x_1 = 3$, $x_2 = 7$, $x_3 = 15$, find the values of $a$, $b$ and $x_6$.

5    At his retirement, aged 65, Jason received a lump sum payment of £350 000 from a life assurance policy. He invested the money in a savings account which earns 0.7 per cent interest per month.

   Jason presently needs £1500 per month to live comfortably and inflation is 0.4 per cent per month. If the interest and inflation rates remain constant will Jason's investment last for 20 years?

## CLASSIFYING RECURRENCE RELATIONS

The recurrence relations for Exploration 10.1 and Example 10.1:

$$P_n = 1.25 P_{n-1} \text{ and } x_n = 1.06 x_{n-1}$$

are examples of **first-order** recurrence relations. A recurrence relation is of $k$th order if the difference between the highest and lowest subscript is $k$.

$x_n = 2x_{n-1} + 3$      is first order because $(n) - (n-1) = 1$
$x_n = 2x_{n-1} - 3x_{n-2}$      is second order because $(n) - (n-2) = 2$
$x_n = 0.5x_{n-1} + 0.7x_{n-4}$      is fourth order because $(n) - (n-4) = 4$

A recurrence relation is linear if it can be written as:
$$x_n = a_{n-1} x_{n-1} + a_{n-2} x_{n-2} \dots + p$$

where $a_{n-1}, a_{n-2}, \dots p$ may depend only on $n$ and not on any $xs$. Otherwise it is **non-linear**. The starting value that generates the sequence from a recurrence relation is called the **initial condition**. The terms in the sequence are called the **iterations** of the recurrence relation.

For some recurrence relations we can find a general formula:
$$x_n = f(n)$$

which satisfies the equation. In Example 10.1, the general term is:
$$x_n = 1.06 x_{n-1}$$

If we look at the first few terms of this sequence we get:
$x_1$
$x_2 = 1.06 x_1$
$x_3 = 1.06 x_2 = 1.06(1.06)x_1 = 1.06^2 x_1$
$x_4 = 1.06 x_3 = 1.06(1.06^2 x_1) = 1.06^3 x_1 \dots$
$x_n = 1.06^{n-1} x_1$

So, for example, if $P_n = 1.25 P_{n-1}$, the general formula will be:

$$P_n = 1.25^n A$$

where $A$ is the value of $P$ when $n = 0$.

Such a formula is called the **general solution** of the recurrence relation. A formula such as:

$$P_n = 1.25^n A$$

is called an **analytical** (or exact) solution. For many recurrence relations no analytical solution exists, so we can only generate the sequence or a numerical solution.

In this chapter we find analytical solutions for first- and second-order linear recurrence relations.

**Exploration 10.2**

## *First-order linear recurrence relations*

Consider the recurrence relation:

$$x_n = kx_{n-1}$$

where $k$ is a constant.

- Generate a sequence of iterations $x_2, x_3 \, x_4, x_5, \ldots$ in terms of the initial condition $x_1$ and $k$.
- From your iterations write down the general solution of the recurrence relation.

## The general solution

You will have found that the general solution of the recurrence relation $x_n = kx_{n-1}$ is:

$$x_n = k^{n-1} x_1$$

If the equation is of the form:

$$x_n = kx_{n-1} + p$$

then a general solution can be found in a similar way by generating a sequence of iterations.

$$x_2 = kx_1 + p$$
$$x_3 = kx_2 + p = k(kx_1 + p) + p = k^2x_1 + kp + p = k^2x_1 + p(k + 1)$$
$$x_4 = kx_3 + p = k^3x_1 + k^2p + kp + p = k^3x_1 + p(k^2 + k + 1)$$
$$x_5 = kx_4 + p = k^4x_1 + k^3p + k^2p + kp + p = k^4x_1 + p(k^3 + k^2 + k + 1) \ldots$$
$$x_n = k^{n-1}x_1 + p(k^{n-2} + k^{n-3} + \ldots + k^2 + k + 1)$$

The expression on the right-hand side $(k^{n-2} + k^{n-3} + \ldots + k^2 + k + 1)$ is the sum of a geometric progression (GP). This has $n - 1$ terms, with the common ratio $k$. So provided $k \neq 1$ the sum of the GP is:

$$S_n = \frac{k^{n-1} - 1}{k - 1}$$

The general solution of the recurrence relation $x_n = kx_{n-1} + p$ where $k \neq 1$ and $n \geq 2$ is:

$$x_n = k^{n-1}x_1 + p\,\frac{k^{n-1} - 1}{k - 1}$$

## Example 10.2

*Find the general solution of the recurrence relation:*
$x_n = 2x_{n-1} + 3, n \geq 2$

*Find the particular solution if $x_1 = 5$.*

### Solution
*For the recurrence relation $x_n = 2x_{n-1} + 3$, $k = 2$ and $p = 3$. The general solution of the recurrence relation is:*

$$x_n = 2^{n-1}x_1 + 3\frac{\left(2^{n-1} - 1\right)}{(2-1)} = 2^{n-1}x_1 + 3(2^{n-1} - 1)$$

*If $x_1 = 5$ then $x_n = 5 \times 2^{n-1} + 3(2^{n-1} - 1) = 8 \times 2^{n-1} - 3$*
$x_n = 2^{n+2} - 3$ *since* $8 = 2^3$.

## Example 10.3

*Find the general solution of the recurrence relation:*
$x_n = 1.5x_{n-1} - p, n \geq 1$

### Solution
*The notation in this recurrence relation is slightly different. The sequence of iterations begins with $x_0$:*

$x_0, x_1, x_2, x_3, \ldots$

*The general solution will look slightly different.*
$x_1 = 1.5x_0 - p$
$x_2 = 1.5x_1 - p = 1.5^2x_0 - 1.5p - p$
$x_3 = 1.5^3x_0 - (1.5^2 + 1.5 + 1)p \ldots$
$x_n = 1.5^nx_0 - (1.5^{n-1} + \ldots + 1.5^2 + 1.5 + 1)p$

$$x_n = 1.5^nx_0 - \frac{\left(1.5^n - 1\right)}{1.5 - 1}p = 1.5^nx_0 - 2(1.5^n - 1)p$$

## General solutions and special cases

In general terms, the solution of the recurrence relation:
$x_n = kx_{n-1} + p \qquad n \geq 1, k \neq 1$

is:
$$x_n = k^nx_0 + p\frac{\left(k^n - 1\right)}{(k-1)}$$

### Special case $k = 1$
In the special case when $k = 1$ the general solution cannot be used because of division by zero. Forming a sequence:
$x_2 = x_1 + p$
$x_3 = x_2 + p = x_1 + 2p$
$x_4 = x_3 + p = x_1 + 3p$

The general solution of the recurrence relation $x_n = x_{n-1} + p$, $n \geq 2$ is:
$x_n = x_1 + (n - 1)p$

## Example 10.4

*Find the general solution of the recurrence relation:*
$$x_n = x_{n-1} + 5, \, n \geq 1$$

### Solution

*The sequence for this equation begins with $x_0$. The general solution is $x_n = x_0 + np$ which in this case is:*
$$x_n = x_0 + 5n$$

## Example 10.5

*Find the monthly repayment on a loan of £1000 at an interest rate of 18 per cent per annum over 18 months.*

### Solution

*Let the amount of the loan owing after $n$ months be $x_n$ so that the initial amount is $x_0 = 1000$.*
*Let $c$ be the monthly repayment.*
*The interest rate per month is assumed to be 1.5 per cent.*
*The model for the monthly repayment is:*

$$x_n = x_{n-1} + \frac{1.5}{100} x_{n-1} - c$$

| amount at end of month | amount at start of month | interest | repayment |

$$x_n = 1.015 x_{n-1} - c \qquad n \geq 1, \, x_0 = 1000$$

*The general solution of this recurrence relation is:*
$$x_n = 1.015^n x_0 - \frac{\left(1.015^n - 1\right)}{\left(1.015 - 1\right)}$$

$$x_n = 1.015^n \times 1000 - \frac{\left(1.015^n - 1\right)}{0.015}$$

*At the end of the repayment period of 18 months the outstanding loan is zero i.e. $n = 18$ and $x_{18} = 0$.*
$$0 = 1.015^{18} \times 1000 - c \frac{\left(1.015^{18} - 1\right)}{0.015}$$

*Solving for $c$ we have:*
$$c = 63.8058 \text{ (to 4 d.p.)}$$

*The constant repayment amount to repay the loan within 18 months is £63.81.*

## Exploration 10.3

### *Investigating convergence*

Consider the recurrence relation:
$$x_n = kx_{n-1} + p \quad n \geq 2$$

■ Find the general solution for $k = -2, -0.5, 0.5$ and 2.
■ Describe what happens to the solution as $n \to \infty$.
■ Deduce a criterion on $k$ for the solution of the recurrence relation to converge.

## EXERCISES

10.2 A

1 Write down the general solution of each of the following recurrence relations.

**a)** $x_n = 4x_{n-1}, n \geq 2$      **b)** $x_n = 4x_{n-1} - 5, n \geq 1$
**c)** $p_n = 0.7p_{n-1} + 100, n \geq 1$      **d)** $u_n = u_{n-1} + 5, n \geq 2$
**e)** $u_n = -2u_{n-1} + 3, n \geq 1$      **f)** $p_{n+1} = 1.5p_n - 50, n \geq 0$
**g)** $x_n = -x_{n-1} + 2, n \geq 2$      **h)** $x_{n+1} = -x_n + 2, n \geq 0$
**i)** $x_n + 0.1x_{n-1} - 1.1 = 0, n \geq 2$      **j)** $u_n - 0.9u_{n-1} - 4 = 0, n \geq 1$

2 Find the particular solution of the following problems.

**a)** $x_n = 3x_{n-1} + 2, n \geq 2, x_1 = 1$      **b)** $u_n = 0.1u_{n-1} + 1, n \geq 1, u_0 = 10$
**c)** $u_{n+1} = u_n + 6, n \geq 0, u_0 = 2$      **d)** $p_n = 1.4p_{n-1} + 60, n \geq 1, p_0 = 100$
**e)** $u_{n+1} = -2u_n + 3, n \geq 2, u_2 = 0$

3 Sally invests £5000 in a building society savings account at an annual interest rate of five per cent compounded monthly. How much will Sally have after 15 years?

4 Find the monthly repayment on a £5000 loan over a term of three years at an interest rate of 24 per cent per annum.

5 Find the monthly repayment on a £300 loan over a term of six months at an interest rate of 15 per cent per annum.

6 By iterating each sequence find the general solution of:

**a)** $x_n = x_{n-1} + n, n \geq 1$      **b)** $u_n = 2u_{n-1} + n^2, n \geq 2$

7 Wesley borrows £1000 to buy a hi-fi system. The annual interest rate on the loan is 18 per cent. Wesley can afford to pay the loan off at £25 per month. How many months will he take to pay off the loan?

8 How long would it take to pay back a loan of £2400 at a rate of £70 per month, if the interest rate is 14 per cent per annum?

9 A population of wild animals is threatened with extinction. The present population is 1000; the numbers of births and deaths per year are 20 per cent and 40 per cent respectively.

**a)** After how many years will the wild animals become extinct?

b) A zoo decides to attempt to save the species. The strategy is to introduce an increasing number of the animals each year. In the first year 20 animals are introduced and in each subsequent year the rate of increase of animals is 20 per cent. Will this strategy save the species from becoming extinct?

**10** This problem investigates the length of a toilet roll. The radius of the inner core is 2.2 cm and the paper thickness is 0.005 cm. When the paper is wrapped $n$ times round the core the radius of the entire toilet roll is $r = 2.2 + 0.005n$.

a) Formulate a recurrence relation model for the length of toilet paper on the roll when another circle of paper is wound onto the roll.
b) Solve the recurrence relation to find the length of the toilet roll.
c) For a new toilet roll its outer radius is 5 cm. Find the total length of paper on the roll.
d) What is the outer radius of the roll when the total length of paper is 200 metres?

## EXERCISES

**10.2 B**

**1** Write down the general solution of each of the following recurrence relations.

a) $x_n = -1.1x_{n-1}, n \geq 2$
b) $x_{n+1} = -1.5x_n + 5, n \geq 1$
c) $u_n = -u_{n-1} + 3, n \geq 2$
d) $u_{n+1} = 0.2u_n, n \geq 0$
e) $p_{n+1} = -0.2p_n, n \geq 0$
f) $x_n = 0.8x_{n-1} - 1, n \geq 2$
g) $x_{n+1} + 0.9x_{n-1} = 1, n \geq 0$
h) $u_n - 2u_{n-1} + 5 = 0, n \geq 1$
i) $2x_{n+1} + 1 = 2x_n - 4, n \geq 1$
j) $2u_n - 3u_{n-1} = 8, n \geq 2$

**2** Find the particular solution of the following problems.

a) $x_{n+1} = -0.2x_n, n \geq 1, x_1 = 1000$
b) $u_n = 1.5u_{n-1} - 5, n \geq 1, u_0 = 4$
c) $x_n = 3x_{n-1} + 1, n \geq 2, x_1 = 0$
d) $p_{n+1} = 1.07p_n + 70, n \geq 1, p_1 = 10$
e) $u_n = -2u_{n-1} + 7, n \geq 2, u_1 = 1$

**3** Consider the recurrence relation:
$$x_{n+1} = 3x_n - 4, n \geq 0$$

a) Find the particular solution given that $x_0 = 3$. For what value of $m$ does $x_m = 245$?
b) Find the value of $x_0$ so that $x_6 = -1456$.

**4** By iterating each sequence find the general solution of:

a) $x_{n+1} = 4x_n + 6n + 5, n \geq 1$
b) $u_n = 2u_{n-1} + 2^n, n \geq 0$
c) $x_{n+1} = x_n + (-1)^n, n \geq 1$

**5** Suppose that you invest £1000 at an interest rate of eight per cent per annum, compounded every six months.

a) What is the value of your investment after ten years?
b) How many years does it take to double your money?

**6** How long will it take to pay back a loan of £3500 at a rate of £80 per month, if the interest rate is 18 per cent per annum?

**7** Find the monthly repayment on a £10 000 loan over a term of five years at an interest rate of 14 per cent per annum.

**8** In a learning experiment, a rat is placed in a starting box at the beginning of a straight runway. Sufficient time is allowed for the rat to run along the runway without dawdling. If he reaches the end of the runway in that time the trial is considered to be a success. When the rat reaches the end of the runway he receives either a reward (food), punishment (a mild electric shock) or nothing.
Let $p_n$ be the probability of success in the $n$th trial. Then from the experiments, it was found that:

$$p_n = p_{n-1} + a(1 - p_{n-1}) - bp_{n-1}$$

where $a$ is the proportion of times success is rewarded and $b$ is the proportion of times success is punished.

**a)** Find the solution of the recurrence relation for the case $a = 0.4$, $b = 0.1$ and $p_1 = 0.2$. What is the probability of success after a very long time?
**b)** When the proportions of reward and punishment are reversed what do you find?
**c)** Investigate the case when $a = b = 0.5$.
**d)** What happens when there is always food at the end of the runway?
**e)** What happens when there is no food and the rat gets an electric shock at the end of the runway?

**9** On 1 January 1996 Hannah invested £2000 in a building society at a fixed interest rate of five per cent compounded annually. At the end of each year (starting on 31 December 1996) she withdraws £200 to pay for Christmas presents. How much will be remaining in her account on 1 January 2005?

**10** Investigate the solution of the recurrence relation:

$$x_{n+1} = nx_n + 2, n \geq 1$$

## Exploration 10.4

### Modelling the spread of dandelions

Airborne seeds from the perennial weed *Taraxacum officinale* (dandelion) landed on an island, previously dandelion-free, in the autumn of 1979. Each dandelion plant sheds seeds every autumn and of these an average of three germinate in the spring of the following year, while an average of twelve germinate in the spring of the year after.

■ Assuming that no dandelions die and that no further airborne seeds arrive, show that the number of dandelions $u_n$ in the summer of year (1980 + $n$) satisfies the recurrence relation
$$u_{r+1} = 4u_r + 12u_{r-1}.$$

■ In the summer of 1980 ($n = 0$) there were four dandelions on the island and one year later there were eight dandelions. Calculate the sequence of iterations to find the number of dandelions growing in the summers of 1985 and 1986.

## SECOND-ORDER RECURRENCE RELATIONS

Exploration 10.4 gives an example of a second-order recurrence relation. In this section we investigate solutions of second-order recurrence relations with constant coefficients which are of the general form:

$$x_n = ax_{n-1} + bx_{n-2} + p, \ n \geq 2$$

where $a$ and $b$ are constants and $p$ may be a function of $n$ only. If $p = 0$ the recurrence relation is said to be **homogeneous**.

**Exploration 10.5**

### *Solving homogeneous recurrence relations*

Consider the recurrence relation:

$$x_n = ax_{n-1} + bx_{n-2}$$

■ Let $x_n = Am^n$ where $A$ and $m$ are constants. Substitute for $x_n$, $x_{n-1}$ and $x_{n-2}$ into the recurrence relation. Show that $m$ satisfies the quadratic equation:
$$m^2 - am - b = 0$$
■ Suggest a form for the general solution of the recurrence relation.

### The auxiliary equation

The quadratic equation for $m$ is called the **auxiliary equation** of the recurrence relation. Its solution is:

$$m_1 = \frac{a + \sqrt{a^2 + 4b}}{2} \text{ or } m_2 = \frac{a - \sqrt{a^2 + 4b}}{2}$$

There are three cases to consider.

■ $m_1 \neq m_2$ and real
■ $m_1 \neq m_2$ and complex
■ $m_1 = m_2$.

Suppose that $m_1 \neq m_2$. Then there are two solutions of the recurrence relation:

$$Am_1^n \text{ and } Bm_2^n$$

where $A$ and $B$ are constants. They both satisfy the recurrence relation, so:

$$Am_1^n = aAm_1^{n-1} + bAm_1^{n-2} \text{ and}$$
$$Bm_2^n = aBm_2^{n-1} + bBm_2^{n-2}$$

Adding these two equations gives:

$$Am_1^n + Bm_2^n = a(Am_1^{n-1} + Bm_2^{n-1}) + b(Am_1^{n-2} + Bm_2^{n-2})$$

From this equation we can see that the sum of the separate solutions is also a solution of the recurrence relation. We deduce that the general solution of the recurrence relation

$$x_n = ax_{n-1} + bx_{n-2}$$

is:  $x_n = Am_1^n + Bm_2^n$

where $m_1$ and $m_2$ are solutions of the auxiliary equation $m^2 - am - b = 0$ and $A$ and $B$ are constants.

## Real roots

**Example 10.6**

*Find the general solution of:*
$$x_n = 5x_{n-1} - 6x_{n-2}, n \geq 2$$

*Find the particular solution given that $x_1 = 1$ and $x_0 = 0$.*

**Solution**
$$x_n - 5x_{n-1} + 6x_{n-2} = 0$$

*So the auxiliary equation is:*
$$m^2 - 5m + 6 = 0$$

*Solving for $m$ we have $m = 3$ or $m = 2$.*
*The general solution is:*
$$x_n = A3^n + B2^n$$

*When $n = 0$, $x_0 = A + B = 0$.*
*When $n = 1$, $x_1 = 3A + 2B = 1$.*
*Solving for $A$ and $B$, $A = 1$ and $B = -1$.*
*The particular solution is $x_n = 3^n - 2^n$.*

## Complex roots

**Example 10.7**

*Find the general solution of:*
$$u_n = -4u_{n-2}, n \geq 2$$

*Find the particular solution given that $u_0 = 1$ and $u_1 = 2$.*

**Solution**
*The auxiliary equation is $m^2 = -4$ and has complex solutions $m = 2i$ and $m = -2i$. The general solution is:*
$$u_n = A(-2i)^n + B(2i)^n$$

*You should ignore this example if you have not studied complex numbers.*

*Now $u_0 = 1$ so $u_0 = A + B = 1$*
*and $u_1 = 2$ so $u_1 = -2iA + 2iB = 2$*
$$\Rightarrow -A + B = -i$$

*Solving for $A$ and $B$ gives:*
$$A = \tfrac{1}{2}(1 + i) \text{ and } B = \tfrac{1}{2}(1 - i)$$

*The particular solution is $u_n = \tfrac{1}{2}(1 + i)(-2i)^n + \tfrac{1}{2}(1 - i)(2i)^n$.*

### Complex and real roots

The solution in Example 10G is given in terms of a complex number. However the solution is always a real number. The first few terms illustrate this.

$$u_0 = 1, u_1 = 2, u_2 = -4, u_3 = -8, u_4 = 16, u_5 = 32, \ldots$$

## The case of equal roots

**Example 10.8**

*Solve $x_n = 6x_{n-1} - 9x_{n-2}$.*

**Solution**

*The auxiliary equation is $m^2 - 6m + 9 = 0$ with solution $m = 3$. There is only one solution, $x_n = A3^n$, where we would expect two solutions because we have a second-order recurrence relation.*

*Consider the expression $x_n = n \times 3^n$, then $x_{n-1} = (n-1)3^{n-1}$ and $x_{n-2} = (n-2)3^{n-2}$.*

*Substituting into the recurrence relation:*

$$\text{LHS} \Rightarrow n \times 3^n$$
$$\text{RHS} \Rightarrow 6(n-1)3^{n-1} - 9(n-2)3^{n-2}$$
$$= 6n \times 3^{n-1} - 9n3^{n-2} - 6 \times 3^{n-1} + 18 \times 3^{n-2}$$
$$= 2n \times 3^n - n \times 3^n = n \times 3^n$$

*The LHS = RHS so that $n \times 3^n$ is also a solution to the recurrence relation. The general solution is:*

$$x_n = A3^n + Bn3^n$$

In general we can show that if there is only one root (sometimes we say two equal roots) $m$ of the auxiliary equation, the general solution is:

$$x_n = Am^n + Bnm^n = (A + Bn)m^n$$

## EXERCISES

**10.3 A**

1 Find the general solution of each of these recurrence relations.

a) $x_n = 7x_{n-1} - 10x_{n-2}, n \geq 2$     b) $x_n = 9x_{n-2}, n \geq 3$
c) $u_n - 3u_{n-1} + 2u_{n-2} = 0, n \geq 2$     d) $u_n - 3u_{n-1} - 2u_{n-2} = 0, n \geq 3$
e) $x_n = 2x_{n-2}, n \geq 2$     f) $u_n = 2u_{n-1} - u_{n-2}, n \geq 2$
g) $u_n + u_{n-1} - 2u_{n-1} = 0, n \geq 2$     h) $x_n - 4x_{n-1} + 4x_{n-2} = 0, n \geq 3$

2 Find the particular solution of each of the following problems.

a) $x_n = 4x_{n-2}, n \geq 2$     if $x_1 = 2$ and $x_0 = 1$
b) $x_n = 4x_{n-1} - 4x_{n-2}, n \geq 3$     if $x_2 = 12$ and $x_1 = 2$
c) $u_{n+1} - 9u_{n-1} = 0, n \geq 1$     if $u_0 = 0$ and $u_1 = 1$
d) $p_n = 6p_{n-1} - 8p_{n-2}, n \geq 2$     if $p_1 = 28$ and $p_0 = 10$

3 Solve $x_n = 2x_{n-1} + 3x_{n-2}, n \geq 3$, if $x_1 = 1$ and $x_2 = 2$. Find the values of $x_4$ and $x_6$.

4 Solve $2u_n + u_{n-1} - u_{n-2} = 0, n \geq 2$, if $u_0 = 1$ and $u_1 = \frac{1}{2}$. Find the values of $u_2$ and $u_5$.

5 Solve $6p_{n+1} - 7p_n + 2p_{n-1} = 0, n \geq 1, p_1 = \frac{1}{6}$ and $p_0 = 0$. Find the value of $p_n$ as $n \to \infty$.

## EXERCISES

**1**  Find the general solution of each of these recurrence relations.

a) $x_n = x_{n-2}, n \geq 2$     b) $x_n = 3x_{n-1} + 4x_{n-2}, n \geq 3$

c) $u_n = 16u_{n-2}, n \geq 2$     d) $u_n = 10u_{n-1} - 25u_{n-2}, n \geq 2$

e) $p_n = p_{n-1} + 2p_{n-2}, n \geq 3$     f) $x_n = 9x_{n-1} - 14x_{n-2}, n \geq 2$

g) $5u_n + 9u_{n-1} = 3u_{n-2}, n \geq 2$     h) $3p_{n+2} + 8p_{n+1} = 3p_n, n \geq 0$

**2**  Find the particular solution of each of these problems.

a) $x_n = 2x_{n-1} + 15x_{n-2}, n \geq 2$, if $x_0 = 1$ and $x_1 = 77$

b) $p_{n+1} - 6p_n + 9p_{n-1} = 0, n \geq 3$, if $p_2 = 3$ and $p_3 = 15$

c) $u_n = 3u_{n-2}, n \geq 3$, if $u_1 = 0$ and $u_2 = 3$

d) $2x_n - x_{n-1} - 3x_{n-2} = 0, n \geq 2$, if $x_0 = 1$ and $x_1 = \frac{1}{4}$

**3**  Solve $x_{n+2} + 2x_{n+1} + x_n = 0, n \geq 0$, when $x_1 = 1$ and $x_2 = 2$. Find the values of $x_4$ and $x_6$.

**4**  Solve $2p_n = 3p_{n-1} - p_{n-2}, n \geq 2$, if $p_0 = 0$ and $p_1 = 1$. Find the value of $p_n$ as $n \to \infty$.

**5**  Elizabeth has a pair of gerbils of opposite sex. She finds that every two months the gerbils produce two pairs of offspring. Each new pair can reproduce four months after birth.

a) Formulate a recurrence relation model for the number of gerbils that Elizabeth has each month. (Assume that no gerbils die and pairs are always one male and one female.)

b) Solve the recurrence relation. How many gerbils does Elizabeth have after two years?

## NON-HOMOGENEOUS RECURRENCE RELATIONS

**Exploration 10.6**

### *Non-homogeneous equations*

Consider the non-homogeneous recurrence relation:

$$x_n + 6x_{n-1} + 8x_{n-2} = n, n \geq 3$$

■ Calculate the first six terms in the sequence of iterations $x_1, x_2, x_3, \ldots$

■ Find the particular solution of the homogeneous recurrence relation:

$$u_n + 6u_{n-1} + 8u_{n-2} = 0$$

and calculate the first six terms $u_1, u_2, u_3, \ldots$.

Calculate the remainder terms

$$r_1 = x_1 - u_1, r_2 = x_2 - u_2, r_3 = x_3 - u_3, \ldots.$$

■ Find a rule between $r_n$ and $n$.

■ Repeat the activity for the initial values:

a) $x_1 = 0$ and $x_2 = 1$     b) your choices of $x_1$ and $x_2$.

■ Deduce a general solution for the recurrence relation:

$$x_n + 6x_{n-1} + 8x_{n-2} = n, n \geq 3.$$

## The general solution

From Exploration 10.6 you should have found that the general solution of the non-homogeneous recurrence relation consists of the general solution of a related homogeneous equation and another expression which is independent of the given conditions for $x_1$ and $x_2$. This remarkable result gives us a method for solving non-homogeneous equations.

The general solution of a non-homogeneous recurrence relation:
$$x_n = ax_{n-1} + bx_{n-2} + f(n)$$

can be expressed as:

$$x_n = \begin{pmatrix} \text{general solution} \\ \text{of associated} \\ \text{homogeneous equation} \end{pmatrix} + \begin{pmatrix} \text{a particular} \\ \text{solution of the} \\ \text{full equation} \end{pmatrix}$$

$$x_n = u_n + r_n$$

$u_n$ is a solution of $u_n = au_{n-1} + bu_{n-2}$. The form of $r_n$ depends on the form of $f(n)$. The following table shows the usual form of the particular solution for some particular cases.

| | **form of $r_n$** |
|---|---|
| constant | $c$ |
| $n$ | $cn + d$ |
| $n^2$ | $cn^2 + dn + e$ |
| $k^n$ | $ck^n$ |
| | (or $cnk^n$ if $k$ is a root of the auxiliary equation) |

where $k, c, d$ and $e$ are constants.

**Example 10.9**

*Find the general solution of:*
$$x_n - 6x_{n-1} + 8x_{n-2} = 7$$

**Solution**

*The associated homogeneous recurrence relation is $x_n + 6x_{n-1} + 8x_{n-2} = 0$ which has auxiliary equation $m^2 + 6m + 8 = 0$ with roots $m_1 = 4$ and $m_2 = 2$. So the general solution of the associated homogeneous recurrence relation is:*
$$u_n = A4^n + B2^n$$

*For a particular solution try $r_n = c$ since here $f(n) = 7$ is constant. On substitution into the left-hand side of the recurrence relation:*
$$c - 6c + 8c = 7$$
$$2c = 7$$
$$c = \frac{7}{2}$$

*The general solution is $x_n = A4^n + B2^n + \dfrac{7}{2}$.*

**Example 10.10**

*Find the general solution of:*
$$x_n - 6x_{n-1} + 8x_{n-2} = 2n - 13$$

**Solution**
*As before:*
$$u_n = A4^n + B2^n$$

*For a particular solution try:*
$$r_n = cn + d$$

*On substitution we get:*
$$(cn + d) - 6(c(n-1) + d) + 8(c(n-2) + d) = 2n + 3$$
$$2cn + 3d - 10c = 2n - 13$$

*Comparing coefficients of n:*
$$2c = 2 \Rightarrow c = 1$$

*Comparing the constant terms:*
$$3d - 10c = -13 \Rightarrow d = -1$$

*The general solution is:*
$$x_n = A4^n + B2^n + n - 1$$

**Example 10.11**

*Find the general solution of:*
$$x_n - 6x_{n-1} + 8x_{n-2} = 3^n$$

**Solution**
*As before:*
$$u_n = A4^n + B2^n$$

*For a particular solution:*
$$r_n = c3^n$$

*on substitution:*
$$c3^n - 6c3^{n-1} + 8c3^{n-2} = 3^n$$
$$\Rightarrow 9c - 18c + 8c = 9$$
$$\Rightarrow c = -9$$

*The general solution is:*
$$x_n = A4^n + B2^n - 3^{n+2}.$$

**Exploration 10.7**

*A special case*

Investigate the general solution of the recurrence relation:
$$x_n - 6x_{n-1} + 8x_{n-2} = 2^n$$

## EXERCISES

**10.4 A**

1   Find the general solution of the recurrence relation:
$$x_n - 5x_{n-1} + 6x_{n-2} = f(n), \ n \geq 2 \text{ when:}$$

   **a)** $f(n) = 3$        **b)** $f(n) = n$        **c)** $f(n) = 2n - 1$
   **d)** $f(n) = 1 + n^2$     **e)** $f(n) = 5^n$       **f)** $f(n) = 3^n$

2   Find the general solution of the recurrence relation:
$$6x_n - 5x_{n-1} + x_{n-2} = f(n), \ n \geq 2 \text{ when:}$$

   **a)** $f(n) = 1$        **b)** $f(n) = n$        **c)** $f(n) = 2n + 3$

   **d)** $f(n) = 1 + n + n^2$    **e)** $f(n) = \left(\dfrac{1}{2}\right)^n$     **f)** $f(n) = 3^n$

3   Find the solution of the problem:
$$x_n - 7x_{n-1} + 12x_{n-2} = 2^n, \ n \geq 3$$

   given that $x_1 = 1$ and $x_2 = 1$.

4   Find the solution of the problem:
$$x_n + x_{n-1} - 2x_{n-2} = n - 3, \ n \geq 3$$

   given that $x_1 = 1$ and $x_2 = 2$.

5   Find the solution of:
$$x_n - 2x_{n-1} - 3x_{n-2} = 4 - 16n, \ n \geq 3$$

   given that $x_3 = -9$ and $x_4 = -57$.

## EXERCISES

**10.4 B**

1   Find the general solution of the recurrence relation:
$$x_n + x_{n-1} - 2x_{n-2} = f(n), \ n \geq 2 \text{ when:}$$

   **a)** $f(n) = 4$        **b)** $f(n) = n$        **c)** $f(n) = 2n + 1$
   **d)** $f(n) = 1 + n^2$     **e)** $f(n) = 5^n$       **f)** $f(n) = 2^n$

2   Find the general solution of the recurrence relation:
$$15x_n - 2x_{n-1} - x_{n-2} = f(n), \ n \geq 2 \text{ when:}$$

   **a)** $f(n) = 2$        **b)** $f(n) = 24n$      **c)** $f(n) = 3n - 1$

   **d)** $f(n) = 1 + n + n^2$    **e)** $f(n) = \left(\dfrac{1}{2}\right)^n$     **f)** $f(n) = \left(\dfrac{1}{3}\right)^n$

3   Find the solution of the problem:
$$x_n - 3x_{n-1} - 10x_{n-2} = 2^n, \ n \geq 3$$

   given that $x_1 = 2$ and $x_2 = 1$.

4   Find the solution of the problem:
$$x_n + 3x_{n-1} - 4x_{n-2} = 4 - n, \ n \geq 2$$

   given that $x_0 = 1$ and $x_1 = 2$.

5   Find the solution of:
$$u_{n+2} - u_{n+1} - 6u_n = \tfrac{1}{2}n, \ n \geq 0$$

   given that $u_0 = \tfrac{3}{2}$ and $u_1 = -\tfrac{29}{2}$.

## MATHEMATICAL MODELLING ACTIVITY
## Mortgage repayments

*Specify the real problem*

### Problem statement

Elizabeth, a 19-year-old student, takes out a loan of £30 000 to buy a flat. The period of the loan is 25 years and the interest rate at the time of taking out the loan was 8.1 per cent per annum. Calculate the monthly repayments.

*Set up a model*

### Set up a model

There are two common ways of repaying a mortgage.

■ An endowment mortgage – a person takes out an endowment insurance policy which should mature to the value of the loan after 25 years. The borrower pays interest on the loan to the building society and an insurance premium. At the end of 25 years the borrower pays off the loan in full by cashing in the insurance policy. There is a small risk that the policy may not have matured to the amount of the loan.

■ A repayment mortgage – a person pays a proportion of the loan, with some interest, to the building society each month. At the end of 25 years the amount owed is zero.

To calculate the repayment amounts, building societies adopt the following method.

■ At the beginning of each year the interest on the outstanding loan is added to the outstanding amount.

■ The twelve monthly repayments $M$ are then subtracted to give the amount owed at the end of the year.

■ This process is repeated 25 times and $M$ is chosen so that the amount owed is then zero.

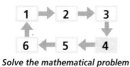

*Formulate the mathematical problem*

### Mathematical problem

If $x_n$ is the amount owed at the end of $n$ years, establish that the amount owed at the end of year $n + 1$ is given by:

$$[x_{n+1}] = [x_n] + \left[\frac{8.1}{100} \times x_n\right] - 12M$$

$$x_{n+1} = 1.081x_n - 12M \qquad\qquad 1$$

The monthly repayment $M$ must now be chosen so that $x_{25} = 0$ when $x_0 = 30\,000$, i.e. the initial mortgage is £30 000 and after 25 years the outstanding loan is zero.

*Solve the mathematical problem*

### Mathematical solution

Equation **1** is a first-order recurrence relation with general solution:

$$x_n = \left(x_0 + \frac{p}{a-1}\right)a^n - \frac{p}{a-1}$$

where $p = -12M$ and $a = 1.081$.

$$x_n = \left(30\,000 - \frac{12M}{0.081}\right)(1.081)^n + \frac{12M}{0.081}$$

Set $n = 25$ and $x_{25} = 0$ and solve for $M$ to establish that the required monthly repayment is $M = £236.20$.

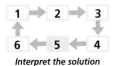

*Interpret the solution*

## Interpretation

For Elizabeth to repay her mortgage within 25 years she must pay £236.20 each month.

Consider what happens with small changes in monthly repayments.

If $M = £236$ then Elizabeth will owe £178.67 after 25 years. However, if she chooses to pay £240 then after 25 years she has paid £3382.12 too much i.e. the loan will be repaid approximately 14 months early. So £4 per month for 24 years saves £3382!!

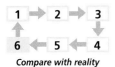

*Compare with reality*

## Refinement of the model

A more worrying situation occurs for Elizabeth if the mortgage rate changes.

Investigate the effects of:
- interest rate increases in steps of 0.1 up to 9.1 per cent
- interest rate decreases in steps of 0.1 down to 7.1 per cent.

The calculations introduce what is called **absolute and relative ill-conditioning** with respect to small changes in the data. Home owners often ask building societies if they can keep their monthly repayment amount fixed as interest rates increase. You can see why building societies need to say 'No'. The ill-conditioning leads to large debts after 25 years.

## CONSOLIDATION EXERCISES FOR CHAPTER 10

1   A public telephone (pay phone) will accept 10p, 20p, 50p and £1 coins only. A woman has a large supply of 10p and 20p coins only. These coins can be put into the pay phone in any order. For instance if she was to make a telephone call costing 50p, she could enter coins in the order:

20p, 20p, 10p    or    10p, 20p, 20p    or    10p, 10p, 20p, 10p

and there are more ways. She wishes to make a telephone call costing exactly £1. How many different ways are there of entering the coins?

Use the information to generate a second-order recurrence relation. Find the general solution and particular solution for this problem.

2   In an experiment the pressure of gas in a container is measured each second and the pressure (in standard units) after $n$ seconds is denoted by $p_n$. The measurements satisfy the recurrence relation $p_0 = 6$, $p_1 = 3$ and:
$$p_{n+2} = \tfrac{1}{2}(p_{n+1} + p_n) \ (n \geq 0).$$

Find the explicit formula for $p_n$ in terms of $n$, and state the value to which the pressure settles down in the long term.

*(AEB Question 5, Specimen Paper, 1996)*

**3**   Ian operates his bank account according to strict rules. Each month he puts £400 into the account and allows himself to spend one-fifth of what was in the account at the end of the previous month.

  **a)** Letting $u_n$ be the amount in the account at the end of month $n$, write down a recurrence relation for $u_{n+1}$ in terms of $u_n$.
  **b)** Find a general expression for $u_n$, given that $u_1 = 400$.
  **c)** Starting with an empty account, how much will Ian have in his account:
  **i)** after 12 months    **ii)** in the long run?

*(UODLE Question 4, Decision Mathematics AS, 1995)*

**4**   A maths teacher intends to set members of her class a problem involving matchsticks. As the first stage, she wishes to have them construct complete $n \times n$ lattices, some examples of which are shown below.

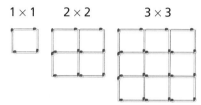

$1 \times 1$    $2 \times 2$    $3 \times 3$

  **a)** if $u_n$ is the number of matchsticks needed to construct an $n \times n$ lattice, explain why another $4n + 4$ matchsticks are needed to extend it to an $(n + 1) \times (n + 1)$ lattice. Express this in the form of a recurrence relation.
  **b)** Solve your recurrence relation to obtain an expression for the number of matchsticks needed to construct an $n \times n$ lattice.
  **c)** How many matchsticks will they need to construct a $6 \times 6$ lattice?
  **d)** The teacher wishes to have them construct a complete series of lattices from $1 \times 1$, through $2 \times 2$, $3 \times 3$, etc., up to $n \times n$. She wishes to find an expression for $s_n$, the number of matchsticks that will be needed. Find a recurrence relation for $s_n$.

*(UODLE Question 3, Decision Mathematics AS, 1994)*

**5**   Sabina is considering taking out a loan of £30 000 to purchase a flat. She will pay interest each month on the outstanding balance at a rate of one per cent per month. She will make monthly repayments. Each month interest will be added, and then her payment will be used to reduce the outstanding balance.

Let her monthly payment be £$X$, and the amount outstanding after $n$ months be £$u_n$. Then the recurrence relation given the amount outstanding after $n + 1$ months is:

$$u_{n+1} = 1.01u_n - X$$

  **a)** Explain the structure of this recurrence relation. Explain the construction and the purpose of the 1.01.
  **b)** Solve the recurrence relation to obtain an expression for the amount outstanding after $n$ months as a function of $n$ and $X$.
  **c)** Given that the answer to **b)** can be expressed as:

$$u_n = 30\,000 \times 1.01^n - 100X(1.01^n - 1)$$

find the value of $X$ required to pay off the loan in 25 years.

*(UODLE Question 6, Decision Mathematics AS, 1993)*

**6** Indiana Smith wants to cross the desert, a 20-day trip. He has a team of 19 porters to help him. All 20 members of the party will carry supplies on day 1, at the end of which a supply camp will be established and one porter will be sent back. The remaining party of 19 will repeat the exercise on day 2, establishing a supply camp and sending one porter back. The returning porter will use the supplies left in the earlier camps on his return journey.

The plan is for this to be repeated until, on the final day, only Indiana will be left to achieve his objective. He and his porters each drink one can of water during each day.

Let $C_k$ be the camp which is $k$ days from the destination. Let $w_k$ be the number of cans of water that have to be transported to camp $C_k$. Thus $w_0 = 0$, since Indiana does not wish to return.

**a)** Show that $w_1 = 2$.

**b)** The water that is transported to camp $C_{k+1}$ has to provide for that required at $C_k$, plus sufficient for the ongoing party to proceed to $C_k$, plus sufficient for all except Indiana to return to $C_{k+2}$ in due course. Show that this implies that:
$$w_{k+1} = w_k + 2(k+1)$$

**c)** Given that the solution to the recurrence relation $w_{k+1} = w_k + 2(k+1)$ is of the form $w_n = an^2 + bn + c$, find the solution of the relation with $w_0 = 0$.

**d)** Use your relation to find out how many cans the party must set out carrying.

Indiana now decides that he will make provision for himself to return as well. Let $y_k$ be the number of cans of water that now have to be transported to $C_k$.

**e)** What is $y_0$? What is $y_1$?

**f)** Find a recurrence relation connecting $y_{k+1}$ and $y_k$.

**g)** By solving the recurrence relation or otherwise, find how many cans the party must set out carrying.

*(UODLE Question 9, Decision Mathematics AS, 1992)*

**7** The cooling system of my car has a capacity of 10 litres. It contains antifreeze at a concentration of 0.25 (2.5 litres of ethylene glycol and 7.5 litres of water). The system leaks 0.1 litres per week. I wish to top it up weekly with 0.1 litres of undiluted ethylene glycol, until the concentration reaches 0.33. Thereafter I shall use a diluted top-up (concentration 0.33) to maintain the concentration.

**a)** Given that $v_k$ is the volume of ethylene glycol in my cooling system immediately after the $k$th top-up, derive a recurrence relation expressing $v_{k+1}$ in terms of $v_k$. Hence show that the recurrence relation for the concentration of ethylene glycol is $c_{k+1} = 0.99c_k + 0.01$, where $c_k$ is the concentration immediately after the $k$th top-up.

**b)** Solve this recurrence relation for $c_k$. Find the concentration after four weeks, and find how many weeks are required for the concentration to reach 0.33.

*(UODLE Question 5, Decision Mathematics AS, 1991)*

**8**    The government of Euroland has a policy of supporting agricultural prices by buying surplus produce in times of plenty. This produce is kept in intervention stores. At the present time there are 250 000 tonnes of soft grain in intervention store.

Each year birds and vermin consume five per cent of the stored grain, another five per cent is disposed of as rotten, and ten per cent is donated to famine relief. The current level of supply and demand for soft grain necessitates the government buying 20 000 tonnes per year.

**a)** Assuming this situation to continue, express the above information as a recurrence relation for $u_n$, where $u_n$ is the quantity of soft grain stored in $n$ years' time.

**b)** How many tonnes should the government expect to have to store in five years' time?
How many tonnes should the government expect to have in store in $n$ years' time?

**c)** In how many years will the government be able to reduce soft grain storage capacity to 105 000 tonnes?

**d)** How much storage capacity will be needed in the long run?
Suppose that three years from now the soft grain crop fails. For that year only, instead of buying in, the government has to hand out 100 000 tonnes from the store. Repeat parts **a)**, **b)** and **d)**, taking account of this new information.

Describe how the yearly change in the quantity of stored grain has been altered by the crop failure.

*(UODLE Question 21, Paper 4, 1990)*

**9**    Susan has £2000 to invest. She puts it into a building society account that pays eight per cent interest at yearly intervals. At the end of each year, immediately after the interest has been paid, she withdraws £250. Let $£u_n$ be the amount in her account immediately after the withdrawal of £250 at the end of the $n$th year. Find a recurrence relation for $u_n$. Hence, show that:
$$u_n = 3125 - (1.08)^n 1125$$

What is the maximum value of $n$ for which the recurrence relation holds if the building society does not allow Susan to overdraw?

In fours years' time Susan will receive a further £2000 from the maturing of savings certificates. She invests this immediately in her building society account and continues to withdraw £250 each year. Find and solve a recurrence relation, for $u_n$ the amount in her account at the end of the $n$th year ($n \geq 4$).

*(UODLE Question 9, Paper 4, 1989)*

**10**    An economic model of the production and selling of melons takes the following form. The demand $x_t$ in millions of melons in year $t$ would be:
$$x_t = 5 - P_t$$

where $P_t$ is the price of a melon. The supply of melons in year $t$ would depend on the price $P_{t-1}$ in the previous year, since melon farmers

would be encouraged to increase or decrease production as $P_{t-1}$ is high or low. Specifically, the supply $Y_t$ in millions of melons would be given by:

$$Y_t = 2 + 0.5\,P_{t-1}$$

The model further postulates that market mechanisms will fix the price $P_t$ such that supply equals demand in each year $t$. Show that this implies a first-order linear recurrence relation of $P_t$.

If $P_1 = 2\frac{1}{4}$ find $P_2$, $P_3$ and $P_4$ and solve the recurrence relation $P_t$ for general $t$.

Hence show that as $t$ becomes very large $P_t$ tends to 2.

*(UODLE Question 9, Paper 4, 1988)*

# Summary

*After working through this chapter you should know that:*

■ a recurrence relation is an equation of the form:
$$x_n + a\,x_{n-1} + bx_{n-2} + \ldots = f(n)$$
where the variable $x_n$ forms a sequence of numbers modelling a discrete situation, and the terms in the sequence are called the iterations of the recurrence relation

■ the order of a recurrence relation is the difference between the highest and lowest subscript in the equation

■ the general solution of a recurrence relation is a formula for $x_n$ that contains the same number of unknown constants as the order of the recurrence relation

■ the general solution of the first order recurrence relation:
$$x_n = kx_{n-1} + P \text{ is}$$
$$x_n = k^{n-1}x_1 + P\left(\frac{k^{n-1}-1}{k-1}\right) \quad k \neq 1$$
$$x_n = x_1 + (n-1)P \qquad k = 1$$

■ the general solution of the second order recurrence relation:
$$x_n + a\,x_{n-1} + bx_{n-2} = f(n) \text{ is}$$

$$x_n = \begin{pmatrix} \text{general solution} \\ \text{of associated} \\ \text{homogeneous equation} \\ x_n + ax_{n-1} + bx_{n-2} = 0 \end{pmatrix} + \begin{pmatrix} \text{a particular} \\ \text{solution of the} \\ \text{full equation} \end{pmatrix}$$

■ the associated homogeneous equation has solution
$Am_1{}^n + Bm_2{}^n$ where $m_1$ and $m_2$ are solutions of the auxiliary equation $m^2 + am + b = 0$

■ the form of the particular solution is the same as the form of f($n$).

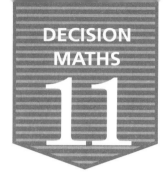

# DECISION MATHS

# 11

# *Coding*

*By the end of this chapter you should:*

- *understand the idea of a code, especially with respect to Morse code*
- *be able to formulate a binary code, which is made up of signals using 0 and 1 only*
- *use methods for detecting and correcting errors in codes*
- *understand the hamming distance as the number of places in which the bits of two codewords differ*
- *use matrices to model codes.*

## INTRODUCTION TO CODING

Most children at some time try out secret codes for communicating with their friends. They may simply replace letters by numbers, or they may experiment, reordering the letters according to some previously agreed pattern. One way of encoding a message is by arranging the letters into a rectangular array.

Suppose we have the message MATHS TEST ON FRIDAY. It has 17 letters, so we could choose to write it in a $3 \times 6$ array, filling the extra space with a null letter – say X, writing the words vertically like this:

| M | H | E | O | R | A |
|---|---|---|---|---|---|
| A | S | S | N | I | Y |
| T | T | T | F | D | X |

We now write the rows in turn, grouping the letters in threes.

MHE  ORA  ASS  NIY  TTT  FDX

This technique can be used to encode short messages in different sizes of array.

## ACTIVITY

11.1

Decode the following messages.

1   DIO ASF ESN TIU CIM HSN

2   MIOH ETEN NTIX EGIA GXTT GTHX

3   CUMPS EALEO PXNET ULXYN ENEXO DNDAX

Use letter arrays to encode some messages of your own.

228

## Exploration 11.1

**Samuel Morse (1791–1872)**

## *Morse code*

### *The Morse code*

| | | | |
|---|---|---|---|
| A • – | G – – • | N – • | U • • – |
| B – • • • | H • • • • | O – – – | V • • • – |
| C – • – • | I • • | P • – – • | W • – – |
| D – • • | J • – – – | Q – – • – | X – • • – |
| E • | K – • – | R • – • | Y – • – – |
| F • • – • | L • – • • | S • • • | Z – – • • |
| | M – – | T – | |

There are two possible signals which use just one beep:
- one short beep (E)
- one long beep (T).

How many possible signals are there which use:
- just two beeps
- just three beeps
- just four beeps?

Are they all used in the Morse code?

How many signals would there be if five beeps were used?

## Codes

The word **code** conjures up images of spies, espionage and sending secret messages. Codes are certainly about messages, but the reality is usually more mundane. We live in an age when the emphasis is on communication of all types; it has been called the 'Information Age'. On desks in offices and in homes there are micro-computers that can communicate to others anywhere in the world through a node to the Internet, the World Wide Web etc. In factories machines controlled by messages sent from a central computer program make cars, TVs, videos or various other everyday items. It is important that all these electronic messages are transmitted efficiently and correctly, and can be received equally efficiently. If there has been an error, either in transmission or reception, or the message has been corrupted it is important that the initial message can be restored. A summary of the process of sending and receiving a message is shown in this diagram.

**Sending and receiving a message**

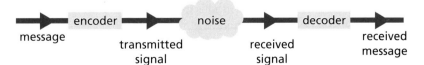

message → encoder → transmitted signal → noise → received signal → decoder → received message

The process of transferring a message into codewords is called **encoding**. The reverse process of transferring the collection of codewords back into a readable message is called **decoding**. Sometimes codes are deliberately obscure and their meaning is only available to those who have the **key** to **decipher** the code. This is the mysterious world of **encryption systems**. These are not included in this book, but behind them are the fundamentals of **binary codes**, which are discussed below.

This chapter is about the way information is transmitted using codes, how they are made up, how much information can be stored in a particular code and how we might detect and correct the errors that arise in transmission.

## Binary codes

*Braille is also an example of a binary code. It uses a 3 × 2 array of dots which are either raised up (on) or flat (off).*

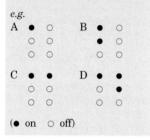

(● on ○ off)

What is a binary code? The Morse code uses two kinds of signal: a short beep and a long beep. Because it uses just two signals it is called a binary code.

Another type of binary code is one that uses the digits 1 and 0 only. A sequence of zeros and ones such as 0100101 is call a **binary codeword** (or simply a **codeword**). Each 0 or 1 is known as a **bit**, short for **binary digit**. A sequence of binary codewords is called a **binary code** (or simply code). For example consider:

{0100, 0000, 0011, 1100, 1111, 1010, 0101, 0010}

This is a code which consists of eight codewords. Each codeword contains four bits. Four is the **length** of each codeword.

## Example 11.1

*Write down all the possible codewords of length 3.*

### Solution
*These are:*

*000, 001, 010, 011, 100, 101, 110 111*

*There are $2^3 = 8$ of them.*

If a code of length $n$, has a maximum of $2^k$ codewords then it is known as an $(n, k)$ code, where the number $k$ is the dimension of the code. In Example 11.1 there are eight codewords with three bits in each code, so it is a (3, 3) code.

The rate, or efficiency, of a code is defined as $\dfrac{k}{n}$, so the code in Example 11.1 has a rate of $\frac{3}{3} = 1$.

## Bar codes

Bar codes are commonly used on goods in supermarkets. They provide a means of identifying goods, so that fewer mistakes should occur at checkouts, and stock levels can be constantly updated. The bar code consists of bands of light (1) and dark (0) that can be translated into a binary number by a laser reader. This unique number corresponds to a specific item, the price of which is automatically tallied by the till.

In reality, straightforward bands of light and dark would make bar codes too long, so different thicknesses are used to enable more complex information to be transmitted with a shorter codeword.

One example of the bar code is the code '3 of 9'. This is used in the health service (e.g. for labelling blood samples). The code itself is made up of letters, numbers and other symbols (the **three**) and each letter, number or other symbol is made from nine elements (the **nine**). Hence code '3 of 9'. The bar code itself is composed of the usual black and white stripes, but these can be either thick or thin. A typical bar code is shown here.

*+A 1 2 3 B 4 C 5 D 6 E T 1 1*

To see how the code is assembled, let us look at the 1. This is shown below.

The black stripes or bars have code 10001 corresponding to thick, thin, thin, thin, thick. The white stripes or spaces have code 0100 corresponding to thin, thick, thin, thin. The symbol A is given by:

which is 10001 (the same as 1) in bars but is 0010 in spaces. Finally, the third type of symbol is + or * (the * is in fact used as a start and a stop for a bar code). The 3 of 9 code for a * is as follows.

The binary code 00110 describes the bars, and the binary code 1000 describes the spaces.

**Exploration 11.2**

### Decoding a bar code

- Look at the complete bar code above and identify the A and 1, then try to deduce the codes for 2 and 3.
- Now try to decode the following bar code.

■ The complete code 3 of 9 is shown in the following table. Use it to decode the bar code below the table.

| Character | Pattern | Bars | Spaces | Character | Pattern | Bars | Spaces |
|---|---|---|---|---|---|---|---|
| 1 | | 10001 | 0100 | M | | 11000 | 0001 |
| 2 | | 01001 | 0100 | N | | 00101 | 0001 |
| 3 | | 11000 | 0100 | O | | 10100 | 0001 |
| 4 | | 00101 | 0100 | P | | 01100 | 0001 |
| 5 | | 10100 | 0100 | Q | | 00011 | 0001 |
| 6 | | 01100 | 0100 | R | | 10010 | 0001 |
| 7 | | 00011 | 0100 | S | | 01010 | 0001 |
| 8 | | 10010 | 0100 | T | | 00110 | 0001 |
| 9 | | 01010 | 0100 | U | | 10001 | 1000 |
| 0 | | 00110 | 0100 | V | | 01001 | 1000 |
| A | | 10001 | 0010 | W | | 11000 | 1000 |
| B | | 01001 | 0010 | X | | 00101 | 1000 |
| C | | 11000 | 0010 | Y | | 10100 | 1000 |
| D | | 00101 | 0010 | Z | | 00110 | 1000 |
| E | | 10100 | 0010 | – | | 00011 | 1000 |
| F | | 01100 | 0010 | . | | 10010 | 1000 |
| G | | 00011 | 0010 | Space | | 01010 | 1000 |
| H | | 10010 | 0010 | * | | 00110 | 1000 |
| I | | 01010 | 0010 | £ | | 00000 | 1110 |
| J | | 00110 | 0010 | / | | 00000 | 1101 |
| K | | 10001 | 0001 | + | | 00000 | 1011 |
| L | | 01001 | 0001 | % | | 00000 | 0111 |

■ Make up your own example of a bar code based on this 3 of 9 code then ask a friend to decode it.

## EXERCISES

**11.1 A**

1  What are the dimensions of the following codes?
{0000, 0101, 1010, 1111}
{0000, 0011, 0101, 0110, 1010, 1100, 1111}

2  Find the rate of the following code.
{0000000, 1101001, 0101010, 1000011, 1001100, 0100101, 1100110, 0001111, 1110000, 0011001, 1011010, 0110011, 0111100, 1010101, 0010110, 1111111}

3  **a)** Write down the possible codewords of length 4.
   **b)** What is the rate of the (4, 4) code?

4  The codewords of length 3 are used to represent letters as shown.

| Space | A | E | H | M | N | R | T |
|---|---|---|---|---|---|---|---|
| 000 | 001 | 010 | 011 | 100 | 101 | 110 | 111 |

Decode the following message.

100010010111000100010000011010110010000001111000111010101

**5**　The following code has eight codewords of which five are given.

0001111, 0110101, 1010011, 1011100, 1100110

The code is such that if $x$ and $y$ are codewords so is $x + y$.
We use the rules $0 + 0 = 0$　$0 + 1 = 1$　$1 + 0 = 1$　$1 + 1 = 0$
to add two codewords together.
**a)** Find the other three codewords.
**b)** Find the length, dimension and rate of the code.

**6**　Send a message in code to a friend using a rectangular array. Tell your friend to reply using the same code.

## EXERCISES

**11.1 B**

**1**　For the codes in questions 1 and 2 of Exercises 11.1A, write down the codewords so that the rate becomes 1.

**2**　Send a message to another person in your Decision maths group using Morse code. In your message, tell them to reply using a rectangular array.

**3**　Determine the dimension and rate of the following code.
{10001, 11010, 00010, 01000, 01110, 10100, 01100, 10110}

**4**　Consider the codewords given in question **3** above. Determine the other codewords of a code so that if $x$ and $y$ are codewords so is $x + y$.

**5**　Explain why in a code of length $n$, if $x$, $y$ and $x + y$ are codewords for all codewords $x$, $y$ in the code, then the codeword consisting of $n$ zeros must also be in the code.

**6**　Give an example of a code of dimension 3 such that for all codewords $x$, $y$ in the code, $x + y$ is in the code but the code does not have rate 1.

## ERROR DETECTION AND ERROR CORRECTION

It is not always possible to tell whether a message that is sent in the form of a binary code and received in the same form is correct or not. For example, consider the code {000, 010, 001, 100}. Suppose that a word is sent and is received as 101. This is obviously incorrect since 101 is not a codeword. However, what was meant to be received? Was it 100 with a corrupted last digit, or 001 with a corrupted first digit? We have no way of knowing. All we can say is that it was more likely to be one of those than 000 which would require *two* errors. One remedy is to repeat each digit. The above code would then be:
{000000, 001100, 000011, 110000}

and if a received word was 010011 then we could correct it to read 000011, as this is the most likely word to have been sent. If it was any other word, at least three bits would have to be in error. This is an example of a **repetition code**.

The original code, with no repetition of digits, has a rate of $\frac{2}{3}$, but the new code has a rate of $\frac{2}{6}$ or $\frac{1}{3}$. By introducing the repeated digits,

the rate has been halved, but it has increased our chances of being able to correct an error in the code.

## Error detection

Once a code is received, it is important that we have a means of detecting and correcting mistakes in transmission. If we consider the code 01001, say, there is no way of telling whether it is correct or not, since we have no idea what it means. On the other hand, a sentence such as 'I enjox most uspecta of matkematirs' that contains five errors can easily be corrected to 'I enjoy most aspects of mathematics'. The original message contained the five errors marked:

'I enjox most uspecta of matkmatirs'.

How are they so easy to detect?

The answer is that they are easy to detect because we match the letter combinations to our vocabulary of known words. The incorrect letters do not prevent us from identifying the words, but, when corrected, they confirm our identification. This concept is called **redundancy**. The letters are, in effect, redundant. They are not essential to our understanding of the words. However, they do help to confirm that our decoding is correct. Without redundancy, errors in a code are difficult to detect, so a numerical code will have redundancy built into it deliberately. The redundancy of a code is calculated by subtracting the rate of the code from 1.

$$redundancy = 1 - rate$$

## Parity check digits

Consider the simple code {0, 1} (possibly the simplest code). Duplication gives the code (00, 11). If the received code is 01, either 00 or 11 is equally likely to have been the intended message. A more usual way of detecting errors, better in the sense of being more efficient, is to use **parity**.

A **parity check digit** is an extra digit (bit) added to a codeword to ensure that the number of 1s in the codeword is even.

So if the codeword has an even number of 1s, a 0 is added to the end of the codeword. If it has an odd number of 1s a 1 is added to make the total number of 1s in the codeword even. So the codeword 101 would have parity check digit 0 to be transmitted as 1010.

The number of 1s in a codeword is called its **weight**, so 101 has weight 2.

**Example 11.2**

*Consider the code {000, 001, 100, 010}. Which of the following codewords with their check digit are correct? Correct the codewords where possible.*

*a) 1101* *b) 1001* *c) 0001*

**Solution**
a) *Is not correct. The check digit 1 tells us that there is an odd number of 1s, so the correct codeword is likely to be 100.*
b) *Is correct.*
c) *Is not correct. However the correct codeword could be 100 or 001 – we cannot determine which is the correct solution from this information.*

### Where the system breaks down
The last example indicates a limitation of this type of simple checking device. The parity check fails if there is more than one error, so the code is called a **single error detecting code**. No method of error correction is entirely foolproof and, in the real world, it inevitably depends on the probability of numbers of errors being made in the transmission of a message.

## Hamming distance

In the previous section we were able to correct some errors, but not others. This is due to the **distance** between two codewords. The definition of the distance between two codewords was introduced in 1950 by R.W. Hamming and is known, therefore, as the **hamming distance**.

Consider all the two-bit codewords {00, 01, 10, 11}. The pair 00 and 01 have the same first digit but different second digits. The pair 01 and 10 have different first and second digits. The hamming distance $d(x, y)$ between two binary codewords $x$ and $y$ is the number of places in which their bits differ. The codewords 00 and 01 differ in only one digit so the hamming distance is $d(00, 01) = 1$; however, the hamming distance between 01 and 10 is $d(01, 10) = 2$.

**Example 11.3**

*Find the hamming distance between the codewords $x = 1010$ and $y = 1100$.*

**Solution**
*Since they differ in both the second and third places, $d(x, y) = 2$.*

The hamming distance $d(x, y)$ has the following properties.
- Symmetry: $\quad\quad\quad\quad\quad d(x, y) = d(y, x)$
- Triangle inequality: $\quad d(x, z) \leq d(x, y) + d(y, z)$
- Positivity: $\quad\quad\quad\quad\quad d(x, y) \geq 0$
- Zero condition: $\quad\quad\quad d(x, y) = 0$ if and only if $x = y$

All of these properties ensure that $d(x, y)$ is a proper measure of distance.

The **minimum distance** $\delta$ of a code is the smallest hamming distance between two distinct codewords of the code.

**Example 11.4**

*Find the minimum distance δ of the (4, 3) code:*
*{0000, 0011, 0101, 0110, 1001, 1010, 1100, 1111}*

**Solution**
*By inspection, δ = 2. The code is often referred to as a (4, 3, 2) code, the last digit referring to the minimum distance.*

Here is an example of a (4, 2, 2) code:
  {0000, 1100, 0011, 1111}

It is not unique. We can write another (4, 2, 2) code:
  {0000, 0101, 1010, 1111}

The following is an important theorem.

**Theorem**
If $C$ is a code with a minimum distance $\delta$ then:

■ if $\delta$ is odd then $C$ can detect and correct up to $\frac{1}{2}(\delta - 1)$ errors

■ if $\delta$ is even, then $C$ can detect and correct up to $\frac{1}{2}(\delta - 2)$ errors.

**Proof**
We shall prove this theorem using some common sense along with properties of the distance function $d(x, y)$. Any decoding process operates by finding the codeword which is as close as possible to the received codeword. If $x$ is the codeword received, the decoding process estimates that the unique codeword $c$ that minimises $d(x, c)$ is the one that is most likely to have been transmitted.

If $t$ errors have affected the transmitted word $c$ then $d(x, c) = t$ from the definition of distance d. Suppose $y$ is any codeword distinct from $c$. Now we must have:
  $$\delta \le d(y, c)$$

since $\delta$ is the minimum distance. Using the 'triangle inequality' property of d:
  $$d(y, c) \le d(y, x) + d(x, c)$$
so    $\delta \le d(y, c) \le d(y, x) + t$
thus  $\delta - t \le d(y, x)$ for any $y \ne c$

This means that if $t < \delta - t$, i.e. $t < \frac{1}{2}\delta$, then $c$ is the codeword that minimises $d(x, c)$, which is another way of saying that the decoding process can recover $c$ from a knowledge of $x$.

If $\delta$ is odd, then:
$t < \frac{1}{2}\delta$ whenever $t \le \frac{1}{2}(\delta - 1)$

If $\delta$ is even, then:
$t < \frac{1}{2}\delta$ whenever $t \le \frac{1}{2}(\delta - 2)$

Thus the code can **correct** up to $\frac{1}{2}(\delta - 1)$ errors when $\delta$ is odd, and up to $\frac{1}{2}\delta$ errors if $\delta$ is even.

### Interpretation

Perhaps a look at the case of $t = \frac{1}{2}\delta$ will help to see what is happening here.

If $t = \frac{1}{2}\delta$ then it is possible for $x$ to be equidistant from two distinct codewords of $c$ (two that are $\delta$ apart with $x$ halfway between them). This is the case when we know there is an error, i.e. an error has been **detected**, but we are unable to say which of the two choices of codewords from $c$ was actually transmitted.

### EXERCISES

**11.2 A**

1  Find the hamming distance between the following pairs of binary words.
   **a)** 10101, 11001   **b)** 110010, 100101   **c)** 11010100, 00101011

2  Find the minimum distance of this code.
   {01010110, 00110011, 01010011, 00110110}

3  How many errors can be detected by a code with minimum distance:
   **a)** 19   **b)** 20?

4  How many errors can be **corrected** by the codes of question **3**?

5  What is the rate of the $(2n, n, n)$ codes?

6  Some of the letters of the alphabet are given by the code:

| A | C | D | E | H | N | S | Space |
|------|------|------|------|------|------|------|------|
| 1011 | 1001 | 0111 | 0010 | 0101 | 1110 | 0000 | 1111 |

The code is used by Elizabeth to send a message to her father. He receives the following signal which includes a parity check on each word.
00000011011110101111111010010101110000011010

What is the message?

### EXERCISES

**11.2 B**

1  Construct a (5, 3, 2) code. How many errors can this code detect and correct?

2  **a)** Find the hamming distance between 110100101 and 001110110.
   **b)** What is the minimum distance $\delta$ of this code?
       {101010101, 110110111, 001010100, 010011001}

3  Consider the code with these codewords.
       {00000000000, 01100100110, 10011011000, 11111111110}
   **a)** How many errors does this code simultaneously correct and detect?
   **b)** If the message is received as 01111110110, which codeword was most likely to have been transmitted?

4  Consider the code with these codewords.
       {0000000000, 0011101010, 1100010101, 1111111111}
   **a)** Determine the rate and minimum distance of the code.
   **b)** Write down the number of errors which the code can correct.

5    A threefold repetition code is a code with codewords 000 and 111. An $n$-fold repetition code $R_n$ is constructed similarly for each positive integer $n > 1$.
Determine the rate of the following repetition codes.
**a)** $R_6$    **b)** $R_7$    **c)** $R_n$
How many errors can each code simultaneously correct and detect?

6    The 2-out-of-5 code is the code of all possible words of length 5 but in which only those with exactly two 1s are used. For example 10010 is a word in the code but 10110 is not.

   **a)** List all the words in this code.
   **b)** Determine the minimum distance for this code and write down the number of errors the code can detect and correct.

## LINEAR CODES

The notation $x = \{$a sequence of 0s and 1s$\}$ is used to denote an arbitrary codeword.

*This property was referred to in Exercises 11.1.*

### Definition

Suppose $x$ and $y$ are codewords of a given code. If $x + y$ is also a codeword of this code, then the code is called a **linear code**.

## Example 11.5

*Check that the code (0000, 0101, 1010, 1111) is linear.*

### Solution

*The checking process is straightforward if a little laborious. We do not have to check the codeword 0000 as adding this to any other codeword leaves it unchanged. Also:*

   $x + y = y + x$

*by the binary addition property.*

   $0101 + 1010 = 1111$
   $0101 + 0101 = 0000$
   $1010 + 1010 = 0000$
   $0101 + 1111 = 1010$
   $1010 + 1111 = 0101$
   $1111 + 1111 = 0000$

*The code is thus linear.*

In fact, $x + x = 0$ for any binary codeword, hence all linear codes must contain the codeword 0 – the codeword that is composed of zeros.

Earlier we introduced the weight of a codeword as the number of 1s in the codeword. If the weight of codeword $x$ is $w(x)$ and $d(x, y)$ is the hamming distance between codewords $x$ and $y$ then $w(x) = d(x, 0)$.

## Generator sets

As the sum of two codewords of a linear code is also a codeword, there must be a minimal set of codewords that can generate the entire linear code. Such a set is called a **generator set**.

- The generator set for the linear code {0000, 0101, 1010, 1111} is <0101, 1111>

- The generator set for the linear code:
  {0000000, 0011101, 0101011, 0110110, 1000111, 1101100, 1110001} is <0011101, 0101011, 1000111>

The convention is to write generator sets between angle brackets < >.

**Example 11.6**

*Find a generator set for the code:*
  {000000, 001110, 010101, 011011, 100011, 101101, 110110, 111000}

*Solution*
*First of all, the code is a (6, 3) code, therefore three codewords will generate the code. Let us try* 001110, 100011, 111000.
  001110 + 100011 = 101101
  001110 + 111000 = 110110
  100011 + 111000 = 011011
  001110 + 100011 + 111000 = 001110 + 011011 = 010101.

*Adding any codeword to itself gives 000000. Thus all are generated and we have a generator G where:*
  *G* = < 001110, 100011, 111000 >

It is not the case that *any* three codewords generate the code. The secret is to pick three codes that between them contain a 1 in every place (column). The three codewords:
  001110, 100011, 111000

fulfil this criterion. The three codewords:
  001110, 010101, 011011

for example, do not because the 1 in the first position cannot be generated. The codewords in the generator set must be what is called a **spanning set** for the code; having a 1 in each position amongst the codewords will ensure success.

## Using matrices

The **generator matrix G** for an $(n, k)$ linear code is a matrix of which the rows are the generator set for the code. So **G** has $k$ rows and $n$ columns. (Recall an $(n, k)$ code has $n$ bits in each codeword and $2^k$ codewords.)

The generator matrix for the code in Example 11.6 is:

$$G = \begin{pmatrix} 0 & 0 & 1 & 1 & 1 & 0 \\ 1 & 0 & 0 & 0 & 1 & 1 \\ 1 & 1 & 1 & 0 & 0 & 0 \end{pmatrix}.$$

## Exploration 11.3

### *Sending messages using the generator matrix*

Suppose $a$ is a binary word of length $k$ written as a row vector. The product $a\mathbf{G}$, where $\mathbf{G}$ is a $(k \times n)$ generator matrix will be the codeword of length $n$ formed by multiplying each binary digit of the codeword $a$ by the corresponding digit in each row of $\mathbf{G}$, then adding. The result will be a codeword of length $n$. For example, if $\mathbf{G}$ is the generator matrix:

$$\mathbf{G} = \begin{pmatrix} 0 & 0 & 1 & 1 & 1 & 0 \\ 1 & 0 & 0 & 0 & 1 & 1 \\ 1 & 1 & 1 & 0 & 0 & 0 \end{pmatrix}$$

then a codeword of length 3, e.g. (110) when multiplied by $\mathbf{G}$ becomes:

$$(110) \begin{pmatrix} 0 & 0 & 1 & 1 & 1 & 0 \\ 1 & 0 & 0 & 0 & 1 & 1 \\ 1 & 1 & 1 & 0 & 0 & 0 \end{pmatrix} = (101101)$$

Let us try another codeword, say (001). This becomes:

$$(001) \begin{pmatrix} 0 & 0 & 1 & 1 & 1 & 0 \\ 1 & 0 & 0 & 0 & 1 & 1 \\ 1 & 1 & 1 & 0 & 0 & 0 \end{pmatrix} = (111000)$$

- Find the new codeword obtained from (100), (010) and (111).
- Find the new codeword obtained from $(x\ y\ z)$.
- Deduce that the original codeword is contained in the fourth, sixth and second positions of the new codeword.
- What do you notice if you write down the matrix from the fourth, sixth and second columns of $\mathbf{G}$?

The following extended example brings together many of the concepts met so far.

## Example 11.7

Let $C$ be the code:
  {000000, 000111, 011100, 011011, 101001, 101110, 110010, 110101}

*Determine the rate of this code. Find a generator set and hence write down the generator matrix. Use the generator matrix to encode the codewords (110, 010, 001).*

### Solution

Since $n = 6$ and there are $8 = 2^3$ codewords, the rate is $\frac{3}{6} = \frac{1}{2}$.
There are plenty of choices for a generator set.
Let us try {101001, 110101, 110010} since all six positions are covered by at least one 1.

To check we see that:
  $110101 + 110010 = 000111$
  $101001 + 110101 = 011100$
  $110010 + 101001 = 011011$

*and finally:*

$$101001 + 110101 + 110010 = 101110$$

*(Remember it is always true that adding a codeword to itself gives the zero codeword.)*

*A generator matrix formed from the above generator set is:*

$$\begin{pmatrix} 1 & 0 & 1 & 0 & 0 & 1 \\ 1 & 1 & 0 & 1 & 0 & 1 \\ 1 & 1 & 0 & 0 & 1 & 0 \end{pmatrix}$$

*and premultiplying this matrix by the three-bit codewords 110, 010 and 001 in turn gives:*

$$(110)\begin{pmatrix} 1 & 0 & 1 & 0 & 0 & 1 \\ 1 & 1 & 0 & 1 & 0 & 1 \\ 1 & 1 & 0 & 0 & 1 & 0 \end{pmatrix} = (011100)$$

$$(010)\begin{pmatrix} 1 & 0 & 1 & 0 & 0 & 1 \\ 1 & 1 & 0 & 1 & 0 & 1 \\ 1 & 1 & 0 & 0 & 1 & 0 \end{pmatrix} = (110101)$$

$$(001)\begin{pmatrix} 1 & 0 & 1 & 0 & 0 & 1 \\ 1 & 1 & 0 & 1 & 0 & 1 \\ 1 & 1 & 0 & 0 & 1 & 0 \end{pmatrix} = (110010)$$

In this encoding, the bits in positions 3, 4, 5 in the new codewords mirror the original codeword. This can be seen as being a direct consequence of the presence of the identity matrix in the generator matrix:

$$\begin{pmatrix} 1 & 0 & 1 & 0 & 0 & 1 \\ 1 & 1 & 0 & 1 & 0 & 1 \\ 1 & 1 & 0 & 0 & 1 & 0 \end{pmatrix}$$

We can also construct parity check matrices if $c$ is a linear $(n, k)$ code. The **parity check matrix** is a matrix $\mathbf{H}$ with $(n - k)$ rows and $n$ columns which has the following property:

if $c$ is a correctly received codeword then $\mathbf{H}c^T = 0$.

## Example 11.8

*Find the parity check matrix for the linear code {0000, 0011, 1100, 1111}.*

**Solution**

*The code is (4, 2) code, thus we look for a matrix with 4 – 2 = 2 rows and 4 columns:*

$$\begin{pmatrix} h_{11} & h_{12} & h_{13} & h_{14} \\ h_{21} & h_{22} & h_{23} & h_{24} \end{pmatrix}$$

*and it must satisfy the conditions:*

$$\begin{pmatrix} h_{11} & h_{12} & h_{13} & h_{14} \\ h_{21} & h_{22} & h_{23} & h_{24} \end{pmatrix} \begin{pmatrix} 0 \\ 0 \\ 1 \\ 1 \end{pmatrix} = 0$$

$$\begin{pmatrix} h_{11} & h_{12} & h_{13} & h_{14} \\ h_{21} & h_{22} & h_{23} & h_{24} \end{pmatrix} \begin{pmatrix} 1 \\ 1 \\ 0 \\ 0 \end{pmatrix} = 0$$

$$\begin{pmatrix} h_{11} & h_{12} & h_{13} & h_{14} \\ h_{21} & h_{22} & h_{23} & h_{24} \end{pmatrix} \begin{pmatrix} 1 \\ 1 \\ 1 \\ 1 \end{pmatrix} = 0$$

*which give rise to these six simultaneous equations:*

$h_{13} + h_{14} = 0$

$h_{23} + h_{24} = 0$

$h_{11} + h_{12} = 0$

$h_{21} + h_{22} = 0$

$h_{11} + h_{12} + h_{13} + h_{14} = 0$

$h_{21} + h_{22} + h_{23} + h_{24} = 0$

*Of course, all the hs are 0 or 1 and the following choice is one of many possibilities.*

$h_{11} = 1, h_{12} = 1, h_{13} = 0, h_{14} = 0$

$h_{21} = 1, h_{22} = 1, h_{23} = 1, h_{24} = 1.$

*All the parity check equations are satisfied, and:*

$$\mathbf{H} = \begin{pmatrix} 1 & 1 & 0 & 0 \\ 1 & 1 & 1 & 1 \end{pmatrix}$$

We can use the parity check matrix to identify errors and possibly to correct them. If we multiply a received codeword by the parity check matrix, the answer should be zero. If the resulting vector is not zero, it is called the **error syndrome** of the received codeword.

**Example 11.9**

*A received word is 011110. If **H** is the parity check matrix given by:*

$$\begin{pmatrix} 1 & 1 & 1 & 0 & 0 & 0 \\ 1 & 0 & 1 & 1 & 1 & 0 \\ 0 & 1 & 1 & 0 & 1 & 1 \end{pmatrix}$$

*find the error syndrome and suggest the correct word.*

### Solution

*First of all, let us find the error syndrome. This is:*

$$\begin{pmatrix} 1 & 1 & 1 & 0 & 0 & 0 \\ 1 & 0 & 1 & 1 & 1 & 0 \\ 0 & 1 & 1 & 0 & 1 & 1 \end{pmatrix} \begin{pmatrix} 0 \\ 1 \\ 1 \\ 1 \\ 1 \\ 0 \end{pmatrix} = \begin{pmatrix} 0 \\ 1 \\ 1 \end{pmatrix}$$

*The error syndrome is thus* $\begin{pmatrix} 0 \\ 1 \\ 1 \end{pmatrix}$. *This is the fifth column of* **H**.

*It therefore suggests that an error has occurred in the fifth place of the received word which should have been* (011100).

Of course if more than one error occurred, this method of error correction would fail. In general, the error would be detected since **H***r* ≠ 0, but it could not be corrected.

## EXERCISES

1   A code has these codewords.
    {0001000000, 000110110, 0010101101, 0110001010, 1000011011, 1100000101, 1110110000, 1110111111}

    a) What is the rate of this code?
    b) What is the minimum distance of this code?
    c) How many errors can this code correct and detect?
    d) Is the code linear?

2   Six of eight codewords of a linear code are:
    {0000000, 0001111, 0110101, 1010011, 1011100, 1100110}

    a) Determine the two missing codewords.
    b) Find the codeword which, taken with the two codewords in a) form the rows of the generator matrix of the linear code.

3   Determine whether any of the codes in Exercises 11.1A are linear. For those which are linear determine a generator set and hence write down the corresponding generator matrix.

4   Consider the linear code with the following eight codewords.
    {0000000000, 0010010111, 1001011100, 1011001011, 0100101110, 0110111001, 1101110010, 1111100101}

    a) Determine a generator matrix and use it to encode the 3-bit codewords: (101), (011), $(xyz)$.
    b) In which positions of the new codeword is the original codeword contained?

**5**    The code $C$ has a parity check matrix:

$$\mathbf{H} = \begin{bmatrix} 1 & 1 & 1 & 0 & 1 & 0 & 0 \\ 1 & 0 & 0 & 0 & 1 & 1 & 1 \\ 1 & 1 & 0 & 1 & 0 & 0 & 1 \end{bmatrix}$$

a) Determine the rate of $C$.

b) A codeword is incorrectly received as 0110000. Assuming that only one error has occurred, find the error syndrome and the transmitted codeword.

c) Find all codewords $x_1 x_2 x_3 x_4 x_5 x_6 x_7$ for which $x_2 = 1$ and $x_5 = 0$.

## EXERCISES

**11.3 B**

**1**    Determine whether the codes in question 3 of Exercises 11.1B and questions 3 and 4 of Exercises 11.2B are linear. If so, determine a generator set and hence write down the corresponding generator matrix.

**2**    Given the generator matrix:

$$\begin{bmatrix} 1 & 1 & 1 & 1 & 1 & 1 & 1 & 1 \\ 0 & 0 & 0 & 0 & 1 & 1 & 1 & 1 \\ 0 & 0 & 1 & 1 & 0 & 0 & 1 & 1 \\ 0 & 1 & 0 & 1 & 0 & 1 & 0 & 1 \end{bmatrix}$$

decode the received word 11100101.

**3**    Let $C$ be the following linear code.

{000000, 000111, 011001, 011110, 101011, 101100, 110010, 110101}

a) What is the rate of this code?

b) Find a parity check matrix for $C$ and use this matrix to correct the received message 010110.

c) Find a generator matrix for $C$ and use it to encode the message 101.

**4**    Let $C$ be the code with parity check matrix:

$$\mathbf{H} = \begin{bmatrix} 1 & 0 & 1 & 1 \\ 0 & 1 & 0 & 1 \end{bmatrix}$$

a) Determine the codewords of $C$.

b) Calculate the minimum distance of $C$.

c) Find a generator matrix of $C$.

**5**    Five codewords of a linear (7, 3) code $D$ are given below.

0000000, 1111000, 1001101, 0000110, 0110101

a) Write down the missing three codewords.

b) Find a generator matrix of $D$ and use it to encode the messages (111), (010) and $(xyz)$.

# CONSOLIDATION EXERCISES FOR CHAPTER 11

1    Nearly all newly-published books include an ISBN (International

Standard Book Numbering) number, printed in the book and on the back cover. The system is used throughout the world in order to identify any book uniquely.

For example, the ISBN number for the Pure Mathematics book in this series is 0-00-322370-1. The number always contains ten digits divided into four parts:

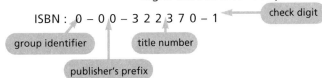

The first three parts of the number must use up the first 9 digits. The tenth digit is the check digit introduced so that any error in the previous nine digits can be found. It is calculated in the following way:

> multiply the first nine digits by 10, 9, 8, ... 2 respectively. The check number is the smallest number that needs to be added to the sum so that it is exactly divisible by 11. (if the number needed is 10, then the symbol X is used).

For example, for the ISBN number above:

$$0 \times 10 + 0 \times 9 + 0 \times 8 + 3 \times 7 + 2 \times 6 + 2 \times 5 + 3 \times 4 + 7 \times 3 + 0 \times 2 = 76$$

So that the check digit is 1, since 76 + 1 = 77 is divisible by 11.

**a)** Do the following ISBN numbers have the correct check digits?
   **i)** 0–00–322372–8          **ii)** 0–00–322371–X
**b)** Find the check digit for the ISBN number 0–86238–468– .
**c)** The following ISBN has an error in it: 0–7135–2272–5. What is the most likely ISBN number for the book?

2    Use the Morse code (page 229) to decode this message.
     • • • – – • – • – – – – – – – • – • • • • • • – – – – • • – – – – – • – • – • –

3    **a)**  Write down the length, dimension and rate of the following code.
        {0000, 0110, 1111, 1001}
     **b)**  Is this code linear?

4    A 3-out-of-7 code is one which consists of all possible codewords of length 7 which contain exactly three 1s.

     **a)**  List all the codewords in the 3-out-of-7 code.
     **b)**  Explain why this code does not need a parity check digit.

**5  a)** Find the hamming distance between 11010100 and 00101011.

    **b)** What is the minimum distance $\delta$ of the following code?

        {01010101, 01101001, 10101010, 10011001}

**6**  A **cyclic code** $C$ has the following property.

        If $x_1, x_2, ..., x_n$ is a codeword of $C$ then so is $x_2, x_3, ..., x_n, x_1$.

        Is this code cyclic?

        {0011101, 0111010, 1110100, 1101001, 1010011, 0100111, 1001110}

**7**  Let $C$ be the code with parity check matrix: $\begin{bmatrix} 1 & 0 & 1 & 1 & 1 & 0 & 0 \\ 1 & 1 & 0 & 1 & 0 & 1 & 0 \\ 1 & 1 & 1 & 0 & 0 & 0 & 1 \end{bmatrix}$

    Determine which two of the following are codewords of $C$.

    {0001111, 0011111, 1000111, 1001111, 1011111, 1100111, 1111111}

**8**  One of the codewords of a cyclic code is 1001110.

    **a)** List the other six words of the code.

    **b)** What is the hamming distance of this code?

    **c)** How many errors in a codeword can be simultaneously detected and corrected? Give a brief reason for your answer.

    **d)** Show that the code is not linear. What is the minimum number of codewords which need to be added to make this code linear?

    **e)** The matrix: $\begin{bmatrix} 1 & 1 & 1 & 0 & 1 & 0 & 0 \\ 0 & 1 & 1 & 1 & 0 & 1 & 0 \\ 1 & 1 & 0 & 1 & 0 & 0 & 1 \end{bmatrix}$

    is a parity check matrix for this code. Show how to use it to correct the received message:

    011011111101000011111.

(AEB Question 8, Discrete Mathematics Specimen Paper 4)

**9**  Let C be the following linear code.

    {000000, 000111, 011001, 011110,1011., 101100, 110010, 110101}

    **a)** What is the rate of this code?

    **b)** Find a parity check matrix for C and use this matrix to correct the received message 010110.

## Summary

*After working through this chapter you should know that:*

■ the Morse code is a system of encoding letters using a system of dots and dashes

■ a binary code is a system of encoding that uses a system of 0s and 1, or 'on's and 'off's

■ a sequence of 0s and 1s in a binary code is a codeword

■ the number of digits in the codeword is its length

■ if a code of length $n$ has a maximum of $2^k$ codewords, then $k$ is its dimension and it is an $(n, k)$ code

■ the rate of efficiency of the code is the ratio of $k : n$

■ a bar code uses bands and spaces to create a unique pattern tha can be read by a laser reader

■ errors can be detected by using a duplication of bits in a codeword or by using a parity check digit, which is an extra digit added to a code to ensure there are an even number of 1s in the codeword

■ the hamming distance is the distance between two codewords

■ a code is linear if the sum of any two codewords is also a codeword

■ a generator set is the minimal set of codewords that will generate an entire linear code

■ generators can be expressed as matrices.

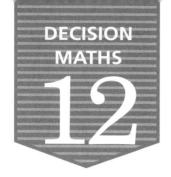
# *Simulation*

*By the end of this chapter you should:*

- *know that a simulation is a mathematical model that can be used to test a real-life situation*

- *be able to carry out simulations using random numbers.*

## Exploration 12.1

### *Queuing at the post office*

A small post office has only one serving counter. Sometimes long queues build up during busy periods, so Mr Simms, the post master, decides to investigate to see whether it is worth installing a second serving counter. How might he go about this?

## SIMULATIONS

One way that the post master could investigate the problem is to set up a **simulation**. A simulation is a mathematical model that can be used to test what might happen in situations where an experiment with real subjects may be too long or too dangerous to be practical, for example:

- What system will minimise the waiting time for customers queuing in a bank or post office?

- How does a disease spread through a population (human or animal)?

It may be theoretically possible to investigate these situations by observation and experimentation, but simulation gives us a feasible alternative approach. A good simulation gives the opportunity to ask 'What happens if ...?' questions and consider the various outcomes of making changes. Therefore it is a very important tool in business and social situations. Setting up a simulation follows a similar procedure to that involved in setting up any mathematical model.

## The modelling loop

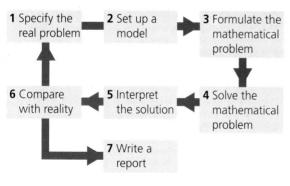

1 Consider the real situation and decide on the important variables and processes (box 1, specify the real problem).

2 Identify the simplifying assumptions that must be made (box 2).

3 Decide on the parameters you need to know. Take a sample of observations in order to estimate the values of these parameters (box 3).

4 Set up the simulation, stating the rules of the simulation clearly, and run it several times (box 4).

5 Compare the results with reality (boxes 5 and 6).

6 If your results are not good enough, amend your simulation to improve results (run the loop again).

7 Interpret the results with reference to the real situation. Use them to predict the likely outcomes if changes are made (box 7).

Notice that point 4 suggests running the simulation several times. This is important in order to generate enough information to give reliable results. However, running a simulation can in itself be a lengthy and repetitive process so computers are normally used. There are two types of simulation model.

## The deterministic model

This model assumes that chance events do not occur. It can only generate one answer to a given situation.

**Exploration 12.2**

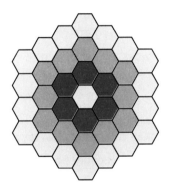

### *Forest fires*

The spread of a forest fire can be modelled as follows. Each tree is represented by a hexagon. The fire is assumed to start at one tree and spreads to all adjacent trees within one minute.

| Time | Trees alight | Total number burning |
|------|-------------|----------------------|
| 0 | 1 | 1 |
| 1 | 6 | 7 |
| 2 | 12 | 19 |
| 3 | 18 | 37 |

● Can you find an equation that will enable you to solve this model to find the number of trees burning after one hour?
● Is the model realistic?

#### *A realistic model?*

This type of model is not always very realistic. For example, an airline may find that a deterministic model suggests that five per cent of passengers do not turn up to claim their seats on flights. On the basis of this information, they may decide to book 105 passengers for every 100 seats available on their flights but, of course, on many flights this assumption will prove to be incorrect.

## The stochastic model

This model allows for an element of chance in the outcome, and will therefore generate a different answer each time the simulation is run. These models use some kind of device for introducing a random element such as a dice or random number generator.

### The collector's problem

Little plastic toys are being given away with breakfast cereal. If there are six different toys, how many boxes of cereal will a collector need to buy on average in order to gain a full set of toys?

This type of problem lends itself to using a die since each toy can be represented by a number on the die, assuming that there is an equally likely chance of getting each of the six toys. One run through would look like this:

2, 1, 4, 4, 6, 3, 4, 1, 2, 3, 5

The collector needs eleven boxes of cereal before she has all six toys.

We should need to run the simulation several times, each time recording how many throws it takes to collect a set of all six numbers. The average of these will give a guide as to how many boxes of cereal the collector would need to buy.

## EXERCISES

**12.1 A**

1   Use the information from the simulation on forest fires to find predictions for the number of trees that will be burning after:

   **a)** 10 minutes    **b)** 30 minutes    **c)** $t$ minutes.

2   Suggest a way in which an element of chance could be introduced into the forest fire simulation.

3   Use a die to perform the collector's simulation ten times. Suggest how many boxes of cereal should be bought to be sure of collecting all six toys.

4   **a)** Suggest how the rules of the collector's simulation could be changed if there were twelve plastic toys instead of six.
   **b)** Run your new simulation ten times and suggest the number of boxes of cereal that would be needed to collect all twelve toys.

5   Calum wants to find the probability that two people in a class of 30 have the same birthday. He decides to use two dice to simulate the day, and sets up a table like this.

First die

| | 1 | 2 | 3 | 4 | 5 | 6 |
|---|---|---|---|---|---|---|
| 1 | 1 | 2 | 3 | 4 | 5 | 6 |
| 2 | 7 | 8 | 9 | 10 | 11 | 12 |
| 3 | 13 | 14 | 15 | 16 | 17 | 18 |
| 4 | 19 | 20 | 21 | 22 | 23 | 24 |
| 5 | 25 | 26 | 27 | 28 | 29 | 30 |
| 6 | 31 | | | | | |

Second die

a) What is his rule for simulation of the day?
b) How might he simulate the month?
c) How do you suggest he deals with dates that do not exist, such as 30 February?
d) Run the simulation for 30 people at least three times. How would you use the results to calculate an estimate for the probability that two people in a class of 30 have the same birthday?

## EXERCISES

**12.1 B**

1 Simulate a gambling game in which you start with £10 and bet £1 at a time on a game. The rules are:

■ you receive £2 back (£1 stake plus £1 winnings) with probability $\frac{1}{3}$

■ you lose your stake of £1 with probability $\frac{2}{3}$.

Run this game ten times, using a six-sided die with each run lasting for 20 throws. (If you run out of money the run finishes.)

2 A mouse in a maze will take a left turn at a T-junction after having taken a left turn at the previous T-junction with probability $\frac{7}{12}$ and a left turn at a T-junction after having taken a right turn at the previous T-junction with probability $\frac{1}{3}$. Fill in the probabilities in each arc in the diagram below.

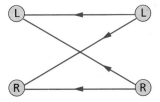

Further moves are to be simulated using a coin and a six-sided die.
a) Give a simulation rule that will generate a right and left turn succeeding:
i) a left turn    ii) a right turn.
b) If the mouse turns right at a junction, use the rules with the coin and die to simulate the next 20 turns of the mouse.

3 In a bagatelle game, a ball is dropped down a chute so that it hits a pin and may be deflected to the left or right with an equal probability of $\frac{1}{2}$. This occurs until the ball enters a slot, as shown in the diagram below. Determine the probability of a ball finishing in each of the five slots.

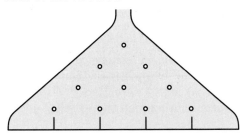

Suggest how this game may be used to simulate the determination of the number of defective items in random sample of four, drawn from a large population, if it is suspected that 50 per cent of this population are defective. Suggest a reason why you may wish to simulate the number of defectives in random samples of four rather than test the whole population.

**4** Elderly people are admitted to a hospital and progress through the diagnostic stages of assessment, rehabilitation and long-stay care, as in the diagram. Most are rehabilitated and discharged but a small proportion become long-stay patients and remain in hospital. The situation can be represented diagrammatically like this.

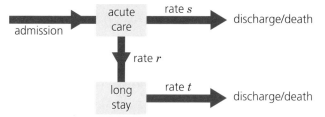

Describe how a simulation could be used by the hospital administration to plan the number of beds needed in this part of the hospital.

**5** Find a situation that can be modelled by a simulation. Devise a method of simulating the situation you are suggesting.

## USING RANDOM NUMBERS

When there are many possible outcomes, or if the outcomes are not all equally likely, then a die, coin or similar method will not be suitable for generating the simulation. When this is the case we often use random numbers.

Most calculators will generate random numbers. A scientific calculator usually has the symbol RAN# or RAND#. Pressing this key will produce a series of three-digit numbers between 0 and 1. Graphic calculators usually have the random number function on the probability menu and most will also have a facility for setting the number of digits required.

To demonstrate a simulation that uses random numbers, let us rejoin the queue in the post office in Exploration 12.1, to see whether it is worth installing a second serving counter. How might Mr Simms proceed?

### 1  Important variables and processes

Since Mr Simms is concerned with the length of queues, he needs to find out how frequently customers arrive at the post office and how long it takes to serve each one. This will enable him to calculate the average waiting time and length of queues.

## 2–3 Simplifying assumptions

First, it is important to establish the parameters that you need to know.

Mr Simms needs to find out how often a new customer comes into the post office. This is called the **inter-arrival time** and is an important feature of simulations of this type. Mr Simms performs a small survey and finds that the inter-arrival times at peak periods are usually between 30 seconds and 3 minutes, with roughly the distribution shown in this table.

| Inter-arrival time (minutes) | 0.5 | 1 | 1.5 | 2 | 2.5 | 3 |
|---|---|---|---|---|---|---|
| Probability (%) | 5 | 10 | 30 | 35 | 15 | 5 |

His counter assistant says that serving times usually vary between one and four minutes, with this pattern.

| Serving time | 1 | 1.5 | 2 | 2.5 | 3 | 3.5 | 4 |
|---|---|---|---|---|---|---|---|
| Probability (%) | 5 | 10 | 25 | 25 | 20 | 10 | 5 |

This provides sufficient information for Mr Simms to set up his simulation.

To avoid the simulation becoming too cumbersome, these simplifying assumptions will be made.

- All times are taken to the nearest 30 seconds.
- Any inter-arrival and serving times which fall outside the normal range will be ignored.
- The simulation will be performed for 25 customers, starting at a time 0.

## 4  Rules of the simulation

The simulation is to be generated using random numbers. There are two variables in this simulation, and so Mr Simms will need to generate two sets of random numbers, one for the inter-arrival times and one for the service times. Since the probabilities are given in percentages, Mr Simms can use random numbers from 0 to 99 (or 0.00 to 0.99 if using a calculator) to represent the probabilities. The numbers will be allocated like this.

### For inter-arrival times

| Inter-arrival time (minutes) | 0.5 | 1 | 1.5 | 2 | 2.5 | 3 |
|---|---|---|---|---|---|---|
| Probability (%) | 5 | 10 | 30 | 35 | 15 | 5 |
| Random numbers | 0–4 | 5–14 | 15–44 | 45–79 | 80–94 | 95–99 |

### For serving times

| Serving time | 1 | 1.5 | 2 | 2.5 | 3 | 3.5 | 4 |
|---|---|---|---|---|---|---|---|
| Probability | 5 | 10 | 25 | 25 | 20 | 10 | 4 |
| Random numbers | 0–4 | 5–14 | 15–39 | 40–64 | 65–84 | 85–94 | 95–99 |

**Note:** it is very important to ensure that you allocate the random numbers correctly – it is easy to make a mistake at this stage of a simulation.

### Solution

Having done this, Mr Simms can now run the simulation. The easiest way to record the results is in a table like this.

| Customer | RND | Inter-arrival time | Arrives (after start time) | Service start (after start time) | RND | Service time (minutes) | Service end (after start time) | Wait time (minutes) | Queue length (maximum) |
|---|---|---|---|---|---|---|---|---|---|
| 1 |  | 0 | 0 | 0 | 72 | 3 | 3 | 0 | 0 |
| 2 | 93 | 2.5 | 2.5 | 3 | 61 | 2.5 | 5.5 | 0.5 | 1 |
| 3 | 14 | 1 | 3.5 | 5.5 | 46 | 2.5 | 8 | 2 | 1 |
| 4 | 66 | 2 | 5.5 | 8 | 77 | 3 | 11 | 2.5 | 1 |
| 5 | 74 | 2 | 7.5 | 11 | 41 | 2.5 | 13.5 | 3.5 | 2 |
| 6 | 82 | 2.5 | 10 | 13.5 | 98 | 4 | 17.5 | 3.5 | 2 |
| 7 | 8 | 1 | 11 | 17.5 | 29 | 2 | 19.5 | 6.5 | 2 |
| 8 | 90 | 2.5 | 13.5 | 19.5 | 97 | 4 | 23.5 | 6 | 2 |
| 9 | 35 | 1.5 | 15 | 23.5 | 0 | 1 | 24.5 | 8.5 | 3 |
| 10 | 92 | 2.5 | 17.5 | 24.5 | 51 | 2.5 | 27 | 7 | 3 |
| 11 | 81 | 2.5 | 20 | 27 | 9 | 1.5 | 28.5 | 7 | 3 |
| 12 | 89 | 2.5 | 22.5 | 28.5 | 57 | 2.5 | 31 | 6 | 4 |
| 13 | 89 | 2.5 | 25 | 31 | 96 | 4 | 35 | 6 | 3 |
| 14 | 17 | 1.5 | 26.5 | 35 | 2 | 1 | 36 | 8.5 | 4 |
| 15 | 63 | 2 | 28.5 | 36 | 43 | 2.5 | 38.5 | 7.5 | 3 |
| 16 | 45 | 2 | 30.5 | 38.5 | 4 | 1 | 39.5 | 8 | 4 |
| 17 | 68 | 2 | 32.5 | 39.5 | 23 | 2 | 41.5 | 7 | 4 |
| 18 | 43 | 1.5 | 34 | 41.5 | 19 | 2 | 43.5 | 7.5 | 5 |
| 19 | 22 | 1.5 | 35.5 | 43.5 | 17 | 2 | 45.5 | 8 | 5 |
| 20 | 72 | 2 | 37.5 | 45.5 | 69 | 3 | 48.5 | 8 | 5 |
| 21 | 5 | 1 | 38.5 | 48.5 | 36 | 2 | 50.5 | 10 | 5 |
| 22 | 41 | 1.5 | 40 | 50.5 | 87 | 3.5 | 53.5 | 10.5 | 5 |
| 23 | 7 | 1 | 41 | 53.5 | 28 | 2 | 55.5 | 12.5 | 6 |
| 24 | 21 | 1.5 | 42.5 | 55.5 | 78 | 3 | 58.5 | 13 | 6 |
| 25 | 57 | 2 | 44.5 | 58.5 | 3 | 1 | 59.5 | 14 | 6 |

| | |
|---|---|
| Average wait time | 6.94 minutes ≈ 7 minutes |
| Maximum wait time | 14 minutes |
| Average queue length | 3.4 |
| Maximum queue length | 6 |

This gives the average waiting time to be around 7 minutes, though it rises to a maximum of 14 minutes and was still rising after 25 customers. The queue length reached a maximum of 6 at the end, with an average of 3.4, showing that an extra service point would definitely be used during this time.

## 5 Compare results with reality

To verify his simulation Mr Simms would then compare his results with the situation he observed in the shop each day.

## 6 Amend your model

In order to improve the model Mr Simms could do a more accurate survey of the inter-arrival times and service times and he may also wish to take smaller time intervals to improve accuracy. Another possibility would be to consider how these factors change once the busiest period is over, to see how quickly the queues decline.

## 7 Interpret the results

On the basis of this simulation, it would appear that queues will inevitably build up during peak periods. If this continues for more than an hour, Mr Simms may well need to install another service point during this time. If, however, the busy period is fairly short, he may feel that it is not necessary as the queues will decline fairly quickly once things slow down again.

**Exploration 12.3**

### *Queuing*

Mr Simms decides to open his new service point. Run another simulation to see which is the most effective method for queuing.

One method is the type usually encountered in a supermarket, where each service point has a separate queue and people join the queue of their choice.

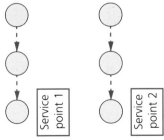

Of course, we all know that, under this system, everyone *feels* that their queue is moving more slowly than others! The alternative is to use the system operated by many post offices and banks, which is to have a single queue feeding into several service points.

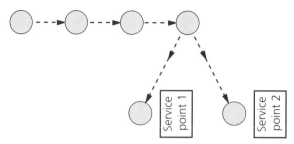

As you try to simulate a queuing procedure, you may well find it helpful to draw the service points on a piece of paper and use counters to represent the people in the queue.

Devise a method for simulating the two queuing procedures given above and state which you think would be the more efficient.

## EXERCISES

**12.2 A**

1   Run the simulation for Mr Simms' queuing problem again, this time for 50 customers. On the basis of this extra information, would you advise him to install a second service point?

2   Quicksnack take-away claims to provide fast food for local workers during lunch time. The inter arrival times of its customers on a typical weekday are summarised here.

| Inter-arrival time (minutes) | 1 | 2 | 3 | 4 | 5 |
|---|---|---|---|---|---|
| Frequency (%) | 55 | 30 | 10 | 4 | 1 |

The average service times are given in this table.

| Service time (minutes) | 1 | 1.5 | 2 | 2.5 | 3 |
|---|---|---|---|---|---|
| Frequency (%) | 20 | 24 | 28 | 18 | 10 |

If the take-away opens at midday, simulate the waiting times for the first 20 customers. Find estimates for the average waiting time and maximum queue length. Does Quicksnack really provide fast food?

3   The time taken to serve a customer at a payment till in a department store varies as shown in the following table.

| Time taken to serve (minutes) | 1 | 2 | 5 |
|---|---|---|---|
| Probability | $\frac{1}{4}$ | $\frac{1}{2}$ | $\frac{1}{4}$ |

Customers arrive at the till according to the following table.

| Time between customers (minutes) | 1 | 2 | 3 | 4 |
|---|---|---|---|---|
| Probability | $\frac{1}{6}$ | $\frac{1}{3}$ | $\frac{1}{3}$ | $\frac{1}{6}$ |

a) Give a rule for generating customer service times from two-digit random numbers.

b) Use your rule and the following random number stream to simulate the service times for the next twelve customers.

| 87 | 64 | 67 | 35 | 15 | 52 | 04 | 57 |
|---|---|---|---|---|---|---|---|
| 91 | 83 | 05 | 23 | 76 | 22 | 41 | |

c) Give a rule for generating times between customers from two-digit random numbers.

d) Use your rule and the following random number stream to simulate the time intervals between the arrivals of the next twelve customers. (The first interval should represent the time interval before the first customer arrives.)

| 37 | 15 | 09 | 35 | 13 | 97 | 65 | 69 |
|---|---|---|---|---|---|---|---|
| 86 | 90 | 85 | 07 | 04 | 55 | 41 | |

e) Using your service times from b) and your inter-arrival times from d), simulate the arrival and serving of twelve customers at the till. Record the events of your simulation in the following way (the numerical entries are for illustration).

| Services times | | | | 5 | 2 | 2 | 2 | | ... |
|---|---|---|---|---|---|---|---|---|---|
| Inter-arrival times | 2 | 1 | 1 | 2 | 1 | 3 | 3 | | ... |

| Time (minutes) | Customer arriving | Time of next arrival | Customer starting service | Time of end of service | Length of queue |
|---|---|---|---|---|---|
| 0 | | 2 | | | 0 |
| 1 | | | | | 0 |
| 2 | first | 3 | first | 7 | 0 |
| 3 | second | 4 | | | 1 |
| 4 | third | 6 | | | 2 |
| 5 | | | | | 2 |
| 6 | fourth | 7 | | | 3 |
| 7 | fifth | 10 | second | 9 | 3 |
| 8 | | | | | 3 |
| 9 | | | third | 11 | 2 |

f) Showing your working, find the mean queue length between the time of the arrival of the fourth customer and the time of the arrival of the ninth customer.
Why is this more representative of the performance of the system than the mean queue length throughout the simulation?

g) Give any times during which the server is not occupied between the time of the arrival of the fourth customer and the time of the arrival of the ninth customer.

Find the percentage of the time between the time of the arrival of the fourth customer and the time of the arrival of the ninth customer for which the server is occupied.

h) What reliance would you place on your simulation for estimating the mean queue length and the server utilisation? How could you improve its reliability?

*(UODLE Question 12, Decision Mathematics AS, 1994)*

4    Twenty companies each have a bay of five reserved spaces in a multi-storey car park (100 spaces in all). Records indicate that at midday, on average, fifteen per cent of spaces are empty.

a) Use the random digits below to simulate the presence (P) or absence (A) of a car at midday in each of the 100 spaces. State your simulation rules clearly.

| | | | |
|---|---|---|---|
| 30 31 02 83 22 | 29 82 73 04 79 | 47 62 79 84 38 | 97 08 59 27 07 |
| 02 28 56 83 24 | 78 07 99 41 32 | 01 87 20 92 75 | 88 00 06 08 68 |
| 84 35 93 65 79 | 13 97 92 79 99 | 01 90 23 41 76 | 77 26 98 00 56 |
| 68 39 19 39 25 | 29 05 01 09 91 | 33 11 71 21 94 | 82 72 42 82 48 |
| 76 33 84 92 71 | 50 10 91 58 66 | 37 76 29 02 51 | 20 01 58 77 04 |

b) Group your results in sets of five, and thus count how many cars are present for each company. Record this number for each company.

c) Using your results from b), estimate:
   i)  the probability that a randomly selected company has all of its spaces occupied
   ii) the mean number of spaces occupied per company.

d) Calculate the theoretical probability that a randomly selected company has at midday:
   i)  all of its spaces occupied
   ii) four of its spaces occupied.

*(UODLE Question 2, Decision Mathematics AS, 1991)*

5    Simulation is often used in population modelling. A pair of birds will produce from four to ten eggs in a clutch each spring, with this distribution.

| **Eggs per clutch** | 4 | 5 | 6 | 7 | 8 | 9 | 10 |
|---|---|---|---|---|---|---|---|
| **Frequency (%)** | 8 | 10 | 16 | 25 | 30 | 7 | 4 |

Of these, 0.4 will either not hatch or die in the first three months. Through the winter 0.5 of the adults and 0.7 of the surviving chicks will also die due to cold and food shortage.

a) Starting with a population of 20 pairs, set up a simulation to estimate the number of birds that will be alive at the beginning of the next spring.

b) Assuming that the young can reproduce after one year, explain how you could continue the simulation. What simplifying assumptions would you need to make?

## EXERCISES

**12.2 B**

1 Allocate random numbers $m$ in the range 00–99 for the following situations.

a) $X$ is distributed as illustrated by the histogram.

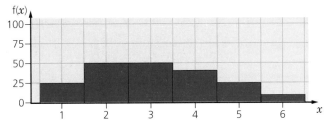

*Parts (b), (c) and (d) should only be attempted by students who have or who are studying A-level Statistics.*

b) The random variable $X$ has a Poisson distribution of mean 2 i.e:

$$P(X = j) = \frac{e^{-2} . 2^j}{j!}, j = 0, 1, 2, 3, \ldots$$

c) The continuous random variable $X$ has a probability density function which is triangular, as illustrated below.

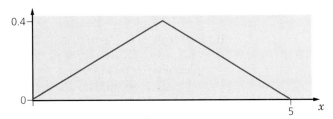

(Divide the range $0 \le x \le 5$ into five classes of equal width.)

d) The continuous random variable $X$ is distributed as a negative exponential distribution of mean 0.5 so its probability density function is $f(x) = 2e^{-2x}$, $x \ge 0$.
(Divide the range $x \ge 0$ into half-unit intervals for five intervals and then consider $x \ge 2.5$. **Note:** The area under the curve $y = 2e^{-2x}$ from $a$ to $b$, where $b > a \ge 0$, is given by $e^{-2a} - e^{-2b}$.)

2 Simulations are usually run on computers and to use random numbers there must be some deterministic way of generating them (i.e. by a fixed rule). Numbers may be generated by the rule:
$$x_{n+1} = ax_n + b \pmod{m}$$

*The numbers generated are called **pseudo-random numbers** and when such generating methods are devised for computers a and m are large, for example one recommended choice is a = 16 807, b = 0, m = 2³¹ – 1.*

i.e. $x_{n+1}$ is the remainder after $ax_n + b$ has been divided by $m$.

To keep the numbers in the range 00–99, we take $m = 100$. Generate ten numbers by taking:
a) $a = 23$, $b = 3$, $x_0 = 7$          b) $a = 30$, $b = 10$, $x_0 = 7$.
Comment on your results.

3 Consider the simulation of a game of tennis between two weak players X and Y. Player X has a 50:50 chance of getting the service in play and only a chance of 0.01 of winning from a valid serve without a rally taking place. Player Y has a 60:40 chance of getting the service in play and a probability of 0.1 of winning directly from a valid serve.

If the game goes to a rally, X has a probability 0.6 of winning and hence Y has a probability 0.4 of winning a rally, independent of who is serving. Draw a flow diagram for the model of this game of tennis.

Simulate a game of tennis in which
**a)** X serves          **b)** Y serves.

4    The local health trust has a short-stay ward for minor illnesses in one of their hospitals. There are five beds in the ward, but they wish to remove one of the beds to cut costs. From previous records, the distribution of the number of patients who became ill each day is summarised in this table.

| Number | 0 | 1 | 2 | Total |
|---|---|---|---|---|
| Frequency | 35 | 50 | 25 | 100 |

The length of stay of patients has the distribution shown in this table.

| Length of stay (days) | 1 | 2 | 3 | 4 | 5 | 6 | 7 | Total |
|---|---|---|---|---|---|---|---|---|
| Frequency | 5 | 10 | 20 | 40 | 16 | 5 | 4 | 100 |

As the local hospital manager, you have to determine the demand on the five beds. Assume that on day 0 two beds are occupied and count day 0 as the first day of these patients' stay; they are in bed for three and four days respectively. Simulate the situation for 28 days and repeat the simulation with four beds. For each simulation run, find the percentage of patients who are accepted and the mean number of empty beds per day. Find the distribution of these quantities over as many runs of the simulation as time allows you.

5    A computer company receives calls for technical support which are dealt with on one direct telephone line. If the line is engaged a customer will try at a later stage. The calls arrive with the following patterns.

| Time between calls (minutes) | 0–4 | 5–9 | 10–14 | 15–19 | 20–39 |
|---|---|---|---|---|---|
| Percentage of calls | 15 | 20 | 25 | 20 | 20 |

The calls are answered as politely as possible and take the following times to be dealt with.

| Time to satisfy a customer (minutes) | 0–4 | 5–9 | 10–19 | 20–29 | 30–89 |
|---|---|---|---|---|---|
| Percentage of calls | 50 | 10 | 15 | 20 | 5 |

A customer who finds the phone engaged will call back. The effect on the frequency of calls has been measured and may be simulated by a ten per cent reduction in the spread of inter-call times for each customer unable to get through.

Simulate the above situation and produce a short report for the management on the suitability of the current position, indicating in particular the number of customers unable to get through when they ring, and the time that the support system is idle.

# MATHEMATICAL MODELLING ACTIVITY
## *Lights or filters*

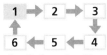
*Specify the real problem*

## Specify the real problem

On the island of Guernsey you would encounter a familiar looking road sign bearing the unfamiliar instruction, '*Filter in turn*'.

Consider the junction below where cars are approaching along roads A and B. They then take it in turns to enter the box and cross the junction, one from A followed by one from B and then one from A again, filtering in turn. Investigate how quickly cars can move through this type of junction.

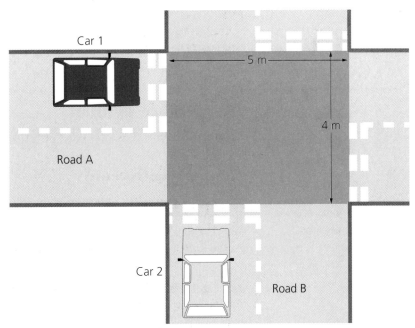

Imagine you are doing a study into whether traffic lights or a 'Filter in turn' would be a more effective way of relieving conjestion at a busy road junction during peak traffic flow times.

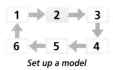
*Set up a model*

# Set up a model

Decide on the simplifying assumptions which must be made. You may wish to consider these ideas.
- All lengths will be measured to the nearest 25 cm.
- All times will be to the nearest second.
- All cars move off from being stationary with the same constant acceleration (e.g. $2\,m\,s^{-2}$).

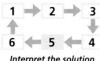
*Formulate the mathematical problem*

# Formulate the mathematical problem

Identify the important variables. These might include:

- the inter-arrival times of vehicles at the junction during a busy time
- the distribution of lengths of vehicles which are likely to use the roads (Do you include buses and lorries? If not you may need to amend your simplifying assumptions.)
- How long the traffic lights stay green.

You may wish to conduct surveys to obtain the information. You will probably end up with tabulated data like this.

| Inter-arrival time (seconds) | 1 | 2 | 3 | 4 | ... |
|---|---|---|---|---|---|
| Percentage of vehicles (%) | 60 | 18 | 12 | ... | ... |

| Length of vehicle (nearest 25 cm) | 2.5 | 2.75 | 3 | ... |
|---|---|---|---|---|
| Percentage of vehicles (%) | 10 | 15 | 20 | ... |

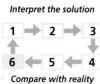
*Solve the mathematical problem*

# Solve the mathematical problem

Run the simulation. State your rules clearly. Run the simulation for both traffic lights and a filter in turn as many times as you think feasible.

# Interpret the solution

Use your simulation to calculate queue length and average waiting times over a given period.

# Compare with reality

See whether your traffic light simulation agrees with observations made at the lights you originally used. You cannot directly check a filter, but if the lights simulation is realistic there is a good chance the filter simulation will also be realistic.

If the simulation does not compare well with reality, you will need to amend your model. Then follow through the procedure again, and interpret your results.

# Write a report

- Which system seems to be more efficient?
- Are your distributions for car lengths and inter-arrival times adequate?
- Did you omit large vehicles from the original model?
- What effect does changing the delay on traffic lights have?

## CONSOLIDATION EXERCISES FOR CHAPTER 12

**1** You are required to simulate a gambler in a casino. The gambler starts with £5 and bets £1 at a time on a game in which he receives £2 back (his £1 stake plus £1 winnings) with probability 0.4, and in which he loses his money with probability 0.6.

Thus, if he starts          loss, loss, loss, win, ...,
then his cash in hand goes   £5 → £4 → £3 → £2 → £3→ ... .

You are to use the following list of random digits in your simulation, reading across the table. For each run of your simulation start with the gambler holding £5, increase his money by £1 if the next digit indicates a win, and decrease his money by £1 if it indicates a loss. Start a new run when he runs out of money.

| | | | | | | | | | | | |
|---|---|---|---|---|---|---|---|---|---|---|---|
| 7 | 6 | 5 | 1 | 5 | 5 | 9 | 9 | 9 | 8 | 6 | 0 |
| 6 | 5 | 2 | 3 | 6 | 8 | 0 | 0 | 0 | 6 | 5 | 6 |
| 7 | 2 | 8 | 2 | 4 | 5 | 4 | 4 | 0 | 4 | 2 | 1 |
| 2 | 8 | 7 | 2 | 8 | 7 | 0 | 1 | 8 | 0 | 5 | 9 |
| 3 | 1 | 6 | 2 | 4 | 6 | 5 | 3 | 2 | 9 | 8 | 1 |
| 5 | 7 | 9 | 4 | 3 | 9 | 9 | 8 | 7 | 4 | 2 | 2 |
| 5 | 6 | 1 | 4 | 8 | 0 | 1 | 0 | 2 | 9 | 5 | 6 |
| 6 | 2 | 7 | 4 | 9 | 3 | 3 | 2 | 5 | 7 | 3 | 8 |
| 9 | 5 | 6 | 9 | 5 | 1 | 5 | 4 | 6 | 5 | 7 | 1 |
| 3 | 2 | 4 | 3 | 9 | 0 | 2 | 7 | 3 | 3 | 4 | 3 |

**a)** Use all of the given digits to perform the required simulation. State the rules which you use for your simulation, and record your results clearly.

**b)** Use your results to find an estimate of how many bets the gambler can expect to make before he runs out of money. Explain how you deal with any unfinished run at the end of your simulation.

**c)** List the possible sums of money which the gambler could be holding after three bets.

**d)** Use your results to estimate the expected amount that the gambler will have left after three bets.

*(UODLE Question 1, Decision Mathematics AS, 1993)*

**2**  The diagram shows part of a pinball machine which has tracks along which balls roll. The tracks divide at points at which separators (labelled 1–6) deflect the balls one way or the other with equal probabilities.

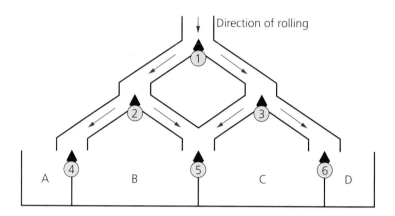

The balls which end up in A or D score five points each. Those which end up in B or C score ten points each.

**a)** You are now to simulate the progress of 24 balls, either by using the random numbers listed beneath this question, or by using the RAN# (or equivalent) key on your calculator.

You must indicate clearly:

■ how many random numbers you use to simulate the progress of each ball

■ the rules which you use to simulate the progress of each ball

■ for each of the first five balls, a list of the random numbers used and a clear description of the path taken.

**b) i)** What was the mean score per ball from your simulation?

**ii)** What would you have expected the mean score per ball to be?

Random numbers:

| | | | |
|---|---|---|---|
| 72824 | 54409 | 53048 | 30383 |
| 04212 | 87273 | 25027 | 43581 |
| 87018 | 05989 | 36415 | 96027 |
| 31624 | 65384 | 40563 | 17496 |

*(UODLE Question 6, Decision Mathematics AS, 1995)*

3   The scoring system for number tennis involves points, games and sets.
    When two players are involved the scoring is as follows.
    Players compete for points until one player has scored *both*:
    i)  at least two points more than the other, and
    ii) at least four points in all.
    The player with the higher points score has then won the game.
    Games are contested until one player has won *both*:
    i)  at least two games more than the other, and
    ii) at least six games in all.
    The player with the higher games score has then won the set.
    When A and B play each other at number tennis, the probability that
    A wins any particular point is 0.56.

   **a)** Using, row by row, the random numbers at the end of this question,
        simulate a game between A and B. State your simulation rules
        carefully. Record full details of the points score throughout the
        game. State the name of the winner and the final score in the
        game. Say how you would estimate the probability of A winning a game.
   **b)** Continue the simulation started in **a)**, and complete the simulation
        of a set of number tennis between A and B. Give details of the
        winner of each game and the games score throughout the set.
        Which player won the set, and what was the final score?
    Number tennis is not restricted to two players. When there are three
    players the rules are amended slightly. The winner of a game must
    have at least two points more than each of the other players, and at
    least four points in all.
    When A, B and C play together at number tennis, the probabilities of
    each winning any particular point are 0.28, 0.22, and 0.5 respectively.
   **c)** Start again at the beginning of the random numbers given.
        Simulate a game of number tennis between A, B and C. State the
        new simulation rules and record full details of the game. What
        was the final score and which player won the game?

    | | | | | | |
    |---|---|---|---|---|---|
    | 4032 | 8635 | 7583 | 2543 | 8594 | 2527 |
    | 1281 | 5109 | 5193 | 6981 | 6755 | 6573 |
    | 0829 | 8002 | 9473 | 6130 | 4798 | 4455 |
    | 9775 | 7884 | 4077 | 4089 | 2931 | 7254 |
    | 9174 | 4993 | 8845 | 0179 | 6859 | 0424 |
    | 8848 | 1143 | 5981 | 4802 | 8706 | 6373 |
    | 4502 | 4802 | 2793 | 2947 | 1716 | 3771 |
    | 9109 | 1853 | 9373 | 3900 | 9143 | 7284 |

    (Use numbers in the order 40, 32, 86, 35, 75, etc.)

    *(UODLE Question 8, Decision Mathematics AS, 1991)*

4   Random number lists for this question are printed at the end of the
    question.
    The random numbers are to be used as two-digit random numbers,
    and are to be read in order from the top left across the page, a row at
    a time, as required. there are separate lists for part **a)** and for part **b)**
    of the question.

**a)** A small post office has one server. Customer inter-arrival times follow the following distribution.

| Inter-arrival time (minutes) | 2 | 3 | 4 |
|---|---|---|---|
| Probability | $\frac{1}{4}$ | $\frac{1}{2}$ | $\frac{1}{4}$ |

Using two-digit random numbers from the list marked A, simulate five inter-arrival times. Describe the rule that you used to generate your inter-arrival times.

**b)** Service times follow the following distribution.

| Length of service (minutes) | 1 | 2 | 5 |
|---|---|---|---|
| Probability | $\frac{1}{3}$ | $\frac{1}{2}$ | $\frac{1}{6}$ |

Using two-digit random numbers from the list marked B, simulate service times for six customers. Describe the rule that you used to generate your service times.

**c)** Assuming that the service of the first customer has just begun, simulate the service of six customers. Number the customers 1 to 6, and record your results in the table.

| Customer number | Arrival time | Inter-arrival time | Start of service | Service time | End of service |
|---|---|---|---|---|---|
| 1 | 0 | | 0 | | |
| 2 | | | | | |
| 3 | | | | | |
| 4 | | | | | |
| 5 | | | | | |
| 6 | | | | | |

**d)** Compute the mean queue time (i.e. the mean of the times for which each customer queues), the mean length of queue and the server utilisation (i.e. the percentage of time for which the server is busy).

**e)** Say how the reliability of the results from your simulation could be improved.

Random number list A
42 22 03 80 19 37 62 93 55 12
12 07 46 91 63 28 20 37 92 83
45 37 64 58 55 07 14 12 75 43
45 94 74 21 56 64 39 54 64 19
23 36 78 53 46 85 10 19 09 05

Random number list B
35 92 78 98 16 06 31 34 82 17
64 54 38 03 17 19 20 91 76 73
36 49 54 84 87 69 46 21 09 70
75 24 15 13 96 50 02 04 56 87
56 62 39 53 81 04 24 25 71 61

*(O & C MEI Question 4, Decision & Discrete Mathematics Paper 19, June 1995)*

5 Genetic theory predicts that the plants of one type of sweet pea should fall into four classes, A, B, C, D in the ratios 9:3:3:1.

a) Use the following random two-digit numbers, reading across the page, to simulate the classes of the next 60 seeds planted by a gardener. State the simulation rules that you use.

| | | | | | | | | | |
|---|---|---|---|---|---|---|---|---|---|
| 84 | 31 | 50 | 71 | 78 | 67 | 43 | 20 | 72 | 12 |
| 60 | 82 | 50 | 78 | 08 | 92 | 15 | 32 | 13 | 53 |
| 29 | 12 | 02 | 27 | 40 | 81 | 91 | 49 | 57 | 77 |
| 14 | 25 | 56 | 97 | 50 | 60 | 22 | 28 | 89 | 14 |
| 11 | 17 | 88 | 62 | 89 | 74 | 90 | 63 | 33 | 08 |
| 78 | 93 | 94 | 20 | 37 | 36 | 18 | 10 | 05 | 90 |
| 75 | 79 | 66 | 96 | 91 | 19 | 54 | 23 | 39 | 01 |

b) When the seeds have grown sufficiently they are set out in groups of three per pot. Use your 60 results from a) to obtain simulation plantings for the next 20 pots. What proportion of these 20 pots contain exactly three plants of class A?

c) Obtain another estimate of the proportion of pots containing three plants of class A using the following algorithm.
Set COUNT = 0
Look at the plants in positions 1, 2 and 3. If they are all of type A, then add 1 to COUNT
Look at the plants in positions 2, 3 and 4. If they are all of type A, then add 1 to COUNT
Look at the plants in position 3, 4 and 5 ... and so on.
Set PROPORTION = COUNT/58
Explain why the division in the final line is by 58.

d) Calculate the theoretical probability that three randomly-selected pea plants are all of class A.

*(UODLE Question 2, Decision Mathematics AS, 1992)*

6 A chocolate factory produces several types of chocolate. The chocolates are mixed together and randomly packed into cartons of 20. Fifteen per cent of the chocolates have nut centres.

a) You are to simulate the process of packing chocolates into cartons of 20, distinguishing between nut centres and other chocolates. You are to use a table of two-digit random numbers from 00 to 99 to generate ten cartons of chocolates. For each carton you are to count the number of nut centres in the carton.
i) Give your simulation rule.
ii) Now apply your rule using the table of two-digit random numbers below. Indicate those numbers which represent nut centres. Count the number of nut centres in each simulated carton.

| | Carton number | | | | | | | | | |
|---|---|---|---|---|---|---|---|---|---|---|
| | **1** | **2** | **3** | **4** | **5** | **6** | **7** | **8** | **9** | **10** |
| **Number of nut centres** | 65 | 06 | 55 | 72 | 04 | 87 | 31 | 29 | 39 | 56 |
| | 29 | 93 | 95 | 65 | 90 | 95 | 99 | 87 | 46 | 66 |
| | 36 | 07 | 93 | 49 | 20 | 02 | 59 | 48 | 54 | 35 |
| | 73 | 34 | 68 | 72 | 44 | 28 | 87 | 44 | 81 | 09 |
| | 77 | 19 | 52 | 52 | 52 | 65 | 29 | 15 | 82 | 81 |
| | 23 | 56 | 99 | 82 | 21 | 01 | 62 | 81 | 98 | 14 |
| | 56 | 32 | 69 | 71 | 27 | 29 | 74 | 87 | 24 | 79 |
| | 42 | 66 | 10 | 50 | 75 | 40 | 87 | 08 | 26 | 35 |
| | 84 | 64 | 56 | 47 | 44 | 11 | 22 | 93 | 84 | 75 |
| | 65 | 06 | 91 | 47 | 67 | 25 | 97 | 25 | 08 | 35 |
| | 68 | 76 | 98 | 45 | 28 | 80 | 46 | 57 | 74 | 80 |
| | 62 | 57 | 51 | 32 | 33 | 42 | 06 | 56 | 17 | 81 |
| | 94 | 25 | 05 | 63 | 58 | 62 | 21 | 99 | 86 | 58 |
| | 90 | 78 | 87 | 05 | 96 | 57 | 38 | 14 | 37 | 35 |
| | 05 | 51 | 87 | 25 | 87 | 71 | 56 | 03 | 65 | 03 |
| | 00 | 51 | 60 | 44 | 72 | 59 | 53 | 94 | 22 | 10 |
| | 74 | 38 | 54 | 43 | 43 | 45 | 29 | 91 | 74 | 43 |
| | 58 | 08 | 72 | 99 | 89 | 09 | 38 | 66 | 75 | 45 |
| | 49 | 00 | 47 | 42 | 75 | 47 | 88 | 59 | 25 | 21 |
| | 04 | 61 | 07 | 14 | 40 | 73 | 42 | 68 | 67 | 25 |

**iii)** Use your *results* to estimate the probabilities of a carton containing:

- exactly three nut centres
- fewer than three nut centres.

**iv)** Suppose that the fifteen per cent figure quoted is only an approximation, and in fact one seventh of the chocolates have nut centres. How would you change your simulation rule to simulate the process using two digit random numbers?

**b)** A quality control manual states that if $p$% of items from a large batch are faulty, then the probability of obtaining fewer than three faulty items in a random sample of size 20 is given by:

$$f(p) = \left(1 - \frac{p}{100}\right)^{20} + 20\frac{p}{100}\left(1 - \frac{p}{100}\right)^{19} + 190\left(\frac{p}{100}\right)^{2}\left(1 - \frac{p}{100}\right)^{18}$$

**i)** Say how this result can be used to compute the probability of obtaining fewer than three nut centres in a carton of 20 chocolates.

ii) Calculate f(15), giving your answer correct to three decimal places.

Compare it with the results of your simulation.

Say why the results might differ, and suggest how to improve the simulation.

*(O & C MEI Question 2, Decision & Discrete Mathematics Paper 19, January 1996)*

7   The weather bureau in a particular country defines each day to be either wet or dry. Records show that if the weather today is dry then the probability that it will be dry tomorrow is $\frac{4}{5}$, and the probability that it will be wet tomorrow is $\frac{1}{5}$. If the weather today is wet then the probability that it will be wet tomorrow is $\frac{2}{7}$, and the probability that it will be dry tomorrow is $\frac{5}{7}$.

Future weather is to be simulated. Each day a two-digit random number is to be used, together with a simulation rule based on that day's weather, to generate the weather for the next day.

When the weather is dry the rule to be used is as follows.

$00 - 79 \Rightarrow$ weather tomorrow is dry

$80 - 99 \Rightarrow$ weather tomorrow is wet.

a)  Give a simulation rule to generate the weather for a day following a wet day.

b)  The weather today is dry. Use the rules, together with the following random numbers to simulate the weather for the next 14 days. Read the two-digit numbers from left to right.
Random numbers:
39  16  44  89  01  56  90  99  11  37  47  84  29  52  21

06  39  43  06  42  82  52  16  39  89  58  61  74  93  82

c)  Use the fourteen results from your simulation to calculate an estimate of the overall proportion of wet days.

d)  Use the fourteen results from your simulation to calculate an estimate of the probability of the weather tomorrow being the same as the weather today.

e)  Give two ways in which the simulation of the weather could be improved.

*(O & C MEI  Question 1, Decision & Discrete Mathematics Paper 19, June 1996)*

8   A circular coin of diameter 3 cm is used in a fairground game. Competitors flick the coin along a board marked across with ten parallel lines at 5 cm intervals, and with end sections of width 4 cm. To be valid the coin must end entirely within the scoring region (shown unshaded in the diagram overleaf), and it will then count as a win if it does not cross a line. Otherwise it will count as a loss. If it does not land entirely within the scoring area then another attempt is made until it does. (The numbers in the diagram indicate the distances in cm between the lines.)

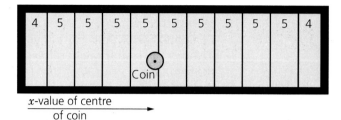

*x*-value of centre
of coin

Assume that a valid coin is equally likely to land in any position within the scoring region. Thus the *x*-value of the centre of a valid coin will be between 1.5 and 51.5.

a) i) Give the ranges of *x*-values which correspond to the centres of winning coins.

ii) Use the following list of five-digit random numbers to simulate the *x*-values of the centres of ten randomly-flicked and valid coins. Explain your method for performing the simulation and give the *x*-value of the centre of each of your coins.

iii) State whether or not each of your simulated coins is in a winning position.

Random numbers:

65236    80077    82581    41085    87273
06567    65104    73943    01674    36415

b) By considering the regions of the board within which the centre of a winning coin must lie, calculate the probability of winning.

c) A new board is to be constructed with ten equally-spaced lines 6 cm apart, but with the width of the end sections each being *e* cm. Find the value of *e* for the board to be fair (i.e. one in which the probability of winning is 0.5).

*(UODLE Question 11, Decision Mathematics AS, 1995)*

9    A manufacturer buys in a component in very large numbers. The nature of the component is such that faulty components are inevitable. The manufacturer is prepared to accept a shipment provided that no more than ten per cent are defective. It is not possible, given the large numbers in a shipment and the testing process, to test each one. It is therefore necessary to select a sample.

Two sampling schemes are under consideration. The first involves selecting a random sample of ten components and accepting the shipment if two or fewer are defective. The second involves taking a sample of size 20 and accepting the shipment if three or fewer are defective.

These sampling schemes have been simulated by repeatedly drawing samples of sizes 10 and 20 respectively from each of a set of nine populations with differing percentages of defective items. The results are summarised in the following table.

**Table showing the number of defectives in each simulated sample**

| Simulation number | Sample size | Percentage of defectives in population (%) | | | | | | | | |
|---|---|---|---|---|---|---|---|---|---|---|
| | | 1 | 5 | 8 | 10 | 12 | 15 | 20 | 25 | 30 |
| 1–9 | 10 | 0 | 1 | 2 | 0 | 1 | 1 | 2 | 2 | 4 |
| 10–18 | 10 | 0 | 0 | 0 | 2 | 0 | 1 | 0 | 0 | 2 |
| 19–27 | 10 | 0 | 2 | 0 | 0 | 2 | 0 | 3 | 3 | 3 |
| 28–36 | 10 | 0 | 0 | 2 | 0 | 0 | 2 | 2 | 1 | 2 |
| 37–45 | 10 | 0 | 0 | 1 | 1 | 0 | 1 | 5 | 1 | 3 |
| 46–54 | 10 | 1 | 1 | 1 | 0 | 1 | 2 | 1 | 2 | 4 |
| 55–63 | 10 | 1 | 0 | 0 | 2 | 2 | 3 | 2 | 2 | 3 |
| 64–72 | 10 | 0 | 0 | 1 | 1 | 2 | 1 | 1 | 4 | 2 |
| 73–81 | 10 | 0 | 0 | 0 | 1 | 1 | 1 | 2 | 1 | 2 |
| 82–90 | 10 | 0 | 1 | 0 | 3 | 1 | 0 | 2 | 3 | 1 |
| 91–99 | 10 | 0 | 2 | 0 | 0 | 1 | 1 | 1 | 4 | 4 |
| 100–108 | 10 | 0 | 2 | 0 | 1 | 0 | 3 | 3 | 3 | 1 |
| 109–117 | 10 | 0 | 1 | 0 | 1 | 0 | 0 | 0 | 3 | 4 |
| 118–126 | 10 | 0 | 0 | 1 | 0 | 1 | 2 | 2 | 3 | 2 |
| 127–135 | 10 | 0 | 0 | 0 | 0 | 2 | 3 | 3 | 3 | 3 |
| 136–144 | 10 | 1 | 0 | 0 | 2 | 0 | 0 | 1 | 3 | 2 |
| 145–153 | 10 | 0 | 1 | 1 | 0 | 1 | 1 | 6 | 1 | 3 |
| 154–162 | 10 | 0 | 0 | 2 | 1 | 2 | 2 | 2 | 2 | 4 |
| 163–171 | 10 | 0 | 1 | 4 | 2 | 1 | 4 | 1 | 2 | 3 |
| 172–180 | 10 | 0 | 0 | 2 | 2 | 2 | 1 | 4 | 3 | 3 |
| 181–189 | 20 | 0 | 2 | 0 | 1 | 2 | 2 | 3 | 4 | 6 |
| 190–198 | 20 | 0 | 2 | 1 | 0 | 5 | 3 | 5 | 8 | 8 |
| 199–207 | 20 | 0 | 1 | 1 | 4 | 3 | 5 | 8 | 7 | 5 |
| 208–216 | 20 | 0 | 0 | 2 | 1 | 3 | 5 | 0 | 3 | 6 |
| 217–225 | 20 | 0 | 0 | 3 | 2 | 2 | 1 | 4 | 7 | 7 |
| 226–234 | 20 | 0 | 1 | 1 | 1 | 1 | 4 | 7 | 6 | 4 |
| 235–243 | 20 | 1 | 0 | 4 | 2 | 3 | 3 | 5 | 4 | 7 |
| 244–252 | 20 | 0 | 2 | 1 | 0 | 2 | 3 | 1 | 7 | 5 |
| 253–261 | 20 | 0 | 1 | 0 | 0 | 0 | 2 | 4 | 3 | 4 |
| 262–270 | 20 | 0 | 0 | 1 | 4 | 1 | 3 | 0 | 6 | 7 |
| 271–279 | 20 | 0 | 5 | 2 | 4 | 3 | 3 | 5 | 7 | 7 |
| 280–288 | 20 | 0 | 2 | 4 | 0 | 4 | 1 | 6 | 3 | 7 |
| 289–297 | 20 | 0 | 2 | 1 | 5 | 2 | 4 | 3 | 2 | 6 |
| 298–306 | 20 | 0 | 3 | 2 | 2 | 3 | 0 | 3 | 2 | 4 |
| 307–315 | 20 | 1 | 2 | 2 | 2 | 1 | 2 | 6 | 5 | 2 |
| 316–324 | 20 | 1 | 0 | 0 | 4 | 2 | 5 | 1 | 5 | 4 |
| 325–333 | 20 | 0 | 2 | 3 | 2 | 4 | 1 | 5 | 5 | 8 |
| 334–342 | 20 | 0 | 0 | 1 | 2 | 4 | 3 | 5 | 6 | 5 |
| 343–351 | 20 | 0 | 0 | 1 | 3 | 1 | 4 | 7 | 5 | 3 |
| 352–360 | 20 | 0 | 2 | 0 | 2 | 5 | 3 | 3 | 3 | 6 |

a) Use the given results to estimate the probability that a shipment containing ten per cent of defective items will be accepted under each of the sampling schemes.

b) Repeat part **a)** for each of the other percentage defectives. Tabulate your results.

c) Produce a graph showing 'Probability of acceptance' against 'Percentage defective' for the first sampling scheme.

d) On the same graph as in part **c)**, but using different symbols, produce a graph of 'Probability of acceptance' against 'Percentage defective' for the second sampling scheme.

e) Which sampling scheme more closely achieves the stated aims of the manufacturer (to accept a shipment if it contains no more than ten per cent defective items)?
Which scheme will be the more expensive to operate?

*(UODLE Question 9, Decision Mathematics AS, 1993 )*

**10** A dentist wishes to analyse the time that her patients spend waiting for their appointments. She begins surgery at 8.30 a.m. and appointments are made every 15 minutes, the last appointment being at 12.15 p.m. She estimates that roughly 33 per cent of her appointments are check-ups which take an average of ten minutes, twelve per cent are extractions which take around fifteen to twenty minutes, and the rest are usually fillings which can take from ten to twenty minutes. Devise a simulation for the dentist's morning surgery, stating your assumptions clearly. Do you think the dentist is correct in making appointments every fifteen minutes?

## Summary

*After working through this chapter you should:*

■ *understand that a simulation is a model set up to represent a real situation*

■ *be able to set up and perform simple simulations, stating the rules you use and the simplifying assumptions, which are the assumptions you make to enable you to carry out the simulation*

■ *be able to interpret the results of your simulation in the context of the original problem.*

# 13

# *Decision-making*

*By the end of this chapter you should:*

■ *know that enumeration methods can be used to find the number of solutions to certain types of decision-making problems*

■ *be able to distinguish between different types of arrangement problems and use the correct formula to solve them*

■ *be able to model decision-making problems by drawing decision trees*

■ *know how to use a decision tree to calculate the expected monetary value*

■ *understand the basic principles of Boolean algebra*

■ *be able to build truth tables and apply these to the output for some types of logic gate in electronics.*

In Chapter 1, *Algorithms*, we considered decision-making problems and we asked the questions:
■ How many solutions (to a problem) are there?
■ Can you list all of the solutions?

There are two problems here. One is a **counting** problem in which we need to know the number of objects or solutions, the other is a **listing** problem in which we try to list all the possible objects or solutions. Together, these types of problems are called **enumeration** problems.

## ENUMERATION

**Exploration 13.1**

### A-level options

Everbrite School uses a blocking system for its sixth-form time table. Students may only take one subject from each of the blocks below. How many different combinations of three A-levels are possible?

| Block A | Block B | Block C | Block D |
|---------|---------|---------|---------|
| Mathematics | Physics | Chemistry | English |
| Art | German | Geography | Mathematics |
| French | Music | English | CDT |
| Biology | Spanish | | Psychology |
| History | | | |

## Choosing

**Example 13.1**

*Julia must choose her subjects from blocks A, B and C. She definitely wants to take French but is unsure which subjects to choose in blocks B and C. In how many ways can she choose her other two subjects?*

**Solution**

*Julia can choose one subject from the four in block B and one from the three in block C, so the number of ways in which she can make her choice is* $3 \times 4 = 12$.

*The combinations of subjects can be shown on a tree diagram which also lists all the possible solutions. Once the number of choices gets larger than this, listing them all becomes more of a problem.*

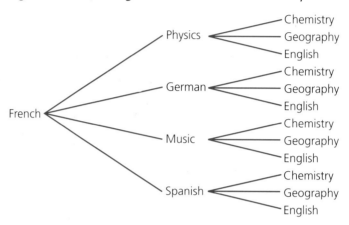

**Example 13.2**

*Ahmed is intending to do three A-levels, one each from blocks A, B and C. In how many ways can he choose three subjects?*

**Solution**

*The subject in block A can be chosen in five ways, the subject in block B in four ways and the one in block C three ways, so the total number of subject combinations is:*

$$5 \times 4 \times 3 = 60 \text{ ways}$$

The solution to the last example would be rather difficult to show in a tree diagram, but we can generalise from these results.

If there are $k$ disjoint sets of items and if item 1 can be chosen in $n_1$ ways, item 2 in $n_2$ ways and so on up to item $k$ in $n_k$ ways, the total number of ways of choosing $k$ distinct objects is:

$$n_1 \times n_2 \times ... \times n_k$$

# The principle of inclusion–exclusion

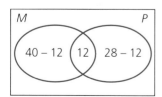

The Head of Science wants to know what proportion of the new Year 12 are taking either Mathematics or Physics. There are 120 students in Year 12; 40 of them have chosen to do Mathematics and 28 have chosen to do Physics. It is known that twelve of these are doing both subjects.

If we add the number of students taking Mathematics to those taking Physics, we will have counted the twelve who take both subjects twice, so the number of students taking one or both of these subjects is:

$$40 + 28 - 12 = 56$$

which gives the fraction of students taking either Mathematics or Physics as $\frac{56}{120}$.

This example demonstrates a method for counting the total number when the sets are not disjoint. It is called the **principle of inclusion–exclusion** and can be formally stated like this.

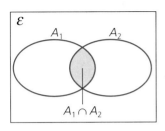

If a set $A$ can be split into two subsets $A_1$ and $A_2$ then:
$$|A| = |A_1| + |A_2| - |A_1 \cap A_2|$$

where $|A|$ is the number of members in set $A$.

## Selections (arrangements)

**Example 13.3**

*Cash cards usually have a four-digit personal identification number (PIN). How many different four-digit numbers can be formed if:*
*a)* *the order does matter and repetition is allowed*
*b)* *the order matters and repetition is not allowed*
*c)* *the order does not matter and repetition is not allowed?*

**Solution**
*The first thing to note is that there are ten digits:*
  *0, 1, 2, 3, 4, 5, 6, 7, 8, 9*
*and that we are to find the ways of choosing four of them.*
*a)* *If the order matters, that means that 1356 is not the same as 6531, even though they contain the same four digits. If we are allowed to repeat digits then we can choose each digit in ten ways, so:*
  *total number of PINs* $= 10 \times 10 \times 10 \times 10 = 10^4 = 10\,000$

> **In general:** If we choose $r$ items from a set of $n$ items in a selection where the order matters and repetition is allowed, the number of arrangements is $n^r$.

*b)* *If the order matters but repetition is not allowed we can choose the first digit in ten ways, the second in nine ways, the third in eight ways and the fourth in seven ways, so:*
  *total number of PINs* $= 10 \times 9 \times 8 \times 7 = 5040$

*which is considerably fewer than we might expect.*
*Another way of writing this would be to use factorial notation:*
$$10 \times 9 \times 8 \times 7 = \frac{10!}{6!} = \frac{10!}{(10-4)!}$$

> **In general:** If we extend this to an ordered selection without repetition of $r$ items from a set of $n$ items, we can choose the first item in $n$ ways, the second in $(n-1)$ ways, and so on until the $r$th item which can be chosen in $(n-r+1)$ ways, so:
> total number of arrangements
> $$= n \times (n-1) \times (n-2) \times \ldots \times (n-r+1)$$
> $$= \frac{n!}{(n-r)!}$$

## Permutations

Ordered selections without repetition are called **permutations**, and the usual notation for a permutation of $r$ items from $n$ is $^nP_r$ or $P_{(n-r)}$.

$$^nP_r = \frac{n!}{(n-r)!}$$

c) *If the order does not matter, then any two numbers with the same four digits will be equivalent, even if the digits are in a different order, so 1356 and 6531 would be the same selection. Therefore, we need to know how many ways any four-digit selection can occur, as these will only count as one selection.*

*If we consider 1356, the arrangements of the digits are:*

| | | | |
|---|---|---|---|
| 1356 | 3156 | 5136 | 6135 |
| 1365 | 3165 | 5163 | 6153 |
| 1536 | 3516 | 5316 | 6315 |
| 1563 | 3561 | 5361 | 6351 |
| 1635 | 3615 | 5613 | 6513 |
| 1653 | 3651 | 5631 | 6531 |

24 in all
$24 = 4 \times 3 \times 2 \times 1 = 4!$

**In general:** There are always $k!$ arrangements of $k$ objects.

*So, the number of selections when order matters and repetition is not allowed will be the number of permutations, divided by the number of arrangements or each four digit combinations* $= \dfrac{5040}{24} = 210.$

**In general:** If we extend this to an unordered selection without repetition of $r$ items from a set of $n$ items, we have:

total number of arrangements $= \dfrac{^nP_r}{r!} = \dfrac{n!}{(n-r)!\,r!}$

*This can be thought of as 'choose r from n'. An alternative notation for $^nC_r$ is $\begin{pmatrix} n \\ r \end{pmatrix}$.*

## Combinations

Unordered selections without repetition are called **combinations** and the usual notation for a combination of $r$ items from $n$ is $^nC_r$.

$$^nC_r = \frac{n!}{(n-r)!\,r!}$$

Returning to Exploration 13.1, we now have to decide on the type of problem we have. Everbrite School offers fourteen different A-level subjects: Mathematics, Physics, Chemistry, English, Art, German, Geography, French, Music, Biology, Spanish, Psychology, History and CDT. If we start by considering the problem as simply being the need to choose three subjects from fourteen then we need to ask two questions.

■ Does the order matter? *No it does not*, since Mathematics, Physics, Chemistry is the same subject combination as Physics, Chemistry, Mathematics.
■ Is repetition allowed? No it is not, students must study three distinct subjects.

So we want an unordered selection without repetition, which is a combination.

$$^{14}C_3 = \frac{14!}{(14-3)!\,3!} = 364$$

The problem becomes more difficult if we take all four blocks into account; we must now ask the question, 'How many different combinations of three A-levels are possible if only one subject can be chosen from each of four blocks?'

| Block A | Block B | Block C | Block D |
|---|---|---|---|
| Mathematics | Physics | Chemistry | English |
| Art | German | Geography | Mathematics |
| French | Music | English | CDT |
| Biology | Spanish | | Psychology |
| History | | | |

Firstly we need to consider how many ways we can choose three blocks from four. Since we want an unordered selection with no repetition, we use the rule for combinations:

$$^{4}C_3 = \frac{4!}{(4-3)!\,3!} = 4$$

We need to list these possibilities, as the combinations available depend upon the blocks: ABC, ABD, ACD, BCD.

Now, for each combination of blocks we must find the possible combinations of subjects.

ABC  These blocks have no subjects in common, so, as we found in Example 13.2, there are $5 \times 4 \times 3 = 60$ possible combinations.

ABD  Since Mathematics appears in two blocks, we need a slightly different strategy. If we consider the total number of subject combinations available if Mathematics can be repeated we have $5 \times 4 \times 4 = 80$.

Now if we consider the combinations which have Mathematics occurring twice in blocks A and D which is $1 \times 4 \times 1 = 4$.

We must then subtract these unacceptable combinations from the total number of combinations, giving $80 - 4 = 76$ possible subject combinations in blocks A, B and D.

ACD  We now have two repeated subjects – Mathematics in blocks A and D and English in blocks C and D. Allowing for repeats we get the total number of subject combinations: $5 \times 3 \times 4 = 60$

The number of combinations in which Mathematics is repeated is $1 \times 3 \times 1 = 3$
and the number of combinations in which English is repeated is $5 \times 1 \times 1 = 5$.

So the number of acceptable combinations from blocks A, C and D is:
$60 - 3 - 5 = 52$

*In question 1 of Exercises 13.1A, you are asked to complete this exploration.*

## EXERCISES

1   Find the number of possible subject combinations in blocks B, C and D and hence complete Exploration 13.1 by finding the total number of different combinations of three A-levels possible within the blocking system.

2   A hotel chain provides each guest room with soap and shampoo. The chain has been offered discounts on five brands of soap and four brands of shampoo. In how many ways can the hotel choose one soap and one shampoo for each guest room?

The hotel chain then decides to provide toothpaste as well and is offered discount on three brands. How many ways can it choose one soap, one shampoo and one tube of toothpaste for each guest room?

3   How many ways can three letters be chosen from the word NETWORK

a) if the order does matter and if repetition is:
   i) allowed   ii) not allowed?
b) if the order does not matter and repetition is not allowed?

4   a) A working party of five people is to be chosen from a team of twelve people. In how many ways can this be done?
   b) The team of twelve is made up of seven women and five men. How many ways can the working party be chosen if it must contain at least two women and two men?

5   A common procedure in quality control is to take a sample of the product and test it for defects. From a batch of 50 personal stereos, the manufacturer wishes to select five for testing. In how many ways can this be done?

The manufacturer believes that six per cent of the stereos may have defects. If he is correct, what is the probability that the sample will contain at least one defective stereo?

## EXERCISES

1   A couple have just had a baby daughter and they have asked their parents for suggestions for names for the child.

| Father 1 | Mother 1 | Father 2 | Mother 2 |
|----------|----------|----------|----------|
| Sharon | Tamsyn | Gillian | Wendy |
| Tracy | Suzanne | Caroline | Joyce |
| Lucy | Samantha | Jane | Gillian |
| Katherine | Katherine | | |

The couple decide to choose two names for their daughter.
a) How many choices do they have if one name is picked from the suggestions of parents 1 and one from parents 2?

**b)** How many choices do they have if a name is chosen from Father 1 and another from Father 2?

**c)** How many different combinations of names are possible if only one name is chosen from each parent?

2    If any element of the set {1, 2, 3, 4} is multiplied by any element of the same set, what is the probability that the product is greater than 8? What is the probability if the case of an element being multiplied by itself is excluded?

In each case list each combination on a tree diagram.

3    Consider six-digit telephone numbers.

**a)** How many six-digit numbers are there?
**b)** How many numbers are there if 0 is not permitted as a first digit?
**c)** How many numbers are there if the combination 999 is not permitted as the last three digits?
**d)** How many six-digit numbers are there if no digit may be repeated?
**e)** How many six-digit numbers are there if no digit is repeated and 0 is not the first digit?

4    **a)** In the British lottery 49 balls are labelled 1 to 49. Initially six balls are selected randomly. A player may choose six different numbers which constitutes a line, and these numbers must correspond to the numbers on the balls.
     **i)** How many different lines may be chosen? Hence, what is the probability of a player playing just one line picking all six numbers?
     **ii)** How many ways can five numbers be chosen from six? How many ways can one number be chosen from 43? Hence, in how many ways can you obtain exactly five numbers in the lottery?

   **b)** A seventh ball is now drawn, called the bonus ball. What is the probability of a winning line consisting of five numbers from the original six plus the bonus ball?

5    **a)** From a group of ten people in how many ways can a chairperson and secretary be chosen?
   **b)** How many four-letter words can be formed from the word COMPANY if repetition is not allowed?
   **c)** A small library has ten books on mathematics, four on statistics and seven on computing. In how many ways can these books be arranged on a shelf:
     **i)** irrespective of subject matter
     **ii)** if all the books on the same subject must be together and the subjects ordered alphabetically starting with the letter 'C'
     **iii)** if all the books on the same subject must be together but order of the subjects is irrelevant?

## DECISION ANALYSIS

### *To develop, or not?*

A kitchenware manufacturer is currently making a profit of £100 000 a year on one of its products. One of the development team comes up with an idea for an improvement to the product and the management team has to decide whether to develop the idea or not. The information available to them suggests that the probability of the idea working is 95 per cent and development costs will be £80 000. If they decide to go ahead they will need to invest a further £25 000 on publicity and there is a 75 per cent chance that the improved product will increase their market share, leading to annual profits of around £220 000. What would you advise the company to do?

## Decision trees

In Chapter 3, *Minimum-connector and the travelling salesman problem*, we defined a tree as a connected graph that contains no cycles. We used an example of a tree in the previous section to help us list the possible outcomes in a counting problem and you will be familiar with the use of trees in other areas, including:

■ family tree
■ classification tree in biology
■ probability tree diagram.

There has been considerable interest in tree structures in the fields of computer science and artificial intelligence, because they are extremely useful when organising data and they can be used to clarify and facilitate complex decision-making strategies.

Decision trees are used extensively in a branch of Decision mathematics known as **game theory** and are at the root of computer programs of two-player strategy games such as Chess and Othello. We can extend the same ideas to help with modelling situations, as in Exploration 13.2, where we are concerned with making the best choice. In order to do this we need to put more information onto the decision tree.

In the same way that, in a flow chart, different-shaped boxes have specific meanings, a decision tree has different-shaped vertices with specific meanings.

□ A rectangle represents a decision vertex (node); a stage at which a player has to make a decision.

○ A circle represents a chance vertex (node); a stage where an element of chance influences the outcome.

△ A triangle represents a pay-off vertex (node); the last stage, giving the final **pay-off** (or win, or successful outcome) to the player.

How can we use decision trees to help the development team in Exploration 13.2?

We can draw a decision tree to represent the manufacturer's dilemma at different stages. As we go through, we can calculate the values at the pay-off vertices by taking account of the expected profits and costs at each stage.

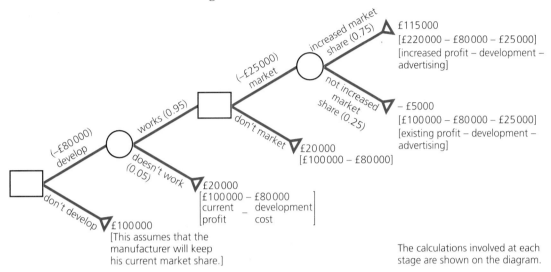

£115000
[£220000 – £80000 – £25000]
[increased profit – development – advertising]

– £5000
[£100000 – £80000 – £25000]
[existing profit – development – advertising]

£20000
[£100000 – £80000]

£20000
[£100000 – £80000]
[current profit – development cost]

£100000
[This assumes that the manufacturer will keep his current market share.]

The calculations involved at each stage are shown on the diagram.

The aim is now to work out which decisions will bring the best return for the manufacturer and so we calculate the **expected monetary value** (EMV) at each stage in the decision-making process. We start at the end of the decision tree and work backwards, calculating the EMV at each chance vertex, then taking the best outcome at each decision vertex to carry back to the next stage.

The figure written in the chance vertex is the EMV at this point. It is found from $(115\,000 \times 0.75) + (-5000 \times 0.25)$.

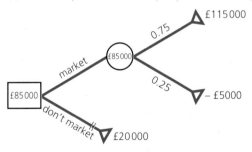

At this stage we choose the best option for the manufacturer. The option you will not choose (in this case 'don't market') is scored through with a double line (//).

We continue by working back to the next vertex, like this.

$(£85\,000 \times 0.95) + (£20\,000 \times 0.05) \longrightarrow$

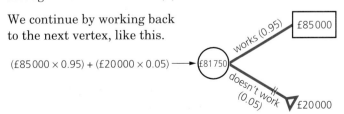

The completed tree looks like this.

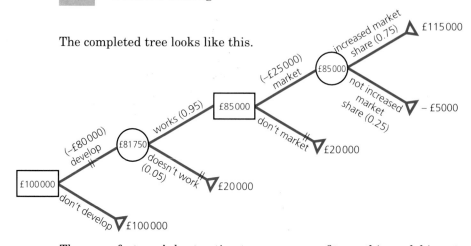

The manufacturer's best option to preserve profits on this model is not to develop the product. Taking costs into account, it would reduce the company's profit from £100 000 to £81 750 if it developed the new idea.

It would appear from our results that it would be better for the manufacturer to continue with the present product rather than take a risk on developing the new product.

This is, of course, very much simplified and there will be other factors to be taken into account, such as the potential of competing companies to develop improvements of their own and thus take some of our manufacturer's existing market share.

## EXERCISES

1   For the situation given in Exploration 13.2, suppose the sales representative points out that their main competitor is also likely to improve their current product and, if this is the case, the company may well lose some of its current market share. He estimates that there is a 60 per cent chance that the profits could be halved. By adding another chance vertex on the 'don't develop' option, calculate the effect of this on the manufacturer's EMV. How will this affect the decision?

2   In a game of chance, the player puts down a £1 stake, then rolls two dice. If he throws a 6 he wins £3 and his stake is returned. If he throws a 1 or a 3, his stake is returned, and if he throws 2, 4 or 5 he loses his stake. Copy and complete this decision tree to find his EMV for the game.

3   In a game, the player rolls two balls down a ramp so that each ball finally settles in one of three slots and scores either three points or eight points as shown.

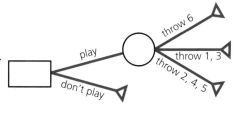

Both balls may settle in one slot, and it is assumed that a ball is equally likely to roll into any of three slots. The player pays 10p to take part in the game, and she wins the total points scored by her two balls.

a) What are the possible totals that can be scored by the two balls?
b) Copy and complete the decision tree to show the EMV of the game.

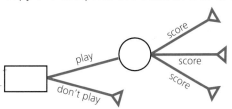

4   The organiser of a golf tournament has to decide whether to take out insurance against bad weather. For a premium of £1000, she will receive a pay-out of £8000 if it rains on the day of the tournament. The weather forecast estimates that there is a 30 per cent chance of rain on the day of the tournament. The organiser calculates that if it rains she will lose £9500, but if it is fine she will make £7500.

Complete this decision tree and work out the best course of action for the organiser.

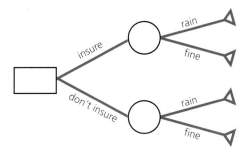

5   A computer software company develops a new game and has to decide whether to launch it immediately or run a trial to see whether it needs further development. If the company decides to launch the game immediately, it may also decide to withdraw one of its older products as previous experience suggests that this will make the new game sell better. The costs for the various options are shown in this table.

| Launch now | Trial |
|---|---|
| **Withdraw old game** | |
| • cost: £25 000 | • chance of success 0.75 |
| • chance of success 0.7 | • estimated profit £130 000 |
| • estimated profit £130 000 | |
| **Don't withdraw old game** | |
| • chance of success 0.5 | |
| • estimated profit £100 000 | |

Use the information to draw a decision tree. What is the maximum amount the company could spend on a trial before this option becomes economically non-viable?

**EXERCISES**

13.2 B

1   **a)** It costs £20 to enter an examination in mathematics and there is a 50:50 chance that Sarah will pass it. If she is successful then she will be given a £36 award. Construct a decision tree and determine Sarah's best course of action.

    **b)** A book of past papers for this examination is now produced which costs £8. If Sarah works diligently through the questions contained in the book her chance of passing is now 80:20. What should Sarah's strategy be now?

2   **a)** A company that mines coal is looking to open new pits in a region which has a number of existing pits. If a coal seam is found at a site the company can expect to make a net profit of £1 000 000 but if no coal is struck it makes a net loss of £400 000. Based on the result that 25 per cent of explorations in the region yield coal, what advice would you give the company on sinking a pit on a new site in the region?

    **b)** The company also has the option of commissioning a geological survey of the site which will cost £200 000. This survey will indicate either probable or improbable conditions. Construct the decision tree diagram to show all the company's options.
A survey indicating the probability of coal will give an 80 per cent chance of finding coal whereas an improbable survey gives only a ten per cent chance of finding coal. A likelihood of a survey indicating the probability of coal is estimated at 50 per cent. Enter the information on your decision tree diagram and determine the company's best strategy.

3   The following game has been devised. You toss two coins and if they both land on heads you win £1.20 plus return of your stake of £1, if they land one head and one tail you lose your stake but if you obtain two tails then you can choose either to have your stake returned or double your stake and toss the coins again. If two heads are tossed you win £2 plus return of your full stake, otherwise you lose your stake.

Construct a decision tree and enter the probabilities and EMVs on the decision tree. What is your strategy when playing this game?

4   Two choices of an electrical item are available, a cheap unbranded one which carries no guarantee at £8 or a named make at £12. This item is less likely to fail and if it does the guarantee means that I have to pay £1 postage to return it to the factory to have it replaced whereas if the cheaper model fails then I have to purchase another at the same price.

    **a)** What strategy should I adopt if records show that 25 per cent of the cheaper brand fail but only two per cent of the more expensive brand do?

    **b)** Determine the probability of the cheaper item failing if my strategy in part **a)** is to be changed.

    **c)** What should be the cost of the cheaper item if my strategy in part **a)** is to be altered, given that the probabilities of failure are the same as in part **a)**?

**5** A company is considering whether to launch a new product. The success of the idea depends on the ability of a competitor also to launch a similar product.

There is a 70 per cent chance that a competitor will do this. The table below gives the profits for each price range that could be set by the company related to the possible competing prices.

| Profits (units of £100 000) | | | | |
|---|---|---|---|---|
| | Competitor's price | | | |
| Company's price | Low | Medium | High | No competitor |
| Low | 3 | 4 | 4.5 | 5 |
| Medium | 4 | 4.5 | 5 | 7 |
| High | 1 | 3 | 5.5 | 9 |

If the company reaches the market first then it must reset the price so that the competitor is able to react to the price. Estimates of the probability of a competitor's price are given below . Recommend what the company's strategy should be.

| | Competitor's price expected to be | | |
|---|---|---|---|
| Company's price | Low | Medium | High |
| Low | 0.75 | 0.2 | 0.05 |
| Medium | 0.2 | 0.6 | 0.2 |
| High | 0.1 | 0.4 | 0.5 |

## BOOLEAN ALGEBRA

When John switches on his graphic calculator, the screen shows the same information that was on it when he switched it off. Electronic circuits inside the machine enable it to retain information (provided the batteries don't go flat).

Throughout this book we have frequently talked of the importance of computer technology in the application and development of algorithms. The branch of Mathematics we now call *Decision mathematics* has developed rapidly since the advent of computers and we could argue that without this development, many of the algorithms that are now widely used in business would have remained little more than recreational maths puzzles, simply because of the time it takes to apply such algorithms by hand. If a technique is to be viable in business, its cost-effectiveness is all important and a method that takes a long time to apply would not be viable.

*'There is no branch of mathematics, however abstract, which may not some day be applied to the phenomena of the real world.'*
*(Nikolai Lobachevsky 1792 – 1856)*

Computers are one of the many electronic devices that we now take for granted, yet their development would not have been possible without one of the most abstract branches of mathematics: logic. A lesson to be learned: never assume a branch of mathematics is useless!

Machines such as computers handle
discrete data and almost always
operate in binary (base 2). The
circuits and networks which
describe their operation are called
**switching circuits**. These were
developed using the work of the
British logician George Boole. He
is considered to have been the
founder of symbolic logic, in which
mathematical notation is applied to
questions of logic, and Boolean algebra
is named after him. **Boolean variables**
can only take two values, often given as
true or false, 0 and 1 or off and on.

***George Boole (1815–64)***

## Logical propositions

A proposition is a statement which is either true or false, but cannot
be both. For example:

$(x + 4)(x - 4) = x^2 - 16$     **true**

$2 + 2 + 2 = 0$     **false**

This means that questions such as 'How are you?' are not acceptable
since no truth value can be assigned to them.

In language we do not construct every sentence as a list of statements,
but link things together using connectives or conjunctions:

■ Jill lives in Birmingham *and* Jenny lives in Leeds.

This contains two propositions and, for the statement to be true, both
of the constituent statements must be true. If we consider the ways of
linking statements that we use in everyday language, we find we use
the same concepts in logic, and we have symbols to represent them.

### Negation

Let $p$ denote a simple proposition, then the negation of $p$ is 'not $p$' and
is written $\sim p$. If $p$ is true, then $\sim p$ will be false, and if $p$ is false, then
$\sim p$ will be true. For example if $p$ denotes the statement 'crocuses
flower in spring' (true) then $\sim p$ denotes 'crocuses do not flower in
spring' (false).

### Conjunction

The most common conjunction used in speech is ***and***. This is
represented in symbolic logic by $\wedge$. If $p$ and $q$ denote two simple
statements then $p \wedge q$ is true if and only if both $p$ and $q$ are true.

This can be represented in a **truth table**, like this.

| $p$ | $q$ | $p \wedge q$ |
|---|---|---|
| T | T | T |
| T | F | F |
| F | T | F |
| F | F | F |

## Disjunction

A disjunction is represented in speech by *or* and in symbolic logic by $\vee$. If $p$ and $q$ denote two simple statements then $p \vee q$ is true if $p$ or $q$ is true. In this case the truth table looks like this. Note the situation where both are true.

| $p$ | $q$ | $p \vee q$ |
|-----|-----|------------|
| T | T | T |
| T | F | T |
| F | T | T |
| F | F | F |

## Condition

A condition is a proposition of the form '$p$ implies $q$' and is represented by the symbol $\Rightarrow$. The truth table for this is like this.

| $p$ | $q$ | $p \Rightarrow q$ |
|-----|-----|-------------------|
| T | T | T |
| T | F | F |
| F | T | T |
| F | F | T |

**Note:** $p \Rightarrow q$ is only false if a false conclusion $q$ is drawn from a true statement $p$.

## Double condition

The double condition $\Leftrightarrow$ is a way of saying '$p$ implies $q$ which implies $p$'. This is rather cumbersome so we usually say 'if and only if' which is abbreviated to *iff*.

Consider these statements.
1   School sports day will be cancelled if it rains.
2   School sports day will be cancelled only if it rains.
3   School sports day will be cancelled if, and only if, it rains.

Let $p$ be the statement 'School sports day will be cancelled.'

Let $q$ be the statement 'It rains.'

Statement **1** says rain will lead to the cancellation of sports day, so we can say $q \Rightarrow p$ (or $p \Leftarrow q$, '$p$ is implied by $q$').

We say rain is a **sufficient condition** for cancellation of sports day.

Statement **2** says if sports day is cancelled, then it must be raining since no other reason for cancellation is given, so $p \Rightarrow q$.

We say rain is a necessary condition for cancellation.

Statement **3** says both of these, so $p \Leftrightarrow q$.

We say rain is a **necessary and sufficient condition** for the cancellation of sports day.

The truth table for ⇔ looks like this.

| p | q | p ⇔ q |
|---|---|-------|
| T | T | T |
| T | F | F |
| F | T | F |
| F | F | T |

### Compound propositions

When more than one connective is needed to express the proposition symbolically we can draw combined truth tables.

**Example 13.4**

*Draw the truth table for $(p \wedge \sim q) \Rightarrow q$.*

### Solution

| p | q | ~q | p ∧ ~q | (p ∧ ~q) ⇒ q |
|---|---|----|--------|--------------|
| T | T | F  | F      | T            |
| T | F | T  | T      | F            |
| F | T | F  | F      | T            |
| F | F | T  | F      | T            |

## Electronics

Logic gates are widely used in computers and other electronic devises. In this case, the variables $p$ and $q$ represent inputs which can be either ON (1) or OFF (0). A simple logic gate is shown in the diagram.

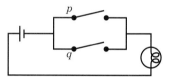

For the bulb to light, both switches must be on, so this is an AND gate which means that $p$ and $q$ must both be on to get an output. We can draw a truth table which describes the behaviour of this system.

| p | q | p ∧ q |
|---|---|-------|
| 0 | 0 | 0 |
| 0 | 1 | 0 |
| 1 | 0 | 0 |
| 1 | 1 | 1 |

The circuit can be represented on a simplified diagram like this.

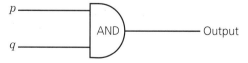

Other types of gates, corresponding to the connectives used in symbolic logic also have circuit symbols.

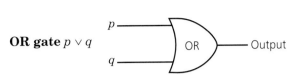

| *p* | *q* | *p* ∨ *q* |
|---|---|---|
| 0 | 0 | 0 |
| 0 | 1 | 1 |
| 1 | 0 | 1 |
| 1 | 1 | 1 |

**OR gate** *p* ∨ *q*

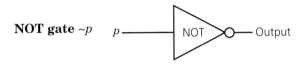

Output

**NOT gate** ~*p*

Output

**NAND gate
(not + and)
~(*p* ∧ *q*)**

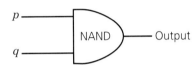

Output

| **p** | **q** | ~(**p** ∧ **q**) |
|---|---|---|
| 0 | 0 | 1 |
| 0 | 1 | 1 |
| 1 | 0 | 1 |
| 1 | 1 | 0 |

**NOR gate
(not + or)
~(*p* ∨ *q*)**

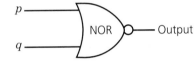

Output

| **p** | **q** | ~(**p** ∨ **q**) |
|---|---|---|
| 0 | 0 | 1 |
| 0 | 1 | 0 |
| 1 | 0 | 0 |
| 1 | 1 | 0 |

*A personal computer
contains about half a
million NOR gates.*

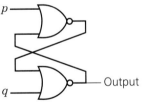

Output

These gates can be connected together in many ways. One of these is called a **bistable**, which can be made of two NOR gates as shown in the diagram. (Remember, NOR is an abbreviation of 'not or', which would be stated symbolically as ~(*p* ∨ *q*). Its output will only be 1 (T) if both inputs are 0).

The bistable forms the basis for computer memories, the sequence of diagrams shows how a bistable based on two NOR gates works.

If the input at *p* is 1 then *x* must be 0 regardless of the value of *y*, so if *q* is 0, the output will be 1 (and hence *y* will also be 1).

If the input at *p* is 0 and the output at *x* is 0, then *y* must be 1 (if *y* were 0, then *x* would also be 0), so *q* must be 0 and the output is also 1.

In other words, changing the state of *p* alone does not alter the output.

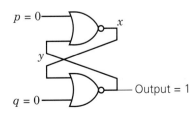

If $p$ is left at 0 and $q$ is now 1, $x$ becomes 1 and $y$ is 0, so the output is 0. The bistable has been cleared.

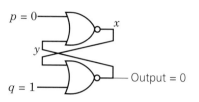

If we switch q back to 0, it will not alter the output which is still 0.

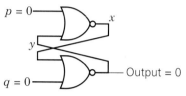

The important point about the bistable is that it 'remembers' whether $p$ or $q$ was last in the logic state 1 and the outcome depends on this. The truth table for a bistable is like this.

| $p$ | $q$ | output |
|-----|-----|--------|
| 1 | 0 | 1 |
| 0 | 0 | 1 |
| 0 | 1 | 0 |
| 0 | 0 | 0 |

The output is stable in two states (hence the name bistable) and we say the bistable remembers one **binary digit**, or **BIT** of information.

John's graphic calculator has several thousand bytes of memory (a byte is 1024, or $2^{10}$ bits), so as long as there is a source of power, it will retain the information he wants it to. A computer has several million bytes of memory which it uses while the machine is on, though it will lose any information not stored when the power is switched off.

## EXERCISES

13.3 A

1   Let $p$ be the proposition 'Ali studies Mathematics'.
    Let $q$ be the proposition 'Ali studies German'.
    Let $r$ be the proposition 'Ali studies Art'.
    Write the following propositions in words.
    **a)** $p \wedge q$          **b)** $p \wedge \sim r$
    **c)** $p \wedge q \wedge \sim r$          **d)** $p \wedge \sim(q \vee r)$

2   Construct a truth table to show $p \Rightarrow \sim q$.

3   Construct a truth table to show $(p \vee q) \Leftrightarrow (\sim p \wedge q)$.

4   The diagram below represents the logic function $F(p, q) = (\sim p \vee q)$.

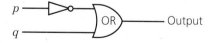

    Draw a table to show the possible outputs for the circuit.

5   A bistable can also be constructed from NAND gates. Draw a bistable with two NAND gates and construct the truth table for it.

**EXERCISES**

**13.3 B**

1   Given that $x$ is a real number, consider the statements
    $r : x = 2$      $s : x^2 = 4$.

a)  Discuss whether the following are true or not.
    $r$ is a necessary condition for $s$.
    $r$ is a sufficient condition for $s$.
b)  Repeat the analysis for new statements
    $r : x < 1$      $s : x^2 < 1$

*$\sim q \Rightarrow \sim p$ is called the*
***contra-positive.***

2   By constructing truth table show that $p \Rightarrow q$ is not equivalent to its converse, $q \Rightarrow p$, but is equivalent to $\sim q \Rightarrow \sim p$.

3   In a switching circuit, $p'$ represents 'not $p$' so when $p$ is open $p'$ is closed and vice versa.

a)  Draw a circuit represented by $(a \vee b \vee c') \wedge (a' \vee b \vee c) \wedge (b' \wedge a) \vee c)$.
b)  Derive an algebraic expression for the switching circuit below. Discuss why it is intuitively acceptable that it could be simplified to $b \wedge c$.

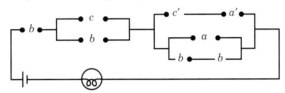

Construct a truth table for the above and show it is logically equivalent to $b \wedge c$.
c)  Using AND, OR and NOT gates, find an equivalent diagram to the switching circuit above.

4   Consider the network below with inputs $x_1$ and $x_2$ and outputs $s$ and $c$.

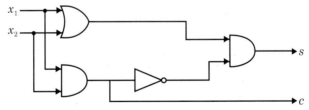

a)  Write down Boolean expressions $c$ and $s$ and use truth tables to show $s$ is equivalent to $x_1' x_2 + x_1 x_2'$.

*The device is called a*
*half-adder.*

b)  If $x_1$ and $x_2$ are binary digits, by considering the truth table for $s$ and $c$, what do you think $s$ and $c$ represent?

5   a)  A corridor is illuminated by a strip lighting element. Design a circuit so that the light is independently controlled by any one of three light switches placed respectively at the two ends and in the middle of the corridor. Write down the equivalent Boolean expression and the corresponding truth table.
    b)  Repeat the above stages for a circuit in which a light is lit when a majority occurs when a panel of three people vote by pressing a switch.

## MATHEMATICAL MODELLING ACTIVITY
### Logic and proof

*Specify the real problem*

## Problem statement

We have the following statements:

If Jones met Smith last night then Smith was the murderer. The murder took place before midnight if Jones is telling the truth. If Jones did not meet Smith last night the inescapable conclusion is that Jones is telling the truth. The murder did take place at or after midnight hence Smith was the murderer.

The problem is to determine whether the above constitutes a valid argument.

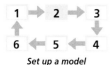

*Set up a model*

## Set up the model

The first step is to identify the simple statements such as:
   $p$ : Jones met Smith last night.

Define each simple statement as above and express the sentences in the introduction in terms of your defined symbol:
   $\Rightarrow \sim$ (not)

## Mathematical solution

A mathematical argument is the process of starting with statements called premises $P$ and deducing another called the conclusion $C$. For an argument to be valid the implication $P \Rightarrow C$ must be true for all possible outcomes. A valid argument is a proof.

A **tautology** is a statement that, if the truth values are always, true shows that:

$$p \wedge (p \Rightarrow q) \Rightarrow q,$$

For example, $(p \Rightarrow q) \wedge (q \Rightarrow r) \Rightarrow (p \Rightarrow r)$ are tautologies.

Two of the more common standard methods of proof are therefore **detachment** where $p$ and $p \Rightarrow q$ are premises and the valid conclusion is $q$, and **syllogism** where $p \Rightarrow q$ and $q \Rightarrow r$ are premises and the valid conclusion is $p \Rightarrow r$.

Using the methods of proof above analyse the statements in the introduction (you may find the result of Question 2 of Exercises 13.3B useful).

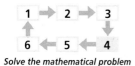

*Solve the mathematical problem*

## Continuation

Consider the following argument by Lord Bridgewater who lived in the 18th century and was a big supporter of canal transport.

> *The development of steam locomotion will surely result in a vast rail system. But for the continued prosperity of the canals we must do without such a rail system. Surely the love of Lord Bridgewater does guarantee the ever-continuing prosperity of the canals. Thus, we may conclude that steam locomotion will not develop further.*

Show that there is nothing invalid in the above argument. Discuss why the above was not patently true.

## CONSOLIDATION EXERCISES FOR CHAPTER 13

1  A car number plate has a letter, followed by three digits and then three more letters e.g. K214 TWU.

   a) Calculate the number of possible arrangements of three digits from ten if repetition is allowed.
   b) Calculate the number of possible arrangements of three letters from 26 if repetition is allowed.
   c) How many number plates of this type are possible?

2  The draw for the quarter finals of the Newshire Sunday league football cup is to be made. Eight counters, numbered 1 to 8, representing the teams are placed in a bag and drawn one at a time.

   a) In how many different orders can the counters be drawn from the bag?
   b) In how many different ways can the counters be drawn so that teams 1 and 2 play each other?

3  a) An examination paper has six questions. The candidate must answer four questions, in any order. In how many ways can the candidate choose four questions?
   b) A second paper is arranged in two sections with three questions in each section. The candidate must answer three questions, choosing at least one question from each section. In how many ways can the candidate choose the questions?

4  There are $n$ ($\geq 4$) identical items in a row and they are to be split into four groups, each consisting of one or more item. The first few in the row will form the first group, the next few will form the second group, and so on, for example:
   00│000│0│0000

   By considering where the breaks between groups must be drawn, explain why the row can be split into four such groups in $\binom{n-1}{3}$ ways.

   Use the result to answer the following questions.
   a) In how many different ways can 30 identical sweets be shared out amongst four children so that each child gets at least one sweet?
   b) By first giving each of the children five sweets, or otherwise, calculate the number of ways in which 50 identical sweets can be shared out amongst four children so that each child gets at least six sweets.
   c) By first taking a sweet from each child, or otherwise, calculate the number of ways in which 30 identical sweets can be shared out amongst four children, where this time some children may get no sweets.

   *(AEB Question 7, Specimen Paper, 1996)*

5 An oil prospector is drilling for oil in an oil-bearing region. So far she has explored ten sites and found oil in three of them. If oil is found at a site she makes a net profit of £5 million, but if no oil is found she makes a net loss of £3 million.

a) The prospector is given the chance to explore a new site. Based on the available evidence, would you advise her to drill or not? Give reasons.

b) The prospector also has the option of ordering a geological survey of the new site. The survey will indicate either promising or unpromising conditions.
Complete the decision tree diagram to show all of the oil prospector's choices.

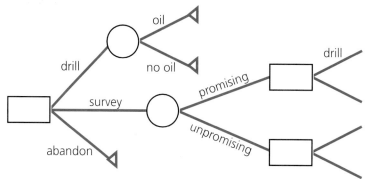

c) The survey will cost £0.5 million. A promising survey will give a 70 per cent chance of finding oil, an unpromising survey a 20 per cent chance of finding oil. The probability of a promising survey is estimated to be 50 per cent. Show the information on the decision tree diagram, and complete the analysis to find the oil prospector's best strategy. State the best strategy, giving reasons.

*(MEI Question 4, D & D, Specimen Paper)*

6 The organiser of a tennis tournament has to decide whether to take out pluvious insurance (that is insurance against lost ticket sales due to rain on the day of the tournament). She estimates that on a fine day the tournament will make a profit of £3000, but on a wet day a loss of £7000. The pluvious insurance will cost £1000 and will pay out £8000 if it rains on the day of the tournament. The probability of rain is estimated to be 0.2.

a) This is illustrated below as a decision tree. Fill in the returns, the probabilities and the expected monetary values (EMVs) on the decision tree.

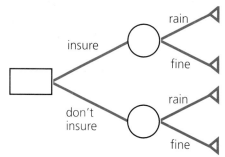

Advise her, with reasons, on whether or not she should take out the insurance.

Madam Tarot, the local fortune teller, has a reputation for success in weather forecasting. In the past if she has said that it would rain then there was a probability of 0.5 of rain, and if she has said that it would not rain, then there has been a probability of 0.1 of rain. A quarter of her forecasts have been for 'rain' and three-quarters for 'no rain'.

In the diagram the decision tree has been extended in order to analyse whether or not the organiser should pay for a consultation with Madame Tarot.

**b)** Enter the probabilities and EMVs on the decision tree.

**c)** How much is it worth paying for a consultation with Madame Tarot?

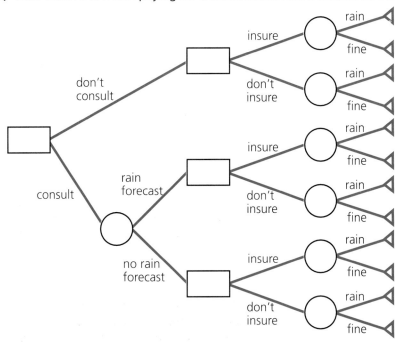

*(MEI Question 4, D & D, January 1995)*

7 In a game of chance, a player pays a £1 stake then picks a card from a shuffled pack. The pay-off is as follows.

- If he picks a jack, his stake is refunded.
- If he picks a queen, his stake is refunded and he wins £1.
- If he picks a king, his stake is refunded and he wins £2.
- If he picks an ace, his stake is refunded or he may choose to roll a dice and if he scores 6 his stake is refunded and he wins £4, otherwise he loses.

Draw a decision tree and use it to work out a strategy for the game.

**8** Let $p$ represent 'I work hard' and $q$ 'I pass my exams'.

   **a)** Express in terms of $p$ and $q$ the proposition 'I work hard so I will pass my exams'.

   **b)** Interpret the proposition $\sim p \wedge q$.

   **c)** If $r$ represents 'I take regular exercise', interpret the proposition:
$(p \wedge \sim r) \Rightarrow \sim q$.

**9** Construct truth tables for the following propositions.

   **a)** $\sim(p \Rightarrow (q \Rightarrow p))$         **b)** $p \Rightarrow (p \wedge \sim(q \wedge r))$

**10** **a)** Show, by constructing truth tables or otherwise, that the following statements are equivalent.
$$p \Rightarrow q \text{ and } \sim(\sim(p \wedge q) \wedge p)$$

   **b)** With the aid of (a), or otherwise, construct combinatorial circuits consisting only of NAND gates to represent the following functions.
$$f(x, y) = x \Rightarrow y \text{ and } g(x, y) = \sim(x \Leftrightarrow y)$$

*(AEB Question 6, Specimen Paper, 1996)*

---

## Summary

*After working through this chapter you should*

- *be able to solve problems including arrangements, permutations and combinations*

- *be able to draw a decision tree and use it to calculate the expected monetary value*

- *appreciate that decision-making problems include an element of uncertainty*

- *understand the ideas of Boolean logic, know the notation and be able to construct simple truth tables*

- *know how the ideas of Boolean algebra can be applied to electronic circuits*

- *know the symbols for AND, OR, NOT, NOR and NAND gates and be able to construct tables of outputs for two inputs.*

# Answers

## CHAPTER 1
### Algorithms

### Exercises 1.1.A (p. 6)

1  $x = \frac{5}{4} \pm \frac{\sqrt{17}}{4}$

2  **a)**  
| | |
|---|---|
| $a$ | 105  75  30  15 |
| $b$ | 180  105  75  30 |
| $R$ | 1  1  2  2 |
| $S$ | 75  30  15  0 |

HCF of 105 and 180 is 15.

**b)**  4 iterations

**c)**  
| | | |
|---|---|---|
| $a$ | 180  105 | 1st iteration exchanges a and b |
| $b$ | 105  180 | thus it gives the solution in 5 |
| $R$ | 0  :  | iterations |
| $S$ | 105  : | |

3  The simplest (though not the most efficient) is to walk on the right-hand side of a passage (or the left) and turn right (or left) at every junction or dead end. This is a quicker route.

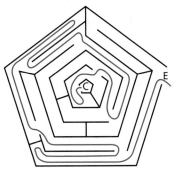

4  **a)**
| A | B | C |
|---|---|---|
| 1 | 1 | 2 |
| 1 | 2 | 3 |
| 2 | 3 | 5 |
| 3 | 5 | 8 |
| 5 | 8 | 13 |
| 8 | 13 | 21 |

**b)** The algorithm generates the first five terms of the Fibonnaci sequence.

### Exercises 1.1B (p. 7)

1  $x = -\frac{1}{5} \pm \frac{\sqrt{61}}{5}$

2  
| | | | | | | |
|---|---|---|---|---|---|---|
| $a$ | 135 | 85 | 50 | 35 | 15 | 5 |
| $b$ | 220 | 135 | 85 | 50 | 35 | 15 |
| $c$ | 1 | 1 | 1 | 1 | 2 | 3 |
| $d$ | 85 | 50 | 35 | 15 | 5 | 0 |

**a)** HCF of 135 and 220 is 5.

**b)** 6 iterations

3  **a)** $x = 0.441$

### Exercises 1.2A (p. 13)

1  Set 1 bubble sort
```
8 7 6 5 4 3 2 1
7 6 5 4 3 2 1 2
6 5 4 3 2 1 3 3
5 4 3 2 1 4 4 4
4 3 2 1 5 5 5 5
3 2 1 6 6 6 6 6
2 1 7 7 7 7 7 7
1 8 8 8 8 8 8 8
```

Set 1 quicksort
```
8 7 6 5 4 3 2 1
7 6 5 4 3 2 1 2
6 5 4 3 2 1 3 3
5 4 3 2 1 4 4 4
4 3 2 1 5 5 5 5
3 2 1 6 6 6 6 6
2 1 7 7 7 7 7 7
1 8 8 8 8 8 8 8
```

Set 2 bubble sort
```
1 1 1 1
3 3 3 2
5 5 2 3
7 2 4 4
2 4 5 5
4 6 6 6
6 7 7 7
8 8 8 8
```

Set 2 quicksort
```
1 1 1 1 1
3 3 2 2 2
5 5 3 3 3
7 7 5 4 4
2 2 7 5 5
4 4 4 7 6
6 6 6 6 7
8 8 8 8 8
```

Set 3 bubble sort
```
5 1 1 1 1
1 4 4 3 2
4 5 3 2 3
7 3 2 4 4
3 2 5 5 5
2 7 6 6 6
8 6 7 7 7
6 8 8 8 8
```

Set 3 quicksort
```
5 1 1 1 1
1 4 4 3 2
4 3 3 2 3
7 2 2 4 4
3 5 5 5 5
2 7 6 6 6
8 8 7 7 7
6 6 8 8 8
```

Summary

| Set | bubblesort | | quicksort | |
|---|---|---|---|---|
| 1 | $c = 28$ | $e = 28$ | $c = 28$ | $e = 28$ |
| 2 | $c = 21$ | $e = 7$ | $c = 19$ | $e = 3$ |
| 3 | $c = 20$ | $e = 11$ | $c = 5$ | $e = 9$ |

2  Bubblesort
```
50 48 48 30 12 4  4  4  4
48 50 30 12 4  12 12 12 12
76 30 12 4  28 28 28 28 21
30 12 4  28 30 30 30 21 28
12 4  28 48 48 48 21 30 29
4  28 50 50 50 21 48 29 30
28 56 56 56 21 50 29 41 41
56 63 63 21 56 29 41 48 48
63 76 21 57 29 41 50 50 50
77 21 57 29 41 56 56 56 56
21 57 29 41 57 57 57 57 57
57 29 41 63 63 63 63 63 63
29 41 76 76 76 76 76 76 76
41 77 77 77 77 77 77 77 77
```

Quicksort

```
50 48 30 12 4  4  4
48 30 12 4  12 12 12
76 12 4  28 28 21 21
30 4  28 21 21 28 28
12 28 21 29 29 29 29
4  21 29 30 30 30 30
28 29 41 41 41 41 41
56 41 48 48 48 48 48
63 50 50 50 50 50 50
77 76 56 56 56 56 56
21 56 63 63 57 57 57
57 63 57 57 63 63 63
29 77 76 76 76 76 76
41 57 77 77 77 77 77
```

**3** Solution by bubblesort

```
14 25 25 25 25 25 25 25
25 14 14 14 14 14 14 14
12 12 12 31 31 31 31 31
31 31 31 12 18 18 18 18
18 18 18 18 12 13 13 13
13 13 13 13 13 12 20 20
20 20 20 20 20 20 12 28
28 28 28 28 28 28 28 12
```

```
25 31 31 31 31 31 31
31 25 25 25 28 28 28
18 18 20 28 25 25 25
14 20 28 20 20 20 20
20 28 18 18 18 18 18
28 14 14 14 14 14 14
13 13 13 13 13 13 13
12 12 12 12 12 12 12
```

**4 a)**

| 28 | 20 | 13 | 18 | 31 | 12 | 25 | 14 | *c* | *e* |
|----|----|----|----|----|----|----|----|-----|-----|
| 12 | 20 | 13 | 18 | 31 | 28 | 25 | 14 | 7 | 1 |
| 12 | 13 | 20 | 18 | 31 | 28 | 25 | 14 | 6 | 1 |
| 12 | 13 | 14 | 18 | 31 | 28 | 25 | 20 | 5 | 1 |
| 12 | 13 | 14 | 18 | 31 | 28 | 25 | 20 | 4 | 0 |
| 12 | 13 | 14 | 18 | 20 | 28 | 25 | 31 | 3 | 1 |
| 12 | 13 | 14 | 18 | 20 | 25 | 28 | 31 | 2 | 1 |
| 12 | 13 | 14 | 18 | 20 | 25 | 28 | 31 | 1 | 0 |

**b)** $c = 28$, $e = 5$

**c)** Comparisons – same as bubble sort but fewer exchanges therefore possibly more efficient on short list.

---

## Exercises **1.2B** *(p. 13)*

**1** Set 1

**a)**
```
4  5  6  7  8  9 10
 5  6  7  8  9 10  9
 6  7  8  9 10  8  8
 7  8  9 10  7  7  7
 8  9 10  6  6  6  6
 9 10  5  5  5  5  5
10  4  4  4  4  4  4
```

**b)**
```
4  5  6  7  8  9 10
 5  6  7  8  9 10  9
 6  7  8  9 10  8  8
 7  8  9 10  7  7  7
 8  9 10  6  6  6  6
 9 10  5  5  5  5  5
10  4  4  4  4  4  4
```

Set 2

**a)** Bubblesort    **b)** Quicksort
```
9 9 9 9        9 9 9 9 9
7 7 7 8        7 7 8 8 8
5 5 8 7        5 5 7 7 7
3 8 6 6        3 3 5 6 6
8 6 5 5        8 8 3 5 5
6 4 4 4        6 6 6 3 4
4 3 3 3        4 4 4 4 3
2 2 2 2        2 2 2 2 2
```

Set 3

**a)** Bubblesort    **b)** Quicksort
```
6 8 8 8 8      6 8 8 8 8
8 6 6 7 7      8 7 7 7 7
2 3 7 6 6      2 6 6 6 6
3 7 4 4 5      3 2 3 4 5
7 4 3 5 4      7 3 4 5 4
4 2 5 3 3      4 4 5 3 3
1 5 2 2 2      1 1 2 2 2
5 1 1 1 1      5 5 1 1 1
```

**2** Bubblesort
```
41 41 57 57 77 77 77 77 77 77 77 77 77 77
29 57 41 77 63 63 63 63 63 63 63 76 76
57 29 77 63 57 57 57 57 57 76 63 63
21 77 63 56 56 56 56 56 76 57 57 57
77 63 56 41 41 41 41 76 56 56 54 54
63 56 29 29 29 30 76 48 48 48 48 48
56 28 28 28 30 76 48 50 50 50 50 50
28 21 21 30 76 48 50 41 41 41 41 41
4  12 30 76 48 50 30 30 30 30 30 30
12 30 76 48 50 29 29 29 29 29 29 29
30 76 48 50 28 28 28 28 28 28 28 28
76 48 50 21 21 21 21 21 21 21 21 21
48 50 12 12 12 12 12 12 12 12 12 12
50 4  4  4  4  4  4  4  4  4  4  4
```

---

Quicksort

```
57 77 77 77 77 77 77
77 63 76 76 76 76 76
63 76 63 63 63 63 63
53 57 57 57 57 57 57
76 53 53 53 53 53 53
48 48 48 50 50 50 50
50 50 50 48 48 48 48
41 41 41 41 41 41 41
29 29 29 29 30 30 30
21 21 21 21 29 29 29
28 28 28 28 21 28 28
4  4  4  4  28 21 21
12 12 12 12 4  4  12
30 30 30 30 12 12 4
```

**3** Bubblesort

```
28 20 13 13 13 12
20 13 18 18 12 13
13 18 20 12 18 14
18 18 12 20 14 18
31 12 25 14 20 20
12 25 14 25 25 25
25 14 28 28 28 28
14 30 30 30 30 30
```

---

## Exercises **1.3A** *(p. 16)*

**1** Full bin

$A + B = 1$ m
$C + E + F = 1$ m
$D + G = 90$ cm
3 racks used

First fit

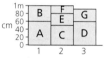

3 racks used
(same solution as full bin)

First fit decreasing

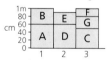

3 racks used
(variation of first two solutions)

**2** Full bin

$I + L = 12$ ft
$J + K = 12$ ft
$A + B + C + D + E = 12$ ft
$F + G + H = 11$ ft
4 lengths of pipe
One 1 ft length left over

First fit

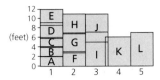

5 lengths of pipe
2 × 1 ft, 1× 6 ft
1 × 5 ft left over

First fit decreasing

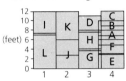

4 lengths of pipe (as full bin)

**3** Full bin e.g.
B + J + P = 800
C + I + E + H + K = 800
L + O + N + G + F = 750
D + M + A = 490
4 discs
It is very hard to find full bin combinations in data such as this.

First fit

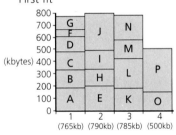

(765kb) (790kb) (785kb) (500kb)

4 discs
Reasonably quick and fairly space efficient.

First decreasing order P J L E N A K M I B O D C H G F

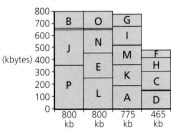

4 discs
Probably best since in terms of space usage, but it takes a long time.

**4** Total breaks time =
180 × 3 = 540 seconds
Total Ads time = 540 seconds
Can be done in theory.
First fit

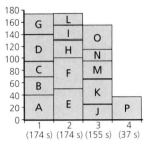

(174 s) (174 s) (155 s)  (37 s)

Comments: Full bin would take ages. Neither FF nor DFG will do it in 3 but DFF could manage if 3rd break was made 5 seconds longer.
First fit decreasing
Reorder: E52, D46, F46, A40, K40, O40, H38, P37, G33, B30, M30, C25, J25, I20, N20, L18

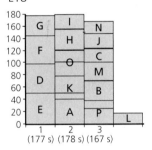

(177 s) (178 s) (167 s)

**5** Sort by size

| Activity | G | J | A | K | B | I | D | F | L | C | E | H |
|---|---|---|---|---|---|---|---|---|---|---|---|---|
| Duration | 9 | 9 | 8 | 8 | 7 | 7 | 6 | 6 | 6 | 5 | 5 | 4 |

Using first fit decreasing, 8 workers are needed.
Using full-bin
A + H = 12
B + C = 12
D + F = 12
E + I = 12      8 workers
G
J
K
L

**Exercise 1.3B** *(p. 18)*

**1** 60 cm : A B C
80 cm : D E F G
120 cm : H
First fit

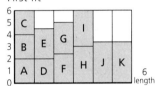

H + A + B = 2.4 m
D + G + F = 2.4 m
G + C = 1.4 m
First fit decreasing

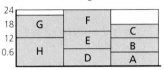

3 racks

**2** 2 m : ABCD
2.5 m : EFG
3 m : HI
3.5 m : JK
First fit

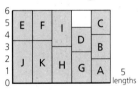

A + B + C = 6
H + I = 6
F + K = 6
D + G = 4.5
First fit decreasing

5 lengths

**3** Too difficult to find full disks.
First fit
A + B + C = 1342
D + E + F + F + H + I = 1312
J + K + L + M + N + O = 947
P + Q = 1204
4 disks – quick and easy

299

First fit decreasing
B + Q = 1351
P + H + L = 1357
A + F + N + E + G = 1363
M + O + D + C + K + J + I = 734
4 disks – more effective use but long time

**4** Total show time = 90 minutes
Total act time = 87 minutes so therefore it is possible to show all acts.
Good use of time
First fit

[bar chart: M W / J H P / D G O / C F L / B E / A K, vertical axis 0, 10, 20, 30]

**5** 5 suitcases

## Consolidation Exercises for Chapter 1 *(p. 20)*

**1** Actual iterations 4, HCF = 15, predicted iterations 10.5
For $a = 233$, $b = 377$, predicted iterations 12, HCF = 1, actual iterations 12

**2**

| 7 | 5 | 2 | 2 | 2 | 1 |
|---|---|---|---|---|---|
| 5 | 2 | 5 | 5 | 1 | 2 |
| 2 | 7 | 6 | 1 | 5 | 5 |
| 8 | 6 | 1 | 6 | 6 | 6 |
| 6 | 1 | 7 | 7 | 7 | 7 |
| 1 | 8 | 8 | 8 | 8 | 8 |

**3** $x = 3$, $y = 41$
**a)**

| $r$ | 41 | 38 | 35 | 32 | 29 | 26 | 23 | 20 | 17 | 14 | 11 | 8 | 5 | 2 |
|---|---|---|---|---|---|---|---|---|---|---|---|---|---|---|
| $q$ | 0 | 1 | 2 | 3 | 4 | 5 | 6 | 7 | 8 | 9 | 10 | 11 | 12 | 13 |

$q = 13$  $r = 2$
repetitions of box 3  13
**b)** $\frac{y}{x} = q$ remainder $r \Rightarrow y = qx + r$
**c)** $x = 3$, $y_1 = 4$, $y_2 = 1$

| $r$ | 4 | 1 | 11 | 8 | 5 | 2 |
|---|---|---|---|---|---|---|
| $q_1$ | 0 | 1 | 1 | 1 | 1 | 1 |
| $q_2$ | | | 0 | 1 | 2 | 3 |

repetitions of box 3  1  $q_1 = 1$, $q_2 = 3$
repetitions of box 6  3  $r = 2$
**d)** First – easier to follow
Second – more efficient (less iterations needed)

**4 a)**

| i = | 1 | 2 | 3 | 4 | 5 |
|---|---|---|---|---|---|
| | 7 | 5 | 1 | 1 | 1 |
| | 5 | 1 | 5 | 3 | 3 |
| | 1 | 7 | 3 | 5 | 5 |
| | 9 | 3 | 7 | 7 | 7 |
| | 3 | 9 | 9 | 9 | 9 |
| | 11 | 11 | 11 | 11 | 11 |
| comparisons | 5 | 4 | 3 | 2 | 1 | $c = 15$ |
| swaps | 3 | 3 | 2 | 1 | 0 | swaps = 9 |

**b)**

| 7 | | 7 | 5 | 1 | | 1 | 1 |
|---|---|---|---|---|---|---|---|
| 9 | | 5 | 7 | 11 | | 3 | 3 |
| 5 | | 9 | 9 | 3 | | 11 | 11 |
| comparisons | | 2 | 1 | | | 2 | 1 |
| swaps | | 1 | 1 | | | 1 | 0 |

Comparisons 6, swaps 3
**c)**

| | i = 1 | 2 | 3 |
|---|---|---|---|
| 5 | 5 | 5 | 1 |
| 7 | 7 | 1 | 3 |
| 9 | 1 | 3 | 5 |
| 1 | 3 | 7 | 7 |
| 3 | 9 | 9 | 9 |
| 11 | 11 | 11 | 11 |
| comparisons | 3 | 3 | 3 |
| swaps | 2 | 2 | 2 |

Total comparisons 15, total swaps 9
There is no advantage to splitting the list.
**d)** You don't need to check the first two pairs on first pass or first pair on second pass. By the third pass the list is guaranteed to be sorted.

**5 (i)** reorder 1.2, 1.1, 0.7, 0.4, 0.4, 0.3, 0.3, 0.2

| 1.2 | | 0.7 | |
|---|---|---|---|
| 1.1 | | 0.4 | 0.4 |
| 0.3 | 0.3 | 0.2 | |

1 m ........... 2 m

**(ii)** Full bin 1.2 + 0.4 + 0.4 = 2 m
0.7 + 0.2 + 1.1 = 2 m
0.3 + 0.3 = 0.6 m

| 1.2 | | 0.4 | 0.4 |
|---|---|---|---|
| 1.1 | | 0.7 | 0.2 |
| 0.3 | 0.3 | | |

1 m ........... 2 m

**6**

| L | U | M | FL | FU | FM |
|---|---|---|---|---|---|
| 0 | 16 | 8 | −60 | 260 | 36 |
| 0 | 8 | 4 | −60 | 36 | −28 |
| 4 | 8 | 6 | −28 | 36 | 0 |

Interval bisection method for finding roots of an equation.

**7 a)** Entries in row $F$ are 18, 20, 16, 22, 16, *.
**b)** The minimum distance is 14.
**8 a)**

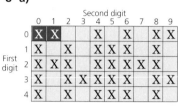

Find the prime numbers up to 49.
**b)**

| | 0 | 1 | 2 | 3 | 4 | 5 | 6 | 7 | 8 | 9 |
|---|---|---|---|---|---|---|---|---|---|---|
| 0 | | | | | 2 | | 2 | | 2 | 3 |
| 1 | 2 | | 2 | | 2 | 3 | 2 | | 2 | |
| 2 | 2 | 3 | 2 | | 2 | 5 | 2 | 3 | 2 | |
| 3 | 2 | | 2 | 3 | 2 | 5 | 2 | | 2 | 3 |
| 4 | 2 | | 2 | | 2 | 3 | 2 | | 2 | 7 |

**9 a) (i)** Choose the middle word each time.
**(ii)** 17
**b) (i)** K4  **(ii)** K6
**(iii)** Choice of either E or F column.  **(iv)** 8
**c)** 9
**10 a)** 3 × 3.5 m
8 × 2 m
2 × 1 m
$4\sqrt{5}$ m (2.3 m)
**b) (i)** 9 × 5 m lengths needed, cost £252
**(ii)** 5 × 5 m @ £28, cost £140; 8 × 2 m @ £12, cost £96; total £236

## CHAPTER 2
# Networks 1: Shortest path

## Exercises 2.1A *(p.32)*

**1 a)** 4 vertices, 5 edges, connected; matches played by football teams
**b)** 10 vertices, 11 edges, not connected; local bus routes
**c)** 7 vertices, 7 edges, connected; teachers and subjects which they teach
**2 a)**

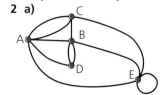

The 2 represents the loop at E which can be travelled in either direction.

**b)**

|   | P | Q | R | S | T |
|---|---|---|---|---|---|
| P | 0 | 1 | 2 | 2 | 1 |
| Q | 1 | 0 | 2 | 0 | 0 |
| R | 2 | 2 | 0 | 0 | 1 |
| S | 2 | 0 | 0 | 2 | 1 |
| T | 1 | 0 | 1 | 1 | 2 |

**3 a)** AZ, ABZ, ABCZ, ACZ, ACBZ
**b)** ABDZ, ABDEZ, ABEZ, ABEDZ, ABCEZ, ABCDZ, ACEZ, ACEDZ, ACDZ, ACDEZ, ACBDZ, ACBEZ, ABCDEZ, ACBEDZ
**c)** ABEH, ABEFH, ABFH, ABCFH, ABCEH, ADGH, ADGFH, ADFH, ADCFH, ADCGH, ABCGH, ACFH, ACEH, ACEFH, ACGH, ACGFH, ADCEH
**5** Number of arcs in $K_n$ is $\frac{1}{2}n(n-1)$.

## Exercises 2.1B *(p.33)*

**1 a)**

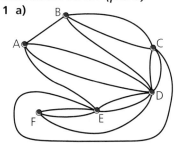

**b)** There are several: ABCEFDA is one example.

**2**

|   | A | B | C | D | E | F |
|---|---|---|---|---|---|---|
| A | 0 | 1 | 0 | 2 | 1 | 0 |
| B | 1 | 0 | 2 | 1 | 0 | 0 |
| C | 0 | 2 | 0 | 2 | 1 | 0 |
| D | 2 | 1 | 2 | 0 | 2 | 1 |
| E | 1 | 0 | 1 | 2 | 0 | 2 |
| F | 0 | 0 | 0 | 1 | 2 | 0 |

**3 a)** ABCS, AS
**b)** ABCS, ABCIS, ABCIJS, AGHKEJS, AGHKEJIS, AGHKEJICS, AFHKEJS, AFHKEJIS, AFHKEJICS, AFEJS, AFEJIS, AFEJICS.
**4 a)** 1 **b)** 4 **c)** $n-1$

## Exercises 2.2A *(p.38)*

**1 a)** SBCT length 20
**b)** SBFT length 93
**c)** 3 possible routes: SADET, SCET, SCFIT all of length 22

---

**2** Route: Bickleigh, Cadbury, Thorverton, Cowley, Exeter; length 24 km
Route: Bickleigh, Silverton, Cowley, Exeter; length 25 km
**3 a)** Ashford – Folkestone – Dover; £12.00 × 4 = £48.00
**b)** Deal – Dover, Canterbury, Maidstone; £21.50 × 4 = £86.00

**4** P × 10
Q × 8
R × 10

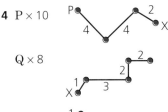

By considering three parts of the network: Use firestation at Q.

**5** Table – shortest distances between pairs of towns.

|   | A | B | C | D | E | F | G | Totals |
|---|---|---|---|---|---|---|---|--------|
| A | – | 10 | 17 | 8 | 13 | 23 | 19 | 90 |
| B | 10 | – | 7 | 9 | 19 | 15 | 20 | 80 |
| C | 17 | 7 | – | 16 | 23 | 6 | 15 | 84 |
| D | 8 | 9 | 16 | – | 10 | 20 | 11 | 74 |
| E | 13 | 19 | 23 | 10 | – | 17 | 11 | 93 |
| F | 23 | 15 | 6 | 20 | 17 | – | 9 | 90 |
| G | 19 | 20 | 15 | 11 | 8 | 9 | – | 82 |

Build sports centre at D, assuming numbers using centre are roughly the same in all villages.

## Exercises 2.2B *(p.39)*

**1 a)** SABT has length 13.
**b)** Three possible routes SBDT, SAEFT, SCFT; all of length 17.
**2** SS–A–B–L–WH 12 miles
**3 a)**

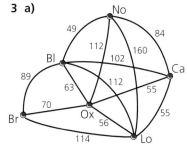

**b) (i)** Bristol – Oxford – Cambridge 125 miles

---

**(ii)** assuming an average speed of 40 mph then Bristol – London – Cambridge = 169 miles might be better assuming no hold-ups round London.
**4 a)** Helsinki to Pori: 280 FIM
**b)** Turku to Joensuu via Tampere and Kuopio: 510 FIM
**5** Lowest costs to:

| Helsinki | £306 |
|----------|------|
| Amsterdam | £68 |
| Copenhagen | £204 |
| Stockholm | £136 |
| Berlin | £272 |
| Frankfurt | £102 |
| Warsaw | £408 |

## Consolidation exercises for Chapter 2 *(p.42)*

**1** Dijkstra: 6 additions, 4 comparisons, 3 subtractions in tracing back
Dynamic programming: 6 additions, 3 comparisons
**2 a)** Quickest route: Manchester, Birmingham, Leicester, Peterborough, Cambridge
Time: 220 minutes
**b)** Manchester, London, Cambridge, 250 minutes
**3** Lowest fare is £1000 along route SBCT.
**4 a)**

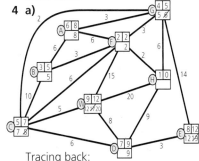

Tracing back:
W — C — G — F — H
Quickest route: HFGCW
Time: 12 mins
**b)** Add 10 to all routes through F and repeat Dijkstra, *or* add 10 to all routes through F – this means that shortest route cannot pass through F so delete F and repeat Dijkstra.
**5 a)** Shortest route ACEJ, length 58
**b)** Add 10 to all routes from E (possibly delete E as this will

mean shortest route does not pass through E) and repeat Dijkstra.

**6 a)** No: Dijkstra's algorithm will not cope with negative values on edges.

**b)** Route SBFT – break even.

**7 a) (i)** Maidstone, Canterbury, Dover: £14 per kg, cost £56
**(ii)** Deal, Dover, Canterbury, Rochester: £19 per kg, cost £76

**b)**

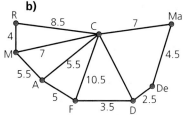

Maidstone, Ashford, Folkestone, Dover: £14 per kg, same price, different route

Deal, Dover, Canterbury, Rochester: £17.50 per kg, cheaper, same route

**8 a)**

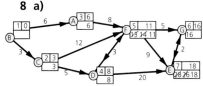

Lowest cost 18
Cheapest route BCDFGE
By tracing back: label – weight on arc = previous label

**b)**

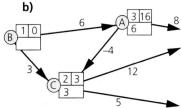

Dijkstra fails because C is labelled at stage 2 so the cheapest route to C (BAC cost 2) is not found.
The algorithm assumes that extra arcs always have extra cost.


**9 a) (i)**
1. label GO
2. T – labels all vertices directly connected to G.
3. P – label the smallest T – label.
4. Consider arcs directly connected to newly P – labelled vertex if weight on arc > P label, T label the vertex with the weight on previous P label.
5. Repeat 3 and 4 until all vertices have been P – labelled (or until B has a P – label).
6. Find minimum time by tracing back.

**(ii)**

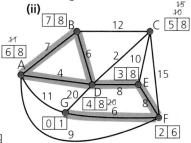

**(iii)** Shortest time 8, route GFEDB *or* GFEDAB
**b)** By selecting slowest edge to be in solution e.g. GDB; time 20

**10** At bottom of B: queue and tow 10 minutes
top of B to bottom of A:
8 minutes       **total** 18 minutes

**b)** Tracing back FGHCA, route ACHGF: time 55 minutes.


# CHAPTER 3
## Networks 2: Minimum-connector

### Exercises 3.1A *(p.50)*

**1 i) a)** e.g. ABCA, ABDCA, ABDECA
**b)** e.g.ABEFCA, ABEGIHFA, ABDEHFCA
**c)** e.g. ABCDEFA, AGHIJKLFA, AGMKEFA

**ii)** There are several answers, here is one tree for each graph.

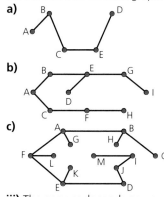

**iii)** The answer depends on your chosen cycles.

**2** For $k_1$ and $k_2$ no cycles exist.

**3 a)** You have to go through vertex F twice.
**b)** Add edge DG or EI.

**4**

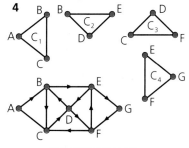

e.g. ABEGFEDFCDBCA
**5** BGHJKFDMLTVWXZQRSNPCB
BGHJKFDCPQRSNMLTVWXZB

### Exercises 3.1B *(p.51)*
**1 a)**

This is a tree.

**b)** e.g.

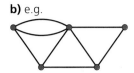

**2** There are no cycles of length 7.

**3**

| Graph | Hamiltonian | Eulerian |
|-------|-------------|----------|
| $G_1$ | No | Yes |
| $G_2$ | No | No |
| $G_3$ | Yes | Yes |
| | (Several H-cycles) | (Several E-cycles) |

**5 a)** $n-1$     **b)** $\frac{1}{2}n(n-1)$

## Exercises 3.2A *(p.55)*

**1**

Length 38

**2**

| | 1 | 2 | 3 | 4 | 6 | 5 | |
|---|---|---|---|---|---|---|---|
| | A | B | C | D | E | F | Start at A |
| A | ∞ | 5 | 7 | ∞ | ∞ | ∞ | order of |
| B | ⑤ | ∞ | 11 | ∞ | 13 | ∞ | connection |
| C | ⑦ | 11 | ∞ | 5 | 18 | ∞ | ABCDFE |
| D | ∞ | ∞ | ⑤ | ∞ | 15 | 12 | length 38 |
| E | ∞ | 13 | 18 | 15 | ∞ | ⑨ | |
| F | ∞ | ∞ | ∞ | ⑫ | 9 | ∞ | |

**3** Order of connection AIEGC
Length 370 miles.
Kruskal's appears quicker to apply once the network is drawn with more vertices. Prim's is quicker to apply as network may be large and difficult to use.

**4**

| | 1 | 6 | 3 | 4 | 5 | 2 | 3 | 7 |
|---|---|---|---|---|---|---|---|---|
| | B | C | D | G | Li | Lon | S | W |
| B | ∞ | 205 | 103 | 196 | 204 | 70 | 126 | 197 |
| C | 205 | ∞ | 154 | 122 | ⑤⑧ | 281 | 200 | 73 |
| D | 103 | 154 | ∞ | 135 | 120 | 146 | 133 | ⑨⑥ |
| G | 196 | 122 | 135 | ∞ | 64 | 176 | ⑨⓪ | 141 |
| Li | 204 | 58 | 120 | ⑥④ | ∞ | 231 | 147 | 77 |
| Lon | ⑦⓪ | 281 | 146 | 176 | 231 | ∞ | 86 | 242 |
| S | 126 | 200 | 133 | 90 | 147 | ⑧⑥ | ∞ | 176 |
| W | 197 | ⑦③ | 96 | 141 | 77 | 242 | 176 | ∞ |

Order of connection Belfast – Londonderry – Sligo – Galway – Limerick – Cork – Waterford – Dublin; length 70 + 73 + 64 + 58 + 86 + 90 + 96 = 537 miles

**5**

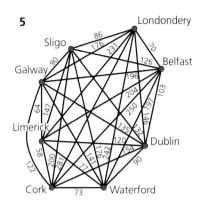

Min spanning tree

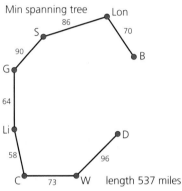

length 537 miles

## Exercises 3.2B *(p.56)*

**1 a)**

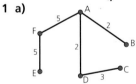

Minimum spanning tree = 17

**b)**

| | A | B | C | D | E | F |
|---|---|---|---|---|---|---|
| A | ∞ | 2 | 8 | ② | 7 | 5 |
| B | ② | ∞ | 4 | ∞ | 8 | ∞ |
| C | 8 | 4 | ∞ | ③ | ∞ | ∞ |
| D | ② | ∞ | 3 | ∞ | 8 | 10 |
| E | 7 | 8 | ∞ | 8 | ∞ | ⑤ |
| F | ⑤ | ∞ | ∞ | 10 | 5 | ∞ |

**2**

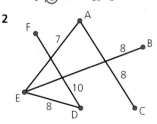

Maximum spanning tree = 41

**3** 363 miles. The introduction of a new centre reduces the shortest distance.

**4**

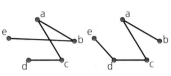

Minimum spanning tree = 94

**5 a) i)** 45     **ii)** $\frac{1}{2}n(n-1)$
**b)** 991 miles

## Exercises 3.3A *(p.64)*

**1 a)**

| | 1 | 2 | 3 | 4 | 6 | 5 | |
|---|---|---|---|---|---|---|---|
| | A | B | C | D | E | F | |
| A | ∞ | 14 | 12 | 18 | ㉟ | 31 | |
| B | 14 | ∞ | ⑨ | 26 | 28 | 28 | |
| C | ⑫ | 9 | ∞ | 17 | 27 | 19 | |
| D | 18 | ㉖ | 17 | ∞ | 40 | 25 | |
| E | ㉟ | 28 | 27 | 40 | ∞ | ⑮ | |
| F | 31 | 28 | 19 | ㉕ | 15 | ∞ | |

Upper bound 12 + 9 + 26 + 25 + 15 + 39 = 126 km
A C B D F E A

**b)**

| Vertex deleted | LB for vertex | LB for network | LB for TSP |
|---|---|---|---|
| A | 26 | 60 | 86 |
| B | 23 | 63 | 86 |
| C | 21 | 72 | 93 |
| D | 35 | 55 | 90 |
| E | 42 | 57 | 99 |
| F | 34 | 65 | 99 |

Best lower bound by deleting E or F = 99

**2**

| | 1 | 5 | 3 | 4 | 2 |
|---|---|---|---|---|---|
| | A | C | E | G | I |
| A | ∞ | ㉒① | 125 | 145 | 105 |
| C | 221 | ∞ | 100 | ㉟ | 263 |
| E | 125 | 100 | ∞ | 40 | ⑯⓪ |
| G | 145 | 95 | ㊵ | ∞ | 168 |
| I | ⑩⑤ | 263 | 160 | 168 | ∞ |

Upper bound (beginning in Aberdeen) A – I – E – G – C – A; length 621 miles
Lower bound by deleting Aberdeen 270 + 295 = 565 miles

**3 a)**

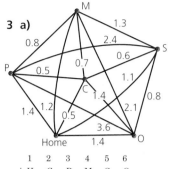

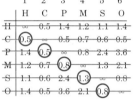

| | 1 | 2 | 3 | 4 | 5 | 6 |
|---|---|---|---|---|---|---|
| | H | C | P | M | S | O |
| H | ∞ | 0.5 | 1.4 | 1.2 | 1.1 | 1.4 |
| C | (0.5) | ∞ | 0.7 | 0.6 | 0.5 | |
| P | 1.4 | (0.5) | ∞ | 0.8 | 2.4 | 3.6 |
| M | 1.2 | 0.7 | (0.8) | ∞ | 1.3 | 2.1 |
| S | 1.1 | 0.6 | 2.4 | (1.3) | ∞ | 0.8 |
| O | 1.4 | 0.5 | 3.6 | 2.1 | (0.8) | ∞ |

H – C – P – M – S – O – H;
distance 5.3 miles
**b)** home – offices – pool – park –
museum – shops – park – home;
length 8 miles

**4**

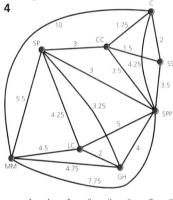

| | 1 | 4 | 5 | 3 | 2 | 8 | 7 | 6 |
|---|---|---|---|---|---|---|---|---|
| | SSP | SS | C | CC | SP | MM | LC | GH |
| SSP | ∞ | 3.5 | 4.25 | 3.5 | 3 | [7.75] | 5 | 4 |
| SS | 3.5 | ∞ | 2 | (1.5) | 4.5 | 10 | 8.5 | 7.5 |
| C | 4.25 | (2) | ∞ | 1.75 | 4.75 | 10 | 9 | 8 |
| CC | 3.5 | 1.5 | 1.75 | ∞ | (3) | 8.5 | 7.25 | 6.25 |
| SP | (3) | 4.5 | 4.75 | 3 | ∞ | 5.5 | 4.25 | 3.25 |
| MM | 7.75 | 10 | 10 | 8.5 | 5.5 | ∞ | (4.5) | 4.75 |
| LC | 5 | 8.5 | 9 | 7.25 | 4.25 | 4.5 | ∞ | (2) |
| GH | 4 | 7.5 | (8) | 6.25 | 3.25 | 4.75 | 2 | ∞ |

by nearest neighbour a possible tour
is: St Peter Port → Saumarez Park →
Craft Centre → St Sampsons →
L'Ancresse common → German
underground hospital → Little Chapel
→ Maritime Museum → St Peter Port;
length 31.75 miles

Improved tour (heuristic)
SPP – SS – C – CC – SP – MM – LC
– GH – SPP
length 26.25 miles

**5** Use a directed network. Decide
which to make last and work
backwards.
Don't forget to add time to
clean the machine before
restarting the cycle.
You must follow on from point
just labelled G – nearest
neighbour algorithm.
**b)** butter – almond – ginger –
chocolate – butter; upper bound
for cleaning time = 71 minutes

## Exercises 3.3B *(p.66)*
**2** a – c – d – b – e – a; 110 seconds
**3** Upper bound 622 miles; lower
bound 467 miles
**5** 72, 78

## Consolidation exercises for Chapter 3 *(p.68)*
**1**

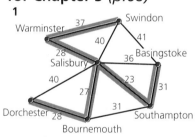

Rank

| | | |
|---|---|---|
| Sals – South | 23 | 1 |
| Sals – Bourn | 27 | 2 |
| Dorch – Bourne | 28 | 3 |
| Warm – Sals | 28 | 4 |
| South – Bas | 31 | 5 |
| Bourn – South | 31 | loop |
| Sals – Basing | 36 | loop |
| Warm – Swin | 37 | 6 |
| Sals – Dorch | 40 | |
| Sals – Swin | 40 | |
| Swin – Bas | 41 | |

| min spanning tree | length |
|---|---|
| 23 + 27 + 28 + 28 + 31 + 37 | = 174 |

**2 a)**

| | 1 | 5 | 7 | 6 | 4 | 3 | 2 |
|---|---|---|---|---|---|---|---|
| | A | B | C | D | E | M | S |
| A | – | – | – | – | – | – | (50) |
| B | 60 | – | – | – | – | (50) | 98.5 |
| C | 98.5 | (50) | – | 50 | 98.5 | 60 | – |
| D | – | – | – | – | 60 | (50) | 98.5 |
| E | – | – | – | – | – | – | (50) |
| M | (50) | – | – | – | 50 | – | 60 |
| S | – | – | – | – | – | – | – |

**b)**

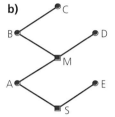

Length 50 × 6 = 300 m
order S A M E B D C
Drains correctly, there is no
house where water
accumulates.

**3 a)**

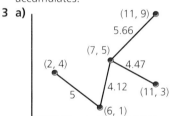

Length = 19.25 (5 + 4.12 +
5.66 + 4.47)
**b) i)** minimum connector = 6
**ii)** minimum connector
= 4 $\sqrt{2}$ = 5.66

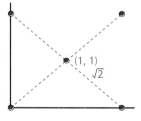

**4 a)** ABDECFA; 45 + 25 + 35 + 70
+ 320 + 145 = 640, e.g.
ABDEFCA costs 537
**b)** 5! = 120

**c)**

|   | A | B | C | D | E | F |
|---|---|---|---|---|---|---|
| A | – | 65 | 80 | 78 | 110 | 165 |
| B | 75 | – | 97 | 55 | 113 | 130 |
| C | 80 | 90 | – | 70 | 90 | 340 |
| D | 90 | 65 | 90 | – | 75 | 250 |
| E | 110 | 90 | 80 | 45 | – | 82 |
| F | 165 | 130 | 320 | 195 | 100 | – |

**d)** B and E;
without tax BDE 25 + 35 = 60;
with tax BDE 55 + 75 = 130,
but BE = 113

**e)** No difference – airport taxes
are paid once at each town so
they add 140 total cost regardless
of which tour you use.

**5 a)**

|   | 1 | 3 | 4 | 2 |   |   |
|---|---|---|---|---|---|---|
|   | B | C | D | E | F | Arcs |
| B | 0.5 | 10 | 31 | 7 | 23 | BE |
| C | (10) | 57 | 14 | 12 | 20 | BC length |
| D | 31 | 14 | 11 | (11) | 9 | ED |
| E | (7) | 12 | 11 | 18 | 12 | DF |
| F | 23 | 20 | (9) | 12 | 43 |  |

length 37

**b)** Add AE and AF, a total of
19, giving a lower bound 56.
**c)** Hamiltonian cycle exists in
this network.
If A is deleted (and arcs to it) we
have a connector for other
vertices which is less than or
equal to a cycle through other
vertices. Since AE + AF are
minimum lengths for
connecting A back into the
network, then AE + AF + min
connector ≤ length of min
Hamiltonian cycle.
**d)** A – E – B – C – D – F – A
        9   7   10  14   9  10
    = 59 > 56
**6** Greedy:
1. Select a starting vertex.
2. Join to nearest vertex.
3. Select shortest edge joining
   vertex not in solution to one
   already in solution.
4 Repeat 3 for all vertices in tree.

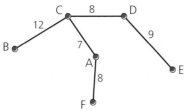

Order
AC
AF
CD
DE
CB
Minimum length of road is 44 km.
Using the nearest neighbour
algorithm gives ACDEFBA and a
distance of 65 km.
An improvement would be
AFEDCBA, a distance of 61 km.
Delete A , then the minimum
connector for the remaining
network is CD, DE, EF, CB
length 40 km.
Reconnect A adding AC + AF =
15 km to give a lower bound of
55 km.

**7 a)**

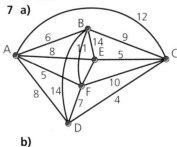

**b)**

|   | A | B | C | D | E | F |
|---|---|---|---|---|---|---|
| A | – | 6 | 12 | 8 | 8 | 5 |
| B | (6) | – | 9 | 14 | 14 | 11 |
| C | 12 | 9 | – | 4 | (5) | 10 |
| D | 8 | (14) | 4 | – | 6 | 7 |
| E | 8 | (14) | 5 | 6 | – | 5 |
| F | (5) | 11 | 10 | 7 | 5 | – |

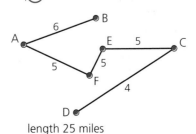

length 25 miles

**c)** A F E C D B A which has
length 39 miles
Interpret this as A → F → E →
C → D → A → B → A
**d)** Revalue DE as 8 and start
algorithm at D → E to include
this edge.
**9 a)** London – Cambridge – Bristol
– Exeter – Manchester – Liverpool
– Leeds – Sheffield – Birmingham
– Oxford – London; 938 miles
**b)** 9 stores ⇒   27 hours
938 miles ⇒ 23 hours 27 minutes
total time of 50 hours 27 minutes
**c)**

Day 1:   7.00 –  8.30
         L – C
         8.30 – 11.30
         Cambridge store
         11.30 – 16.00
         C – Br
         Overnight in Bristol

Day 2:   8.30 – 11.30
         Bristol store
         11.30 – 13.36
         Br – Ex
         13.36 – 16.36
         Exeter store
         16.36 – 21.00
         Ex ⇒ 184 miles ⇒
         Overnight stop

Day 3:   7.00 –  8.51
         Overnight stop –
         Manchester
         8.51 – 11.51
         Manchester store
         11.51 – 12.44
         Man – Liv
         12.44 – 15.44
         Liverpool store
         15.44 – 17.39
         Liv – Leeds
         Overnight in Leeds

Day 4:   8.30 – 11.30
         Leeds store
         11.30 – 12.24
         Leeds – Sheff
         12.24 – 15.24
         Sheffield store
         15.24 – 17.33
         Sheff – Birm
         Birmingham overnight

Day 5:  8.30 – 11 30
Birmingham store
11.30 – 13.12
Birm – Oxford
13.12 – 16.12
Oxford store
16.12 – 17.36
Oxford – London and
home

## CHAPTER 4
## Networks 3: Route inspection problems

### Exercises 4.1A *(p.78)*

**1 a)** Planar

$v = 5, e = 9, f = 6$
$v - e + f = 2$

**b)** Planar

$v = 6, e = 12, f = 8$
$v - e + f = 2$

**c)** Not planar
$v = 6, e = 15$
For Euler to hold
$6 - 15 + f = 2 \Rightarrow f = 11$
$2e \geq 3f \Rightarrow 30 \geq 3f$
$\Rightarrow 10 \geq f$
$\therefore$ cannot be planar.

**3 a)** Not Eulerian,
(B and D odd)
**b)** Eulerian,
all vertices even
ADBAFDCFECBEA
**c)** Not Eulerian,
all vertices odd

**4** Graph 2 and graph 3 have the same sequence of vertex degrees and so are isomorphic.

**5** Traversable graphs are not necessarily Eulerian – they can have two odd vertices, provided you do not end at the same vertex as you began.

### Exercises 4.1B *(p.79)*

**1 b)** and **c)** are planar graphs, **a)** $K_{3,\,3}$ and **d)** are non-planar graphs.

**2 b)** is Eulerian.

**3** There are several edge-disjoint cycles. Here are two for vertex a.

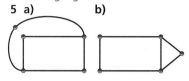

They give the Eulerian cycle abcdbgaegdefa.

**5 a)**      **b)**

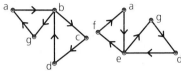

### Exercises 4.2A *(p.81)*

**1**

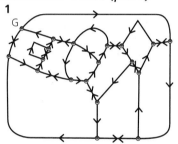

It is not possible without repeating paths since there are lots of odd vertices.

**2**

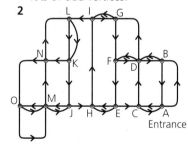

Graph vertices are at junctions of paths.
12 odd vertices – repeat 6 paths
e.g. AC, GI, EH, FB(FDB), OJ (OMJ), KJ
Possible route:
ABACDBDGFDFEHIGILKJMNLKN OMOMJHECA

**3** Odd vertices S, T $\Rightarrow$ repeat ST length 6; distance along all edges 111 km; total to travel $111 + 6 = 117$ km.

**4** Odd vertices A, C, D, G; distance along all edges 148 km; total travel $148 + 35 = 183$ km by repeating ABEG and CD.

**5** A route of minimum length, 98 miles, is ABEGFEBACBFGDCDA If road between E and F is closed then the route ABEGFBCADGDCA is 87 miles (note that these answers are not unique).

### Exercises 4.2B *(p.83)*

**1**

No Eulerian cycle is possible. Each door may be used at least once but you do not end where you started.

**2** Graph is $K_6$ and each vertex is order 6.
No change by adding the doubles, as each vertex then has a loop.

**3** 61 miles

**4 a)** 911 miles
**b)** KF length 101

**5 a)** 460 mm      **b)** 548 mm

## Consolidation exercises for Chapter 4 *(P.85)*

**1 a)** 18
**b)**

degrees of vertices 4, 4, 3, 3, 2, 2
**c) i)** planar – yes
**ii)** Eulerian – no. It has two odd vertices.

**2 a)** 0, 2, 4 ($\sum$degv must be even)
**b)** 2, 4   **c)** 2, 4   **d)** 2

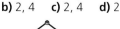

**3 a)** needs two, **b)** needs three;
**c)** $\frac{1}{2}k$

**4** Minimum length is 63 repeating QT, RS and UV.

**5** Route S A B S D E G F B E F S C G C B D A S with cost £18.51

**6 a)** Minimum length is 48 cm repeating AF and CG.
**b)** Upper bound is 28 cm (AEFDBCGFA)
1. A E F D B C G A  28
2. A E F D C G B A  24
3. A E F D G C B A  23

**7 a)** Even if even number of edges incidental.
**b)** Odd if odd number of edges incidental.
Total distance travelled is 8300 m and a possible route is DEFCGFGHBAHCDAED.
Every edge is duplicated hence no need for repeats; ⇒ total distance = 7600 × 2 = 15 200 m

**8 a) i)** odd vertices A, C, F, I; CPP route must duplicate an edge at every odd vertex since it must enter and leave the vertex.
**ii)** The minimum distance is 2620 m by repeating ABC and FI.
**iii)** B  C  D  E  F  G  H  I
       3  2  2  2  3  2  2  2
**b) i)** Redraw the network with two edges for every road.
**ii)** All vertices are now even ⇒ traversable distance = 2300 × 2 = 4600 m.
**c)** He can only go one way along each side of the road, so edges will need to be directed.

**9 a)**

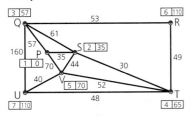

**b)** Minimum cost route is 110p
**c)** Odd vertices P + U cost 110; possible route V P Q U T V S T R Q S P V U V; total cost 699 + 110 = 809 = £8.09
**d)** Odd vertices P R S U; cost 699 + 132 = 831p by repeating PS and RTU.

**10 a)** 4 odd vertices ⇒ network not transversable.
**b)** By Dijkstra, shortest path A H K J or A H I J , 230 mm order of labelling vertices A G H B K I F C J D E.
**c)** Shortest distance is 2220 mm by repeating AB and EJ. Route
A B̲A̲ H B C H G K H I C D I J K F J D E J̲E̲F G A
 ←—— no icing ——→
**d)** 14 odd vertices; no of pairings 13 × 11 × 9 × 7 × 5 × 3 × 1 = 135 135

# CHAPTER 5
## Network flows

### Exercises 5.1A (p.92)

**1 a)** A one-way system of roads.
**b)** A food web – the arrow indicates 'derives food energy from'.
**c)** Any 'tree' situation e.g. W is a manufacturer, R, S & T are warehouses and A – G are retail outlets.

**2 a)** & **b)** are isomorphic. In **c)** arcs PR and RQ are opposite direction.

**3**

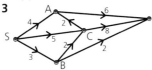

**4** Maximum flow is 9 (but any feasible flow is acceptable)

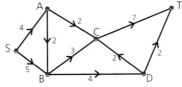

This saturates **all** arcs.

**5** Maximum flow is 13.

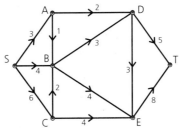

Saturated arcs are SA, AD, DT, SB, BC, BE, SC, DT.

### Exercises 5.1B (p.93)

**1 a)**

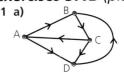

Arrows indicate 'has beaten'.

**b)**

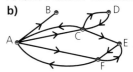

Arrows indicate 'talks to'.

**c)**  Information from

| | a | b | c | d | e |
|---|---|---|---|---|---|
| a | 0 | 0 | 1 | 0 | 1 |
| Information b | 0 | 0 | 0 | 0 | 1 |
| to   c | 0 | 1 | 0 | 0 | 0 |
| d | 1 | 1 | 0 | 0 | 0 |
| e | 0 | 0 | 0 | 1 | 0 |

**2**

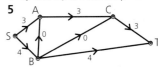

**3**

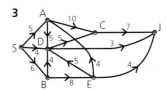

**4 c)** and **d)** are isomorphic.

**5**

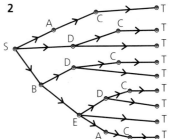

### Exercises 5.2A (p.99)

**1**

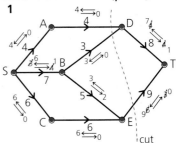

| Path | Flow | Saturated |
|------|------|-----------|
| SADT | 4 | SA, AD |
| SCET | 6 | SC, CE |
| SBDT | 3 | BD |
| SBET | 3 | ET |

Maximum flow 16.

**2 a)** 15 000 tonnes with flows
ABEF 6 000 tonnes, ADF 5000
tonnes, ACF 4000 tonnes

**b)**

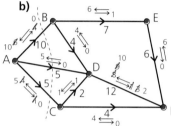

| Path | Flow | Saturated |
|------|------|-----------|
| ABEF | 6 | EF |
| ACF | 4 | CF |
| ADF | 5 | AD |

Flow augmenting paths

| | | |
|------|------|------|
| ABDF | 4 | AB, BD |
| ACDF | 1 | AC |

Maximum flow 20 000 tonnes
**c)** If AC and CD are upgraded
to 8 and 6 the flow increases by
2 000 tonnes since 2 units are
available in DF (CF is saturated).
Improve either DF or CF to
increase flow by a further 2000
tonnes per week.

**3** 12 to A but only 11 can leave,
14 to B but only 12 can leave,
maximum flow 23 units.

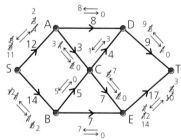

| Path | Flow | Saturated |
|------|------|-----------|
| SADT | 8 | AD |
| SBET | 7 | BC |
| SACDT | 1 | DT |
| SBCET | 5 | BC |
| SACET | 2 | AC, CT |

Maximum flow 23 units.

**4** Maximum flow 27 000 gallons
per hour.

**5**

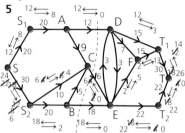

| Path | Flow | Saturated |
|------|------|-----------|
| $SS_1$ $ADT_1$ T | 12 | AD |
| $SS_2$ $BET_2$ T | 18 | BE |
| $SS_2$ $CET_2$ T | 4 | $ET_2$, $T_2$T |
| $SS_2$ $CEFT_1$ T | 2 | CE |

Maximum flow is 3600 vehicles
per hour.

The problem at F has no effect
as the solution has less than 8
units of flow passing through F.

## Exercises 5.2B *(p.100)*

**1**

| Path | Flow | Saturated |
|------|------|-----------|
| SADT | 2 | AD |
| SBET | 4 | BE, SB |
| SCET | 4 | CE |
| SCBDT | 2 | CB, SC |
| SABDT | 1 | AS, DT |

Maximum flow of 13.

**2**

| Path | Flow | Saturated |
|------|------|-----------|
| SAT | 4 | SA |
| SCT | 5 | SC |
| SBCT | 2 | |
| SABDT | 1 | SB |

Maximum flow of 12.

**3** Maximum flow of 1000
containers.

**4 a)** maximum flow of 180 units

**5** Reduces maximum flow by 900
vehicles per hour.
Maximum flow is now 2700
vehicles per hour.

## Exercises 5.3A *(p.103)*

**1** Cut    AD, BD, ET
Capacity   4 + 3 + 9 = 16

**2** Cut    AB, AD, AC
Capacity   10 + 5 + 5 = 20

**3** Cut    $R_1A$, $R_1B$, $R_2B$, $R_2C$
Capacity   7 + 6 + 6 + 8 = 27 000
OR $RR_1$, $RR_2$

**4** Cut    AD, CE, BE
Capacity   12 + 6 + 18 = 36

**5**

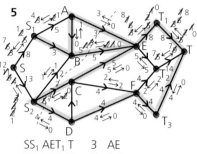

| | | | |
|------|------|---|---|
| $SS_1$ $AET_1$ T | 3 | AE | |
| $SS_1$ $BET_1$ T | 3 | $S_1B$ | |
| $SS_1$ $ABET_1$ T | 1 | AB | |
| $SS_2$ $DFT_3$ T | 4 | $S_2D$, DF, $FT_3$, $T_3T$ | |
| $SS_2$ $CET_2$ T | 5 | CE | |
| $SS_2$ $CET_2$ T | 2 | $S_2C$ | |
| $SS_2$ $BET_1$ T | 1 | BE, ET, $T_1T$ | |

Maximum flow 19 (other flows
through paths also give 19).
Cut    AE, BE, $S_2C$, $S_2D$
Capacity   3 + 5 + 7 + 4 = 19

## Exercises 5.3B *(p.104)*

**1 a)** Value of minimum cut ≥ 10
**b)** Maximum flow ≤ 10
**c)** 7 ≤ maximum flow ≤ 10
**d)** Maximum flow = 10
**e)** You have made a mistake!
Either the flow is too large or
the cut is too small.

**2**

| Cut | Arcs in cut | Capacity of cut |
|-----|-------------|-----------------|
| $c_1$ | SA, SC, SB | 120 |
| $c_2$ | SA, SC, BC, BT | 100 |
| $c_3$ | SA, AC, CA, CT, BT | 170 |
| $c_4$ | AT, CT, BT | 80 |
| $c_5$ | AT, AC, CA, SC, SB | 160 ← minimum cut |
| $c_6$ | AT, AC, CA, SC, BC, BT | 140 |
| $c_7$ | AT, CT, BC, SB | 110 |
| $c_8$ | SA, CA, AC, CT, BC, SB | 200 |

**4 a)** The following flow is an
example of one that satisfies
the maximum and minimum
conditions.

| Path | Flow |
|------|------|
| SAT | 2 |
| SCAT | 2 |
| SBCT | 1 |
| SBT | 1 |

**b)** Maximum flow is 16.
**5 a)** Yes conditions can be satisfied.
**b)** No, cannot achieve a
minimum flow through AT.

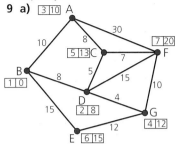

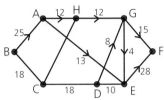

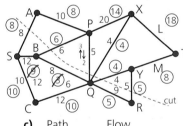

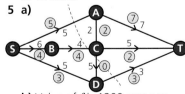

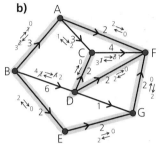

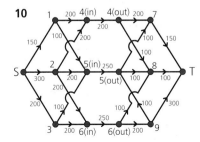

## Consolidation Exercises for Chapter 5 *(p.105)*

**1 a)** Maximum flow ≤ 9
**b)** Minimum cut ≥ 11
**c)** Maximum flow and minimum cut most lie between 28 and 35 inclusive.

**2 a)** Minimum cut of 15 is through AD, BD and ET.
**b)** Minimum cut of 14 is through AD, BD, BE and CE.
**c)** Statement false. It should be: a minimum cut will consist of saturated arcs in direction S to T and arcs with zero flow in direction T to S.

**3** An initial feasible flow that satisfies the minimum conditions is 31 units.

| | |
|---|---|
| $SADT_1 T$ | 10 |
| $SACT_1 T$ | 6 |
| $SBT_2 T$ | 9 |
| $SBCT_2 T$ | 6 |

Flow augmenting gives:

| Path | Flow | Saturated |
|---|---|---|
| $SADT_1 T$ | 2 | $DT_1$ |
| $SACT_1 T$ | 2 | SA, AC |
| $SBT_2 T$ | 3 | SB |

The maximum flow is 38 units.
**c)** The minimum of 38 units is through SA, SB.

**4 a)** Note that in this problem the network is undirected so you can send flow in either direction.

| Path | Flow |
|---|---|
| SACT | 10 |
| SBDT | 5 |
| SBDACT | 4 |

This gives a maximum flow of 19 units.
**b)** The cut SA, SB has a capacity of 19 units.

**5 a)**

**b)** Value of $f$ is 1200 cars per hour.
**c)** Cuts SA, BC, CD, DT has a capacity of 1200.

**6 a)** Source B, sink F.
**b)** Cut partitions network into 2 parts, one containing source and one containing sink.
Flow across cut
$12 + 13 + 18 - 5 = 38$
Not max because capacity of cut $12 + 13 + 18 + 0 = 43$
**c)** BCHGF flow augmentation 4
**d)** Maximal flow is 43.
**e)** SAHGF 12 – then no flow in HC

Maximum flow – minimum cut theorem.

**7 a)** The cut SA, CB, CT, DT gives the minimum cut of 750 vehicles per hour.
The flows are BT 300, CT 350, DT 100.
**b)** 550
**c)** Increase DT by 70 (to 170).

**8 a), b)**

**c)**

| Path | Flow |
|---|---|
| SAPXT | 8 |
| SBPXT | 6 |
| SCQYT | 4 |
| CCQRYT | 5 |
| SCQXT | 1 |
| SBQXT | 3 |

The maximum flow is 27.
**d)** The cut SA, BP, QX, RY gives a capacity of $8 + 6 + 4 + 2 + 5 = 27$.
**e)**

| Path | Flow | Path | Flow |
|---|---|---|---|
| SA | 8 | XT | 18 |
| SB | 9 | YT | 9 |
| SC | 10 | | |
| total | 27 | total | 27 |

**f)** Flow augmenting paths
SBQPXT   3

**g)** Introduce RT with a large capacity.

**9 a)**

**b)**

Tracing back gives route BDCF length 20.

| Path | Flow | Saturated |
|---|---|---|
| BAF | 2 | AF |
| BACF | 1 | BA |
| BDCF | 2 | DC |
| BDF | 2 | DF |
| BEFG | 2 | BE, EG, GF |
| Total flow | 9 | |

The cut BA, DC, DF, GF has capacity $3 + 2 + 2 + 2 = 9$.
Value of flow = capacity of cut; therefore by 'maximum flow – minimum cut theorem' this flow is a maximum.
**c)** Dijkstra on network above.
BDCF   flow   2

% flow = $\frac{2}{9} \times 100 = 22\%$

**10**

## Maximum flow

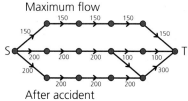

## After accident

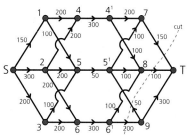

Minimum cut (look for cut < 550)

7T, 8T, 5'9, 6'9

Capacity 150 + 100 + 0 + 200 = 450 < 550

Maximum flow ≤ capacity of any cut, therefore the demand cannot be satisfied.

## CHAPTER 6
## Critical path analysis

### Exercises **6.1A** *(p.113)*

**1**
| Activity | Duration | Preceding |
|----------|----------|-----------|
| A | 2 | – |
| B | 4 | A |
| C | 5 | A |
| D | 2 | C |
| E | 3 | B |
| F | 8 | C |
| G | 4 | D, E |
| H | 1 | G, F |

**2** Activity on vertex

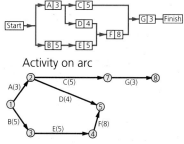

Activity on arc

**3** Activity on vertex

### Activity on arc

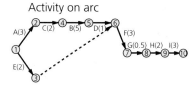

**4**
| Activity | Duration | Preceding act |
|----------|----------|---------------|
| A | 5 | – |
| B | 3 | – |
| C | 5 | – |
| D | 20 | C |
| E | 1 | B |
| F | 30 | E |
| G | 10 | C, E |
| H | 15 | A, D, F, E |
| I | 1 | – |
| J | 7 | I, H |
| L | 8 | J |

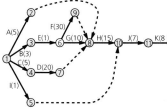

### Exercises **6.1B** *(p.114)*

**1** Activity on vertex

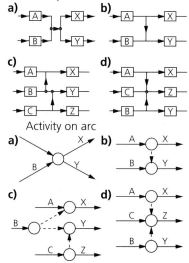

Activity on arc

**2 a)** Remove (3, 4)
**b)** Remove (1, 4)
**c)** Remove (3, 5) and (4, 6)

The activity of vertex networks are drawn for questions 3 – 5.

**3**

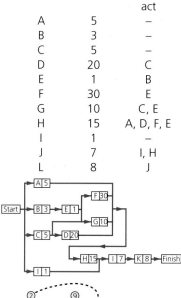

**4**

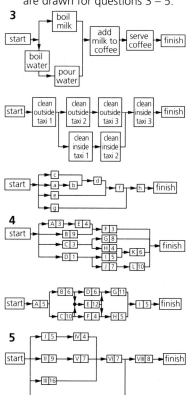

**5**

### Exercises **6.2A** *(p.121)*

**1** Critical path ACFH
**2** Critical path BEFG
**3** Critical path ACBDFGHI
**4** Critical path BEFHJK

### Exercises **6.2B** *(p.121)*

**1** Critical path CGIL; minimum project time 47.
**2** Critical path KFGM; minimum completion time 16.
**3** Exercise 6.1B question 4:
**a)** Critical path DJ L, time 18 weeks
**b)** Critical path ACG EI, time 43
Exercise 6.1B question 5:
Make test rig, test and dispatch; 33 weeks.

## Exercises **6.3A** *(p.128)*

**1**

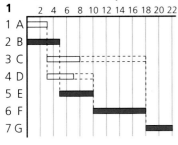

| Can | Activity |
|-----|----------|
| 1 | A |
| 2 | C |
| 3 | D |
| 4 | B |
| 5 | E |
| 6 | F |
| 7 | G |

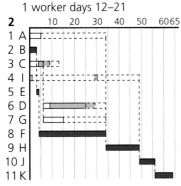

2 workers days 1–12
1 worker days 12–21

**2**

| Can | Activity |
|-----|----------|
| 1 | A |
| 2 | B |
| 3 | C |
| 4 | D |
| 5 | I |
| 6 | E |
| 7 | D |
| 8 | G |
| 9 | F |
| 10 | H |
| 11 | J |

Although it appears that 3 people are needed for a minimum time, if Marc times his activities correctly, he can do it alone since he doesn't need to work while the pie is cooking for example, so he can do another task. There is no time saved if his girlfriend helps.

**3**

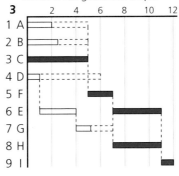

| Can | Activity |
|-----|----------|
| 1 | A |
| 2 | B |
| 3 | C |
| 4 | D |
| 5 | F |
| 6 | H |
| 7 | E |
| 8 | G |
| 9 | I |

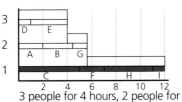

3 people for 4 hours, 2 people for $1\frac{1}{2}$ hours, 1 person for 6 hours.

**4 a)** Critical path BDHGI
**b)** Need 3 workers for 12 days and 2 workers for 7 days.

**5 a)** Critical path ABCDHIJ
**c)** For E:
**i)** independent float 23 days
**ii)** interfering float 4 days
**d)** Reduce time on a critical activity. If you choose the cheapest i.e. H, then the increase in cost is £1000.
Normal cost £89 000
Percentage increase 1.12%

## Exercises **6.3B** *(p.130)*

**1** Critical path: CEH length 51

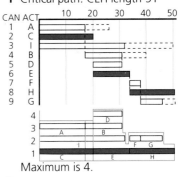

Maximum is 4.

**2**

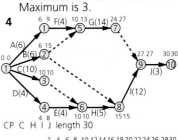

Maximum is 3.

**4**

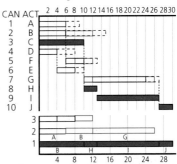

CP C H I J length 30

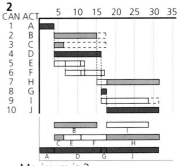

3 persons for 12 days; 2 persons for 14 days; 1 person for 4 days.

## Consolidation exercises for Chapter 6 *(p.132)*

**1 a)**

| Task | A | B | C | D | E | F | G |
|------|---|---|---|---|---|---|---|
| Time | 2 | 4 | 4 | 3 | 4 | 6 | 3 |
| Predecessors | – | – | A | A | D | CE | B |

**b)**

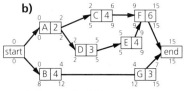

Critical path A D E F;
time 15 minutes

**c)** Float for C 3 minutes

**d)**

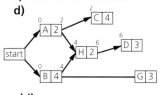

**e) i)**

| Task | A B C D E F G H |
|---|---|
| Time | 2 4 4 3 4 6 3 |
| Predecessors | – – A H D CE B AB |

**ii)** H lies on critical path – it
cannot start until 4 mins into
project. Critical path B H D E F
add 4 mins to completion time
(19 mins)

**2**

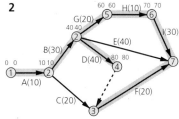

**c)** Length of critical path 100.

**d)** ABGHI

**3 a) i)** 3 → 4 → 5 so 3 → 5 is not
needed

**ii)**

| Activity | A B C D E F G |
|---|---|
| Predecessors | – A – ABDCCDE |
| Event | 1 2 3 4 5 6 |
| Early | 0 2 6 6 8 9 |
| Late | 0 2 6 6 8 9 |

**iv)**

| Activity | A B C D E F G |
|---|---|
| Float | 0 1 3 0 0 1 0 |

**v)** Minimum completion time 9

Critical activities  A D E G

**b)** Interfering float will affect the
scheduling of other activities.
Independent float in amount C
can be changed without affecting
the scheduling of other activities.

**c)** Interfering float = 3
Independent float = 0

**4 a)**

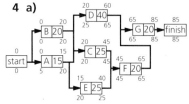

Critical path BCFG
Start 4.30, time 85 minutes,
ready at 5.55 p.m.

**b)**

| Float | A B C D E F G |
|---|---|
| | 5 0 0 5 5 0 0 |
| Reporter | 1 B D |
| | 2 A C F |
| | 3 E |

No activity has more than
5 minutes float therefore you
cannot re-schedule to fit E in for
Reporters 1 and 2

**5 a)**

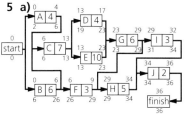

**b)** For G earliest start 23
(from E 13 + 10 = 23)
latest start 23
(from H 29 – 6 = 23)

**c)** Critical acitivities BCEGHJ
early start = late start for critical
activities

**d)** 36 hours

**e)**

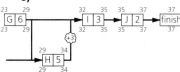

I becomes critical.
Final completion time 37 hours.

**6**

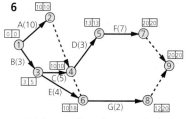

Minimum time for completion
20 days.

Critical path ADF

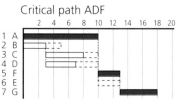

| Can | Activity |
|---|---|
| 1 | A |
| 2 | B |
| 3 | C |
| 4 | E |
| 5 | D |
| 6 | G |
| 7 | F |

**7 a), b)**

| Event | 1 2 3 4 5 6 7 8 9 10 11 |
|---|---|
| Early time | 0 7 4 3 9 15 19 19 24 27 35 |
| Late time | 0 7 6 10 11 17 21 21 26 31 35 |

**c)** Independent float for X =
19 – 17 – 2 = 0
Total float for X = 21 – 15 – 2
= 4
Independent – doesn't affect
scheduling of other activities.
Total float – can be moved but
affects other activities.

**d)** Early time for 52   19
Early time for 58   can't do
Late time for 50   17
Late time for 35   can't do

**8**

| Activity | | Preceeding activities | Duration (days) |
|---|---|---|---|
| A | Sand Hull | – | 2 |
| B | Fit doors | – | 0.5 |
| C | Cut and fit windows | – | 1 |
| D | Fit sea cocks and toilet | A, B | 1 |
| E | Fit galley | B | 0.25 |
| F | Paint hull and deck | A, D | 3 |
| G | Apply anti-fouling | F | 1 |
| H | Fit winches | C | 0.5 |
| I | Step mast | H | 0.25 |
| J | Fit internal lining | C | 2 |
| K | Fit upholstery | D E J | 0.5 |
| L | Launch | G I K | 0.5 |

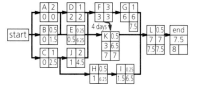

For D:
Forward pass
A → D    0 + 2 = 2
maximum {2, 0.5} = 2
B → D    0 + 0.5 = 0.5

Backward pass
D → F    3 − 1 = 2
minimum {2, 5} = 2
D → K    6.5 → 0.5 = 6
**c)** Critical path A D F G L
where earliest start = latest start.
**d)** 7.5 days
**c)** Critical path now A D F K L,
length 8 days.

**9** Minimum completion time 9
days. Critical path BEG
For C – at time = 3
For F – at time = 4

**10** Critical activities BEGCF
Time to completion 13 days
13 day cost £30 200
In 12 days – choose to cut one
day of cheaper activity on
critical path
C (on C, F path) £700
E (on B E G path) £600
So extra cost is £1300.
% increase is 4.3%.

# CHAPTER 7
## Matchings

## Exercises 7.1A *(p.144)*

**1 a) i)**

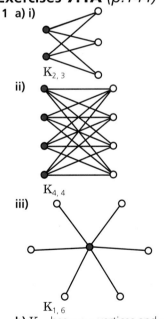

$K_{2, 3}$

**ii)**

$K_{4, 4}$

**iii)**

$K_{1, 6}$

**b)** $K_{r,s}$ has $r + s$ vertices and $rs$
edges. All vertices, $r$, in A have
degree $s$.
All vertices, $s$, in B have degree $r$.

**2 a)** non-planar – subdivision $K_5$
**b)** non-planar – contains $K_{3,3}$

**c)** planar

**3** A 1, 3
   B 1, 3, 4
   C 1, 2, 5
   D 3, 5
   E 4

**4**  Girls           Boys

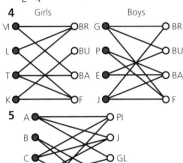

**5**

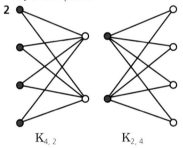

## Exercises 7.1B *(p.145)*

**1 a)** and **c)** are planar
   **b)** is non planar

**2**

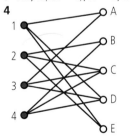

$K_{4, 2}$        $K_{2, 4}$

**3** $2n$, $n(2n − 1)$, $n = 1, 2$

**4**

1 ────── A
        B
2        C
3        D
4        E

## Exercises 7.2A *(p.150)*

**1** Miss Abraham – Environmental
                     studies
   Mr Bowen – Science
   Mrs Carter – Geography
   Ms Delgardo – Maths
   Mr Eggham – Graphics

**2** A – 1, B – 3, C – 2, D – 5, E – 4
   (**or** A – 3, B – 1, C – 2, D – 5,
   E – 4 etc)

**3** There are several solutions.
   **i)** Meena – backstroke
   Lauren – breaststroke
   Tara – freestyle
   Kerry – butterfly
   **ii)** Meena – breaststroke
   Lauren – freestyle
   Tara – backstroke
   Kerry – butterfly
   **iii)** Meena – backstroke
   Lauren – freestyle
   Tara – breaststroke
   Kerry – butterfly
   **i)** Gary– breaststroke
   Paul – backstroke
   Eddie – butterfly
   James – freestyle
   **ii)** Gary– breaststroke
   Paul – freestyle
   Eddie – backstroke
   James – backstroke
   **iii)** Gary – butterfly
   Paul– freestyle
   Eddie – butterfly
   James – breaststroke
   **iv)** Gary – freestyle
   Paul – backstroke
   Eddie – butterfly
   James – breastroke
   **v)** Gary – freestyle
   Paul – butterfly
   Eddie – butterfly
   James – breastroke

**4** Anne – plastering
   Brian – bricklaying
   Carl – joinery
   Daljit – glazing
   Errol – roofing

**5** Annabel – ladies fashions
   Bettina – furnishing
   Chris – electrical goods
   Doreen – hardware
   Imran – menswear

## Exercises 7.2B *(p.150)*

**1** $x_1 − y_3$, $x_2 − y_2$, $x_3 − y_4$, $x_4 − y_1$
**2** $1 − b$, $2 − a$, $3 − d$, $4 − c$
**3** $a − E$, $b − G$, $c − D$, $d − C$,
   $e − F$, $f − B$, $g − A$; This is a
   unique solution
**5** $1a$ – Doing, $16a$ – Going, $20a$ –
   Frame, $8di$– draft, $12d$ – Grant

## Consolidation exercises for Chapter 7 *(p.152)*

**1**

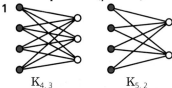

$K_{4, 3}$       $K_{5, 2}$

**2 i)** Non-planar (contains $K_3$, 3)

  **ii)** Planar $v = 8$, $e = 15$, $f = 9$

       $v - e + f = 2$

**3** A – 1, B – 4, C – 2, D – 3, E – 5, F – 6

**4**

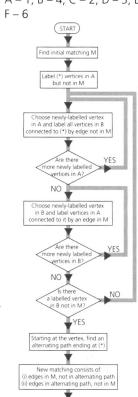

**5** A – carrots, B – tomatoes, C – peas, D – courgettes, E – herbs

**6** There are 2 solutions:  A – W, B – M, C – Tu, D – Thu, E – F **or** A – W, B – M, C – F, D – Thu, E – Tu

**7 a)**

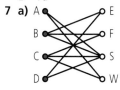

    Ahmed – watches, Bridget – stationery, Chris – food, Diane – electrical

**8** Mrs Frawt – publicity, Miss Delgardo – tickets, Miss Perry – costumes, Mr Quince – lights, Ms Rouse – make-up, Mrs Standish – scenery, Mr Taylor – producer, Mr Unwin – sound

**9** Couple – twin room
Family 4 – family room
Family 5 – appartment 6
Family 7 – flat 8
3 friends – appartment 4
Five friends will be disappointed.

**10 a)** 4 ways

    **b)** Mike – butterfly
        Ned – freestyle
        Oliver – breaststroke
        Pete – backstroke

# CHAPTER 8
# Linear programming 1

## Exercises 8.1A *(p.157)*

**1** Maximise $x + y$
Where $2x + 3y \leq 36$ (eggs)
$250x + 150y \geq 3000$ (margarine)
$x \geq 0, y \geq 0$
and $x$ is the number of chocolate cakes and $y$ is the number of coffee cakes.

**2** Maximise $p = 15x + 11y$
Where $x \leq 3$    (moulder)
$3x + y \leq 12$    (lathe)
$x + y \leq 7$    (assembler)
$x \geq 0, y \geq 0$
and $x$ is the number of bicycles and $y$ is the number of trucks.

**3** Minimise $x + y$
Where $6x + 18 y \geq 54$ (distance covered)
$4x + 3y \leq 12$    (aerobic points)
$y \geq 2x$
$x \geq 0, y \geq 0$
and $x$ is time spent jogging and $y$ is time spent cycling.

**4** maximise $P = 8000x + 11500y$
Where $150x + 200y \leq 24000$ (labour)
$x + 2y \leq 150$            (plots)
$40x + 105y \leq 9600$    (capital)
and $x$ is the number of 'one plot houses' and $y$ is the number of 'two plot houses'.

**5** Maximise $P = 1.55X - (0.7x_1 + 1.05x_2 + 1.05x_3 + 1.3x_4 + 1.1x_5)$
where $X = x_1 + x_2 + x_3 + x_4 + x_5$ = amount of Alma margarine.
$5.6X < 1.2x_1 + 3.4x_2 + 8x_3 + 10.8x_4 + 8.3x_5 < 7.4X$
$x_1 + x_2 + x_3 \leq 40\,000$
$x_4 + x_5 \leq 32000$
and $x_1, x_2, x_3, x_4, x_5$ are the weights of groundnut, soya bean, palm, lard and fish respectively.

## Exercises 8.1B *(p.159)*

**1** $\frac{5}{x} + \frac{3}{y} \leq \frac{4}{3}$, $x > 0$, $y > 0$

**2** Minimise £$C$ where
$C = 0.9x + 0.25y$
subject to $5x + 20y \geq 2$, $0.5x + 0.2y \geq 0.3$, $0.04x + 0.08y \leq 0.04$
$x \geq 0, y \geq 0$ where $x$ kg of meat and $y$ kg of biscuits are purchased.

**3** If $x$ denotes the number of thousand kg of XTRA and y the number of thousand kg of YTER produced per week, $x \geq 0$, $y \geq 0$, $5 \leq 0.9x + 0.5y \geq 20$, $2 \geq 0.1x + 0.5y \geq 7$, $x + y \geq 25$ and profit $P = 1.5x + 2.1y$ (in thousand pounds).

**4** Minimise £$C$ where $C = 20x + 16y$ subject to $x \geq 0$, $y \geq 0$, $6x + 2y \geq 12$, $2x + 2y \geq 8$, $4x + 12y \geq 24$, $x + y \leq 8$
where $x$ hours is spent in Orchard A and $y$ hours in Orchard B.
New cost function $C = 44 + 8x$.

**5** If $x, y$ in 1000 litres are the capacities of CAH and BBM respectively $x, y \geq 0$, $\frac{1}{3}x \leq y \leq 4x$, $x + y \leq 5$, $y \geq 1 - \frac{x}{2}$
Cost (\$) $C = 300x + 220y$, profit (\$) $P = 1700x + 1780y$

## Exercises 8.2A *(p.164)*

**1 a)** $f = 132$ at (12, 36)
    **b)** $f = 190$ at (10, 90)
    **c)** $f = 4$ at (2, 2)
    **d)** $f = 47$ at (7, 6)
    **e)** $f = 38\frac{1}{3}$ at (20, 25)

**2** $c = 12360$ at (1080, 180)

**3** $f = 20$ at (5, 10)

**4** $f = 18$ on line $2x + y = 36$ for $9 \leq x \leq 16.5$ and $3 \leq y \leq 18$

**5 a)** $f$ = 1550 at (90, 65)
**b)** $c$ = 60 at (60, 0)
**6** 'Mathematical solution' is 0.86 cups of rice and 3.43 cups of soya beans: cost 95.14p
In terms of easily measurable quantities, 1 cup of rice and $3\frac{1}{2}$ cups of soya beans costs 100.5p.
**7** 4 days for Refinery 1 and 3 days for Refinery 2.
**8** Mathgem should publish 28 sixth-form books, 8 undergraduate books and 4 graduate books to give a profit of £296 000.
**9** 1.25 ounces of wheatgerm and 1.5 ounces of enriched oat flour.
**10** 1) $14\frac{2}{3}$ at $(8, \frac{20}{3})$
2) 89 at (3, 4)
3) $\frac{11}{3}$ at $(1, \frac{8}{3})$
4) Build 150 'one-plot houses'
5) Too many variables.

## Exercises 8.2B *(p.166)*

**1** (0, 0), (0, −1), (2, 2)
**2** (−1, −1) and $z$ = 5
**3** (15, 0), −30
**4** $(\frac{13}{3}, \frac{5}{3})$ $P = \frac{62}{3}$
**5 a)** −2 at (1, 2) **b)** 10 at (−1, 2)
**6 a)** $\frac{1}{2} < \frac{m}{n} < 1$, **b)** $1 < \frac{m}{n}$, **c)** $\frac{m}{n} > 2$
Any point on $2y + x = 10$ from (0, 5) to (2, 4) will maximise $f$.
**7** $x$ = .5, $y$ = .25; $C$ = 0.5125
**8** $x$ = 13.75, $y$ = 11.25; $P$ = 44250
**9** $x$ = 1, $y$ = 3,   $C$ = 68;
$x$ = 0, $6 \le y \le 8$, $C$ = 44
**10** $x = \frac{2}{9}$, $y = \frac{8}{9}$, $C$ = 262.22;
$x$ = 1, $y$ = 4, $P$ = 8820.

## Consolidation exercises for Chapter 8 *(p.169)*

**1 a)** $P = 55\frac{1}{12}$ at $(\frac{17}{4}, \frac{19}{6})$
**b)** $P$ = 53 at (3, 4)
**2 a)** $3x + y \ge 9$
$x + y \ge 7$
$x + 2y \ge 8$
**b)**

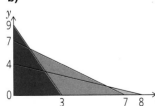

**c)** 2 Xtravit tablets and 3 Yeastalife tablets
**3** 16 000 biscuits of both types. For whole packets of biscuits then make 1333 packets of type A biscuits (15 996 biscuits) and 800 packets of type B biscuits (16 000 biscuits).
**4 b)** $P = 7x + 8y$
**c)** 3 cushions and 4 ragdolls
**5 a)** $4x + 3y \le 5100$
**b)** $4x + 5y \le 5500$
$2x + 3y \le 3000$
**c)** $P = 5x + 7y$
**d)** Maximise $P = 5x + 7y$ subject to $4x + 3y \le 5100$, $4x + 5y \le 5500$, $2x + 3y \le 3000$, $x, y \ge 0$
Bake 750 biscuits and 500 buns. Income £72.50
**6 a)** Maximise $f = 3a + 4b + 1600$ subject to $a + b \le 200$, $5a + 7b \le 1200$, $a, b \ge 0$
**b)** Produce 100 m of plan A and 100 m of plan B
Income £23.00
**7 a), b)** and **c)** $3x + y \ge 240$
$3x + y \le 480$
**d)** Cheapest 100 ml gin, 100 ml martini
Most expensive 140 ml gin, 60 ml martini
**e)** $\frac{1}{2} \le p \le \frac{7}{10}$ corresponding to $100 \le x \le 140$
**8** Produce 4 Alpha and 4 Beta spaceships.

## CHAPTER 9
## Linear programming 2

## Exercises 9.1A *(p.175)*

**1 a)** (2, 3), $r$ = 3
**b)** $(\frac{31}{5}, -\frac{9}{5})$, $r$ = 4
**c)** $(\frac{94}{11}, \frac{23}{11})$, $r = \frac{1}{2}$
**d)** $(\frac{111}{41}, \frac{1}{41})$, $r = \frac{4}{3}$
**e)** (1, 1), $r = \frac{3}{2}$
**f)** (0.92, 3.2), $r = \frac{1}{5}$
**2 a)** (−3, 0, 1)
**b)** (0.8, −1.8, 1.8)
**c)** $(-\frac{8}{7}, -\frac{9}{7}, -\frac{3}{7})$
**d)** (1, 2, 3)

**3 a)** (5, 2, 1)
**b)** (1, −1, 2, 3)
**4 a)** $(-\frac{23}{3}, \frac{25}{3})$ **b)** No solution, parallel lines
**5** $y = x^4 + 2x^3 - 3x^2 - 4x + 4$
**6** $h = 2 + 30t - 5t^2$
Launched at $h$ = 2 metres

## Exercises 9.1B *(p.176)*

**1 a)** $x$ = 1, $y$ = 2
**b)** $x$ = 2, $y$ = 0
**c)** $x$ = 1, $y$ = 3
**d)** $x$ = 2, $y$ = −1
**e)** $x$ = 1, $y$ = −1
**f)** $x = 2a + b$, $y = a$
**2 a)** Infinity of solutions
**b)** No solutions
**3 a)** $x$ = 1.7, $y$ = −0.2
**b)** No solutions
**4 a)** $x$ = 1, $y$ = 1, $z$ = −1
**b)** $x1$ = −2, $x_2$ = 0, $x_3$ = −1
**5 a)** $0x + 0y + 0z = 0$ infinity of solutions
**b)** $0x + 0y + 0z = 1$ no solutions
**6** $d$ = 8, $a$ = 2, $b$ = −5, $c$ = 1

## Exercises 9.2B *(p.179)*

**3 a)** Unique solution $z$ = 2, $y$ = 4, $x$ = 2
**b)** No solutions
**c)** Infinity of solutions, $z = t$, $y = 2 - 3t$, $x = 1 + 2t$
**4 a) i)** $\lambda \ne -1$, $\lambda \ne 1$, $x = \frac{\lambda}{\lambda + 1}$, $y = \frac{1}{\lambda + 1}$
**ii)** $\lambda = 1$, $y = t$, $x = 1 - t$, for any $t$
**iii)** $\lambda = -1$
**b) i)** $\lambda \ne 0$, $\lambda \ne -2$, $z = \frac{2}{\lambda}$, $y = \frac{4 - \lambda}{\lambda}$, $x = \frac{4\lambda - 6}{\lambda}$
**ii)** $\lambda = -2$, $z = t$, $y = 2t - 1$, $4 - 3t$, for any $t$
**iii)** $\lambda = 0$

## Exercises 9.3A *(p.188)*

**1 a)** $f = \frac{2502}{19}$ at $(\frac{222}{19}, \frac{686}{19})$
**b)** $c$ = 15 at (15, 0)
**c)** $f$ = 1457.69 at (88.46, 57.69)
**d)** $g$ = 65 at (15, 5)
**e)** $M$ = 806 at (81.5, 241.5)
**2 a)** $f$ = 150 at (0, 30)
**b)** $f$ = 4.5 at (1.5, 3)
**c)** $c$ = 12 on line $2x + y = 24$ between (9, 6) and (10, 2)

**d)** $c = 10\,560$ at (1080, 180)

**3 a)** $f = \frac{144}{5}$ at $\left(0, \frac{38}{5}, \frac{6}{5}\right)$

**b)** $c = 51$ at (0, 12, 0)

**c)** $M = 45$ at (0, 10, 0)

**d)** $f = 68.2$ at (3.8, 0, 0.6)

**4** 3 bicycles and 3 trucks (assuming whole number of toys; if not, $2\frac{1}{2}$ bicycles and $4\frac{1}{2}$ trucks.

**5** 8 chocolate cakes and 6 coffee cakes.

**6** 28 sixth-form books, 8 undergraduate books, 4 graduate books.

## Exercises 9.3B (p.190)

**1 a)** $x = 10, y = 1, f = 230$

**b)** $x = 2, y = 0, g = 2$

**c)** $x = 0, y = \frac{35}{3}, h = \frac{245}{3}$

**d)** $x = 5, y = 15, M = 440$

**e)** $x = 10, y = 10, N = 200$

**2 a)** $x = 0, y = 20, f = 60$

**b)** $x = 80, y = 0, g = 640$

**c)** $x = 2, y = 32, h = 182$

**d)** $x = 8, y = 14, m = 46$

**3 a)** $x = 4, y = 0, z = 0, f = 28$

**b)** $x = 0, y = \frac{14}{31}, z = \frac{8}{3}, g = 34$

**c)** $x = 5, y = 0, z = 2, h = 19$

**d)** $x = 3, y = 0, z = 0, m = 50$

**4** 40 kg of A, 470 kg of B, income = £3060

**5** 120 large, 260 small containers, income = £315

**6** 150 16-bit, 250 32-bit, profit = £145 000

**7** 240 kg of cake A, 120 kg of cake B, profit = £2280, buy 96 kg flour, 90 kg sugar and 840 eggs

## Exercises 9.4A (p.194)

**1 a)** $f = 15.6$ at (0.6, 4.4)

**b)** $c = 150$ at (50, 0)

**c)** $M = 24$ at (4, 0)

**2 a)** $f = 16$ at $\left(\frac{8}{3}, 2\right)$

**b)** $M = 8$ at $\left(\frac{5}{2}, \frac{1}{2}\right)$

**c)** $f = 4.5$ at (1.5, 3)

**d)** $f = 20$ at (5, 10)

**3** 1.25 ounces of wheatgerm and 1.5 ounces of enriched oat flour

**4** Karen should cycle for 3 hours and not jog at all.

**5** Use 100% of grade C and no grades A and B.

**6** 'Mathematical solution' is 0.86 cups of rice and 3.43 cups of soya beans, in terms of easily measurable quantities, 1 cup of rice and $3\frac{1}{2}$ cups of soya beans.

## Exercises 9.4B (p.196)

**1 a)** $x = 3, y = 2, f = 40$

**b)** $x = 25, y = 0, g = 500$

**c)** $x = 20, y = 0, h = 400$

**d)** $x = 2, y = 6, m = 18$

**2 a)** $x = 3, y = 4.5, f = 7.5$

**b)** $x = 3.75, y = 2.25, g = 12.75$

**c)** $x = 0, y = 7, h = -7$

**d)** $x = 2, y = \frac{8}{3}, m = \frac{98}{3}$

**3** Send 20 from A to I, 10 from A to III and 15 from B to II. Cost £170.

**4** 500 kg of A, 125 kg of B, income = £3750.

**5** Minimum cost £6321.43, commercial run for 4 minutes 17 seconds, the trio 17 minutes 8 seconds, variety act 8 minutes 35 seconds.

**6** 16 desks, $P = £800$
$66\frac{2}{3}$ tables on average, $P = £666.67$

## Consolidation exercises for Chapter 9 (p.199)

**1** $P = 140$ at (0, 22.5, 5)

**2** $f = 200$ at (30, 0)

**3** Minimise $z = x_1 + 2x_2 + 3x_3$ subject to $x_1 + 3x_2 + 2x_3 > 300$
$2x_1 + x_2 + 3x_3 > 150$
$2x_1 + 2x_2 + 5x_3 > 200$
$x_1, x_2, x_3 > 0$.

**4 a)** $c = 3.8x + 4.8y$

**b)** $3x + 0.2y > 32$
$300x + 40y > 5500$
$x + 10y > 30$
Use integer solutions, the parents should serve 19 measures of milk and 2 measures of juice.

**5** Ben should spend 6 hours per week mowing lawns and 9 hours per week cleaning cars.

**6** Build 150 'one plot houses'.

**8 a)** $2x + y \le 10, x + y \le 6,$
$x + 3y \le 15, x, y \ge 0$

**b)** $P = 160x + 120y$

**c)** 4 tonnes of product $X$ and 2 tonnes of product $Y$ gives a profit of £880.

**9** Maximise $D = m + c + l$
subject to $2m + 3c + 6l \le 300$
$6m + 3c + 4l \le 360$
$m \le 55$
$c + l \le 55$
$m, c, l \ge 0$

**b)**

| | $m$ | $c$ | $l$ | $u$ | $v$ | $w$ | $x$ | $P$ | |
|---|---|---|---|---|---|---|---|---|---|
| $R_1$ | 2 | 3 | 6 | 1 | 0 | 0 | 0 | 0 | 300 |
| $R_2$ | 6 | 3 | 4 | 0 | 1 | 0 | 0 | 0 | 360 |
| $R_3$ | 1 | 0 | 0 | 0 | 0 | 1 | 0 | 0 | 55 |
| $R_4$ | 0 | 1 | 1 | 0 | 0 | 0 | 1 | 0 | 55 |
| $R_5$ | −1 | −1 | −1 | 0 | 0 | 0 | 0 | 1 | 0 |

Increase $m$ $R_3$

| | | | | | | | | | | |
|---|---|---|---|---|---|---|---|---|---|---|
| $R_1A = R_1 - 2R_3$ | 0 | 3 | 6 | 1 | 0 | −2 | 0 | 0 | 190 |
| $R_2A = R_2 - 6R_3$ | 0 | 3 | 4 | 0 | 1 | −6 | 0 | 0 | 30 |
| $R_3A = R_3$ | 1 | 0 | 0 | 0 | 0 | 1 | 0 | 0 | 55 |
| $R_4A = R_4$ | 0 | 1 | 1 | 0 | 0 | 0 | 1 | 0 | 55 |
| $R_5A = R_5 + R_3$ | 0 | −1 | −1 | 0 | 0 | 1 | 0 | 1 | 55 |

Increase $c$ $R_2$

| | | | | | | | | | | |
|---|---|---|---|---|---|---|---|---|---|---|
| $R_1B = R_1A - 3R_2B$ | 0 | 0 | 2 | 1 | −1 | 4 | 0 | 0 | 160 |
| $R_2B = 3R_2A$ | 0 | 1 | $\frac{4}{3}$ | 0 | $\frac{1}{3}$ | −2 | 0 | 0 | 10 |
| $R_3B = R_3A$ | 1 | 0 | 0 | 0 | 0 | 1 | 0 | 0 | 55 |
| $R_4B = R_4A - R_2A$ | 0 | 0 | $-\frac{1}{3}$ | 0 | $-\frac{1}{3}$ | 2 | 1 | 0 | 45 |
| $R_5B = R_5A - R_2B$ | 0 | 0 | $\frac{1}{3}$ | 0 | $\frac{1}{3}$ | −1 | 0 | 1 | 65 |

**d)** Not optimal (objective function line not all positive.) Travel 10 miles by car and 55 miles by moped; a total distance of 65 miles. All time is used and $1\frac{1}{3}$ gallons of fuel left.

**10 b)** Violates trivial constraints

**c) i)** $x = 0, y = 8, c = 8$

**ii)** $s_1 = 0, s_2 = 2, s_3 = 10$ which satisfies trivial constraints.

**d)** $x = 4, y = 1, s_1 = 1, s_2 = 0, s_3 = 0$ and $c = 5$

## CHAPTER 10
## Recurrence relations

## Exercises 10.1A (p.206)

**1 a)** £7764.85

**b)** £6518.67

**c)** Pasi has £57.03 more than Claire.

**2 a)** $u_{n+1} = 0.95u_n + 800$

**b)** 18 395 trees

**c)** Forest tends to 16 000 trees.

**3** £140.42

**4 a)** 4, 1, −0.5, −1.25, −1.625, −1.825, −1.906 25, −1.953 125, −1.976 5625, −1.988 28125
$x_n \to -2$

**b)** 0, 3, 9, 21, 45, 93, 189, 381, 765, 1533; $x_n \to \infty$

**c)** 2, –1, 2, –1, 2, –1, 2, –1, 2, –1   $x_n$ alternates.

**d)** 1, 4, 3.8125, 3.806 65, 3.806 307, 3.806 301, 3.806 300 724, 3.806 300 717
$x_n \to 3.806\,300\,716\,741$ (to 12 dp)

**5 a) i)** 7   **ii)** 15   **iii)** 31

**b)** $u_{n+1} = 2u_n + 1$

**c)** 1023

**d)** After 585 billion years.

## Exercises **10.1B** *(p.207)*

**1 a)** $P_{n+1} = 1.01P_n + 20\,000$
$P_1 = 5.2$ million

**b)** 6.7 million

**2 a)** $u_n = 0.995\,u_{n-1}$, $u_1 = 1000$

**b)** £547.99

**c)** £1221.20

**3 a)** 14 years

**b)** 13 years 11 months

**c)** 9 years

**4** $a = 2$, $b = 1$, $x_6 = 127$

**5** Yes; investment continues to increase.

## Exercises **10.2A** *(p.212)*

**1 a)** $x_n = 4^{n-1}\,x_1$

**b)** $x_n = 4^n\,x_0 - 5\left(\frac{4^n-1}{3}\right)$

**c)** $p_n = 0.7^n\,P_0 + 100\left(\frac{1-0.7^n}{0.3}\right)$

**d)** $u_n = u_1 + 5(n-1)$

**e)** $u_n = (-2)^n\,u_0 + 1 - (-2)^n$

**f)** $p_n = 1.5^n\,p_0 - 100\,(1.5^n - 1)$

**g)** $x_n = (-1)^{n-1}\,x_1 - ((-1)^{n-1} - 1)$
or $x_n = x_1$ for $n$ odd and
$x_n = 2 - x_1$ for $n$ even

**h)** $x_n = x_0 + 2n$

**i)** $x_n = (-0.1)^{n-1}\,x_1 - ((-0.1)^{n-1} - 1)$

**j)** $u_n = 0.9^n\,u_0 - 40(0.9^n - 1)$

**2 a)** $x_n = 2 \times 3^{n-1} - 1$

**b)** $u_n = 0.1^{n-2} + \left(\frac{1-0.1^{n-1}}{0.9}\right)$

**c)** $u_n = 2 + 6n$

**d)** $p_n = 250 \times 1.4^{n-1} - 150$

**e)** $u_n = 1 - (-2)^{n-1}$

**3** £10 568.52

**4** £200.01

**5** £62.26

**6 a)** $x_n = x_1 + \frac{1}{2}(n^2 + n - 2)$

**b)** $u_n = 2^{n-1}\,u_1 - n^2 - 4n - 6$

**7** 63 months

**8** 45 months

**9 a)** 30 years

**b)** Yes; population falls for 10 years and then starts to increase again.

**10 a)** $x_{n+1} = x_n + 2\pi(2.2 + 0.005n)$; $x_0 = 0$

**b)** $x_n = 4.4\pi(n - 1)$
$+ 0.01\pi\,\frac{(n-1)(n-2)}{2}$

**c)** 126.27 metres

**d)** 6.06 cm

## Exercises **10.2B** *(p.213)*

**1 a)** $x_n = (-1.1)^{n-1}\,x_1$

**b)** $x_n = (-1.5)^{n-1}\,x_1$
$- 2((-1.5)^{n-1} - 1)$

**c)** $x_n = (-1)^{n-1}\,x_1 + 3\,\frac{((-1)^{n-1}-1)}{-2}$
$x_n = x_1$ when $x$ is odd;
$x_n = 3 - x_1$ when $n$ is even.

**d)** $u_n = 0.2^n\,u_0$

**e)** $p_n = (-0.2)^n\,p_0$

**f)** $x_n = (0.8)^{n-1}\,x_1 + 5\,(0.8^{n-1} - 1)$

**g)** $x_n = (-0.9)^n\,x_0 + 1\,\frac{((-0.9)^n-1)}{-1.9}$

**h)** $u_n = 2^n\,u_0 - 5(2^n - 1)$

**i)** $x_n = x_1 - \frac{5}{2}(n - 1)$

**j)** $u_n = \left(\frac{3}{2}\right)^{n-1}\,u_1 + 8\left(\left(\frac{3}{2}\right)^{n-1} - 1\right)$

**2 a)** $x_n = 1000 \times (-0.2)^{n-1}$

**b)** $u_n = 4 \times (1.5)^n - 10(1.5^n - 1)$

**c)** $x_n = 0.5(3^{n-1} - 1)$

**d)** $p_n = 10 \times 1.07^{n-1} + 1000$
$(1.07^{n-1} - 1) = 1010 \times 1.07^{n-1}$
$- 1000$

**e)** $u_n = -\frac{4}{3} \times (-2)^{n-1} + \frac{7}{3}$
$= -\frac{1}{3}(-2)^{n+1} + \frac{7}{3}$

**3 a)** $x_n = 3^{n+1} - 2(3^n - 1) = 3^n + 2$;
$m = 5$

**b)** $x_0 = 0$

**4 a)** $x_n = 4^{n-1}\,x_1 - 2n - \frac{7}{3}$

**b)** $u_n = 2^n\,u_0 + n2^n$

**c)** $x_n = x_1$ for $n$ odd and
$x_n = x_1 - 1$ for $n$ even

**5 a)** £2191.12

**b)** 9 years

**6** 73 months

**7** £235.42.

**8 a)** $P_n = 0.5^{n-1} \times 0.2 - 0.8$
$(0.5^{n-1} - 1) = 0.8 - 0.6 \times (0.5)^{n-1}$
$P_n \to 0.8$

**b)** $P_n \to 0.2$

**c)** $P_n \to 0.5$

**d)** $P_n \to 1$

**e)** $P_n \to 0$

**9** £616.43

## Exercises **10.3A** *(p.217)*

**1 a)** $x_n = 5^n A + 2^n B$

**b)** $x_n = 3^n A + (-3)^n B$

**c)** $u_n = 2^n A + B$

**d)** $u_n = \left(\frac{3+\sqrt{17}}{2}\right)^n A + \left(\frac{3-\sqrt{17}}{2}\right)^n B$

**e)** $x_n = \left(\sqrt{2}\right)^n A + \left(-\sqrt{2}\right)^n B$

**f)** $u_n = A + Bn$

**g)** $u_n = (-2)^n A + B$

**h)** $x_n = 2^n A + n2^n B$

**2 a)** $x_n = 2^n$

**b)** $x_n = -2^n + n2^{n+1}$

**c)** $u_n = \frac{1}{2}(3^{n-1} + (-3)^{n-1})$

**d)** $p_n = 4^{n+1} + 3 \times 2^{n+1}$

**3** $x_n = \frac{1}{4}(3^n + (-1)^{n+1})$
$x_4 = 20$
$x_6 = 182$

**4** $u_n = \left(\frac{1}{2}\right)^n$
$u_2 = \frac{1}{4}$
$u_5 = \frac{1}{32}$

**5** $p_n = \left(\frac{2}{3}\right)^n - \left(\frac{1}{2}\right)^n$
$p_n \to 0$ as $n \to \infty$

## Exercises **10.3B** *(p.218)*

**1 a)** $x_n = A + (-1)^n B$

**b)** $x_n = 4^n A + (-1)^n B$

**c)** $u_n = 4^n A + (-4)^n B$

**d)** $u_n = 5^n (A + Bn)$

**e)** $p_n = 2^n A + (-1)^n B$

**f)** $x_n = 7^n A + 2^n B$

**g)** $u_n = \left(\frac{-9+\sqrt{141}}{10}\right)^n A + \left(\frac{-9-\sqrt{141}}{10}\right)^n B$

**h)** $p_n = \left(\frac{1}{3}\right)^n A + (-3)^n B$

**2 a)** $x_n = 2 \times 5^{n+1} - (-3)^{n+2}$

**b)** $p_n = 3^{n-2}(-1 + 2n)$

**c)** $u_n = \frac{1}{2}\left(\left(\sqrt{3}\right)^n + \left(-\sqrt{3}\right)^n\right)$

**d)** $x_n = \frac{1}{2}\left(\left(\frac{3}{2}\right)^n + (-1)^n\right)$

**3** $x_n = (-4 + 3n)(-1)^n$
$x_4 = 8$
$x_6 = 14$

**4** $p_n = -\left(\frac{1}{2}\right)^{n-1} + 2$
$p_n \to 2$ as $n \to \infty$

**5 a)** $u_{n+2} = 3u_n + 2(2n - 3)$

**b)** 2485 pairs of gerbils

## Exercises **10.4A** *(p.221)*

**1 a)** $x_n = 2^n A + 2^n B + \frac{3}{2}$

**b)** $x_n = 3^n A + 2^n B + \frac{1}{2}n + \frac{7}{4}$

**c)** $x_n = 3^n A + 2^n B + n + 3$

**d)** $x_n = 3^n A + 2^n B$
$+ \frac{1}{2} n^2 + \frac{7}{2} n + 8$

**e)** $x_n = 3^n A + 2^n B + \dfrac{5^{n+2}}{6}$

**f)** $x_n = 3^n A + 2^n B + n3^{n+1}$

**2 a)** $x_n = \left(\frac{1}{3}\right)^n A + \left(\frac{1}{2}\right)^n B + \frac{1}{2}$

**b)** $x_n = \left(\frac{1}{3}\right)^n A + \left(\frac{1}{2}\right)^n B + \frac{1}{2} n + \frac{3}{4}$

**c)** $x_n = \left(\frac{1}{3}\right)^n A + \left(\frac{1}{2}\right)^n B + n + 3$

**d)** $x_n = \left(\frac{1}{3}\right)^n A + \left(\frac{1}{2}\right)^n B + \frac{1}{2} n^2$
$- n + \frac{9}{4}$

**e)** $x_n = \left(\frac{1}{3}\right)^n A + \left(\frac{1}{2}\right)^n B + \left(\frac{1}{2}\right)^{n+1} n$

**f)** $x_n = \left(\frac{1}{3}\right)^n A + \left(\frac{1}{2}\right)^n B + \dfrac{3^{n+2}}{40}$

**3** $x_n = \frac{1}{2}\left(4^n\right) - \frac{5}{3}\left(3^n\right) + 2^{n+1}$

**4** $x_n = \frac{5}{27}(-2)^n + \frac{49}{27} + \frac{1}{6} n^2 - \frac{11}{18} n$

**5** $x_n = -3^n + (-1)^n + 4n + 7$

## Exercises **10.4B** *(p.221)*

**1 a)** $x_n = (-2)^n A + B + \frac{4}{3} n$

**b)** $x_n = (-2)^n A + B + \frac{1}{6} n^2 + \frac{7}{18} n$

**c)** $x_n = (-2)^n A + B + \frac{1}{3} n^2 + \frac{10}{9} n$

**d)** $x_n = (-2)^n A + B + \frac{1}{3} n^3$
$+ \frac{7}{18} n^2 - \frac{23}{18} n$

**e)** $x_n = (-2)^n A + B + \dfrac{5^{n+2}}{28}$

**f)** $x_n = (-2)^n A + B + 2^n$

**2 a)** $x_n = \left(-\frac{1}{5}\right)^n A + \left(\frac{1}{3}\right)^n B + 1$

**b)** $x_n = \left(-\frac{1}{5}\right)^n A + \left(\frac{1}{3}\right)^n B + 2n - \frac{2}{3}$

**c)** $x_n = \left(-\frac{1}{5}\right)^n A + \left(\frac{1}{3}\right)^n B + \frac{1}{4} n - \frac{1}{6}$

**d)** $x_n = \left(-\frac{1}{5}\right)^n A + \left(\frac{1}{3}\right)^n B + \frac{1}{12} n^2$
$+ \frac{1}{36} n + \frac{25}{216}$

**e)** $x_n = \left(-\frac{1}{5}\right)^n A + \left(\frac{1}{3}\right)^n B + \frac{1}{7}\left(\frac{1}{2}\right)^n$

**f)** $x_n = \left(-\frac{1}{5}\right)^n A + \left(\frac{1}{3}\right)^n B + \frac{n}{8}\left(\frac{1}{3}\right)^{n+1}$

**3** $x_n = \frac{23}{25}(5)^{n-1} + \frac{11}{7}(-2)^{n-1} - \frac{1}{3}(2^n)$

**4** $x_n = -\frac{14}{125}(-4)^n + \frac{139}{125} - \frac{1}{10} n^2 + \frac{9}{10} n$

**5** $u_n = -\frac{87}{40}(3^n) + \frac{347}{90}(-2)^n$
$- \frac{1}{12} n - \frac{13}{72}$

## Consolidation exercises for Chapter 10 *(p.223)*

**1** $u_n - u_{n-1} - u_{n-2} = 0$
$u_n = \left(\frac{5+\sqrt{5}}{10}\right)\left(\frac{1+\sqrt{5}}{2}\right)^n + \left(\frac{5-\sqrt{5}}{10}\right)\left(\frac{1-\sqrt{5}}{2}\right)^n$

**2** $p_n = 4 - 4\left(-\frac{1}{2}\right)^n$
$p_n \to 4$

**3 a)** $u_{n+1} = 0.8u_n + 400$
**b)** $u_n = 2000(1 - 0.8^n)$
**c)** $u_{12} = £1862.50$
$u_n \to £2000$

**4 a)** $u_{n+1} = u_n + 4n + 4$
**b)** $u_n = 2n(n + 1)$
**c)** 84
**d)** $S_{n+1} = S_n + 2(n + 1)^2 + 2(n + 1)$

**5 b)** $u_n = 1.01^n(30\,000 - 100X)$
$+ 100X$
**c)** $X = £315.97$

**6 a)** $w_n = n(n + 1)$
**d)** 400 cans needed
**e)** $y_0 = 1, y_1 = 4$
**f)** $y_{k+1} = y_k + (k + 1) + 1 + (k + 1)$
$= y_k + 2k + 3$
**g)** 420 cans

**7 b)** $C_k = 1 - 0.75(0.99)^k$
12 weeks required

**8 a)** $u_{n+1} = 0.8u_n + 20\,000$
$u_0 = 250\,000$
**b)** $u_5 = 149\,152$
**c)** 16 years
**d)** $u_n \to 100\,000$
Quantity stored is increasing.

**9** valid for $n \le 13$.
$u_n = 3125 + 345.05 (1.08)^n, n \ge 4$

**10** $p_t = (-0.5)^{t-1} p_1 - 2(-0.5)^{t-1} + 2$

## CHAPTER 11
## Coding

## Exercises **11.1A** *(p.232)*

**1** 2, 3

**2** $\frac{4}{7}$

**3 a)** 0000 0001 0010 0100 1000
0011 0101 0110 1010 1100
1001 0111 1011 1101 1110 1111
**b)** 1

**4** Meet me here at ten.

**5 a)** 1101001, 0111010, 0000000
**b)** Length = 7, dimension = 3,
rate of code = $\frac{3}{7}$

## Exercises **11.1B** *(p.233)*

**1** {0001, 0010, 0011, 0100,
0110, 0111, 1000, 1001, 1011,
1100, 1101, 1110}
{0001, 0010, 0100, 0111,
1000, 1001, 1011, 1101, 1110}

**3** Dimension = 3, rate = $\frac{3}{5}$

**4** Remaining 5-bit codewords;
there are 24 of them!

**5** If $x$ is a codeword then $x + x = 0$
for any $x$.
So 0 must be in the code.

**6** There are many possible
answers. The code in question 5
of Exercises 11.1A is an example.

## Exercises **11.2A** *(p.237)*

**1 a)** 2    **b)** 4    **c)** 8

**2** 2

**3 a)** 9   **b)** 10

**4 a)** 9   **b)** 9

**5** $\frac{1}{2}$

**6** SEND CASH
(note the errors in E and H).

## Exercises **11.2B** *(p.237)*

**1** Code can detect 1 error and
correct 0 errors.

**2 a)** 6
**b)** 8 = 2

**3 a)** 2   **b)** 11111111110

**4 a)** rate = $\frac{1}{5}$, $\delta = 5$
**b)** 2

**5 a)** $\frac{1}{6}$, 2   **b)** $\frac{1}{7}$, 3
**c)** $\frac{1}{n}$, $\frac{1}{2}(n - 1)$ for $n$ odd,
for $n$ even

**6 a)** {11000, 10100, 10010,
10001, 01100, 01010, 01001,
00110, 00101, 00011}
**b)** Minimum distance = 2, can
detect and correct 0 errors!

## Exercises **11.3A** *(p.243)*

**1 a)** $\frac{3}{10}$   **b)** 4
**c)** Detect 2 errors and correct 1
error.
**d)** No; if example it does not
contain 0.

**2 a)** {0111010, 1101001}
**b)** 1100110

**3** For Exercises 11.1A:
*Question 1* {0101, 1010}
$$\begin{pmatrix} 0 & 1 & 0 & 1 \\ 1 & 0 & 1 & 0 \end{pmatrix}$$
Second code is not linear since 1001 is missing
*Question 2* {1101001, 0101010, 0010110, 1001100}
$$\begin{pmatrix} 1 & 1 & 0 & 1 & 0 & 0 & 1 \\ 0 & 1 & 0 & 1 & 0 & 1 & 0 \\ 0 & 0 & 1 & 0 & 1 & 1 & 0 \\ 1 & 0 & 0 & 1 & 1 & 0 & 0 \end{pmatrix}$$
*Question 3* {0110101, 1010011, 1011100}
$$\begin{pmatrix} 0 & 1 & 1 & 0 & 1 & 0 & 1 \\ 1 & 0 & 1 & 0 & 0 & 1 & 1 \\ 1 & 0 & 1 & 1 & 1 & 0 & 0 \end{pmatrix}$$

**4 a)**
$$G = \begin{pmatrix} 0 & 0 & 1 & 0 & 0 & 1 & 0 & 1 & 1 & 1 \\ 1 & 1 & 0 & 1 & 1 & 1 & 0 & 0 & 1 & 0 \\ 0 & 1 & 0 & 0 & 1 & 0 & 1 & 1 & 1 & 0 \end{pmatrix}$$
is one example
$(101) G = (0110111001)$
$(011) G = (1001011100)$
$(xyz) G = (y\,y + z\,x\,y\,y + z\,x + y\,z\,x + z\,x + y + z\,x)$
**b)** 3rd, 4th, 7th columns.

**5 a)** $n = 7, n - k = 3 \Rightarrow k = 4$ and rate $= \frac{4}{7}$

**b)** Error syndrome $\begin{pmatrix} 0 \\ 0 \\ 1 \end{pmatrix}$

Transmitted codeword (0111000)
**c)** 0111000, 0110011, 1100010, 1101001

## Exercises 11.3B *(p.244)*

**1** Exercises 11.1B question 3 is not linear, 0 is missing.
For Exercise 11.2B:
*Question 3* is linear
$$\begin{pmatrix} 0 & 1 & 1 & 0 & 0 & 1 & 0 & 0 & 1 & 1 & 0 \\ 1 & 0 & 0 & 1 & 1 & 0 & 1 & 1 & 0 & 0 & 0 \end{pmatrix}$$
*Question 4* is linear
$$\begin{pmatrix} 0 & 0 & 1 & 1 & 1 & 0 & 1 & 0 & 1 & 0 \\ 1 & 1 & 0 & 0 & 0 & 1 & 0 & 1 & 0 & 1 \end{pmatrix}$$
**2** {1101}
**3 a)** $\frac{1}{2}$
**b)** For example
$$\begin{pmatrix} 1 & 1 & 1 & 0 & 0 & 0 \\ 1 & 0 & 0 & 1 & 1 & 0 \\ 0 & 1 & 0 & 0 & 1 & 1 \end{pmatrix}$$ {011110} is the correct message.
**c)** For example
$$\begin{pmatrix} 1 & 1 & 0 & 0 & 1 & 0 \\ 1 & 0 & 1 & 1 & 0 & 0 \\ 0 & 1 & 1 & 0 & 0 & 1 \end{pmatrix}$$
{101} → {101011}

**4 a)** {0000, 0111, 1010, 1101}
**b)** 2
**c)** For example $\begin{pmatrix} 0 & 1 & 1 & 1 \\ 1 & 0 & 1 & 0 \end{pmatrix}$

**5 a)** 0110011, 1001011, 1111110
**b)** For example
$$\begin{pmatrix} 1 & 1 & 1 & 1 & 0 & 0 & 0 \\ 1 & 0 & 0 & 1 & 1 & 0 & 1 \\ 0 & 0 & 0 & 0 & 1 & 1 & 0 \end{pmatrix}$$
$(111) \to (0110011)$
$(010) \to (1001101)$
$(x, y, z) \to (x + y\,x\,x\,x + y\,y + z\,z\,y)$

## Consolidation exercises for Chapter 11 *(p.245)*

**1 b)** 0
**c)** Several answers are possible. The total comes to 170. Possible ISBN's with one wrong digit are 0–7135–1272–5 or 0–7136–2272–5
**2** I want more homework.
**3 a)** Length 4, dimension 2, rate $\frac{1}{2}$
**b)** Yes, the code is linear.
**4 a)** 10001 01001 00110
   10010 01010 00011
   10100 01100
   11000 00101
**b)** There are always 2 ones and 3 zeros, so any codeword not containing these must be incorrect.
**5 a)** 8
**b)** $\delta = 4$
**6** Yes, the code is cyclic
**7** 1000111 1111111
**8 a)** 0011101, 0111010, 1110100, 1101001, 1010011, 0100111,
**b)** 4
**c)** 2 detected and 1 corrected
**d)** 0000000 is not a codeword. Include this and the code is linear.
**d)** Message should read 0100111 1110110 0011101
   ↑        ↑        ↑
↑ shows corrected bits
**9** $P_{n+2} = \frac{1}{2}(P_{n+1} + P_n)\ n \geq 0$
$P_0 = 6, P_1 = 3$
$P_n = 4 - 4(-\frac{1}{2})^n$
as $n \to \infty$   $P_n \to 4$

## CHAPTER 12
## Simulation

## Exercises 12.1A *(p.250)*

**1 a)** 331
**b)** 2791
**c)** $3t^2 + 3t + 1$
**2** Using a dice to simulate spread to a tree rather than all at once.
e.g.

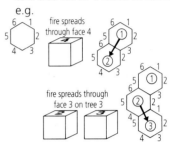

**4** Use a dice and coin and the following table.

| | H | T |
|---|---|---|
| 1 | TOY 1 | TOY 7 |
| 2 | TOY 2 | TOY 8 |
| 3 | TOY 3 | TOY 9 |
| 4 | TOY 4 | TOY 10 |
| 5 | TOY 5 | TOY 11 |
| 6 | TOY 6 | TOY 12 |

Toy 6 would correspond to a 6 and a head.
**5 a)** Use numbers on dice from $(1, 1) \to$.
Ignore (2, 6) to (6, 6) inclusive if they come up.
**b)** Use a dice and coin to generate 12 possibilities (or 2 dice: 1–6 and even/odd)
**c)** If you generate a non-existent date, ignore it and continue.

## Exercises 12.1B *(p.251)*
**3** $\frac{1}{16}, \frac{1}{4}, \frac{3}{8}, \frac{1}{4}, \frac{1}{16}$

## Exercises 12.2A *(p.256)*
**1** Queueing    Use counters to record.
Situation 1 – Each new customer chooses to join shorter queue
Situation 2 – Record which service point is free hence where next customer goes.

**2** Rules:

| Interarrival time | 1 | 2 | 3 | 4 | 5 |
|---|---|---|---|---|---|
| RND | 0–54 | 55–84 | 85–94 | 95–98 | 99 |

| Service time | 1 | 1.5 | 2 | 2.5 | 3 |
|---|---|---|---|---|---|
| RND | 0–19 | 20–43 | 44–71 | 72–89 | 90–99 |

[My simulation
Average wait – 4 minutes,
Maximum wait – 6.5 minutes
Average queue – 2.5
Maximum queue 4
Not very fast – would be fast
food with 2 service points]

**3 a)**

| Time to serve | 1 | 2 | 5 |
|---|---|---|---|
| RND | 0–24 | 25–74 | 75–99 |

**b)**

| Customer | 1 | 2 | 3 | 4 | 5 | 6 | 7 | 8 | 9 | 10 | 11 | 12 |
|---|---|---|---|---|---|---|---|---|---|---|---|---|
| Service time | 5 | 2 | 2 | 2 | 1 | 2 | 1 | 2 | 5 | 5 | 1 | 1 |

**c)**

| Time between customers | 1 | 2 | 3 | 4 |
|---|---|---|---|---|
| RND | 1–15 | 16–45 | 46–75 | 76–90 |

Using 01–90 ($\frac{1}{6}$ is 15 numbers,

$\frac{1}{3}$ is 30 numbers)
Ignore 91–99.

**d)**

| Customer | 1 | 2 | 3 | 4 | 5 | 6 | 7 | 8 | 9 | 10 | 11 | 12 |
|---|---|---|---|---|---|---|---|---|---|---|---|---|
| Time (between) | 2 | 1 | 1 | 2 | 1 | 3 | 3 | 4 | 4 | 1 | 1 | |

**e)**

| Time (minutes) | Customer arriving | Time of next arrival | Customer Start service | Time end service | Length of queue |
|---|---|---|---|---|---|
| 0 | | 2 | | | 0 |
| 1 | | | | | 0 |
| 2 | 1st | 3 | 1st | 7 | 0 |
| 3 | 2nd | 4 | | | 1 |
| 4 | 3rd | 6 | | | 2 |
| 5 | | | | | 2 |
| 6 | 4th | 7 | | | 3 |
| 7 | 5th | 10 | 2nd | 9 | 3 |
| 8 | | | | | 3 |
| 9 | | | 3rd | 11 | 2 |
| 10 | 6th | 13 | | | 3 |
| 11 | | | 4th | 13 | 2 |
| 12 | | | | | 2 |
| 13 | 7th | 17 | 5th | 14 | 2 |
| 14 | | | 6th | 16 | 1 |
| 15 | | | | | 1 |
| 16 | | | 7th | 17 | 0 |
| 17 | 8th | 21 | 8th | 19 | 0 |
| 18 | | | | | 0 |
| 19 | | | | | 0 |
| 20 | | | | | 0 |
| 21 | 9th | 25 | 9th | 26 | 0 |
| 22 | | | | | 0 |
| 23 | | | | | 0 |
| 24 | | | | | 0 |
| 25 | 10th | 26 | | | 1 |
| 26 | 11th | 27 | 10th | 31 | 1 |
| 27 | 12th | | | | 2 |
| 28 | | | | | 2 |
| 29 | | | | | 2 |
| 30 | | | | | 2 |
| 31 | | | 11th | 32 | 1 |
| 32 | | | 12th | 33 | 0 |

**f)** Average queue length

$$= \frac{3+3+3+2+3+2+2+2+1+1+0}{15} \text{ (minutes)}$$

= 1.47 (people each minute).
Neglects zeros at the beginning
and any build up at the end.

**g)** Not occupied for 2 minutes
(19 + 20 minutes).
Percentage occupation 87%

**h)** Too small – needs a longer
run *and* several repeats in order
to be confident of reliability.

**4 a)** Present 0–84 Absent 85–99

| 1–4 | PPPP | PPPPP | PPPPP | APPPP |
|---|---|---|---|---|
| 5–6 | PPPPP | PPAPP | PAPAP | APPPP |
| 9–12 | PPAPP | PAAPA | PAPPP | PPAPP |
| 13–16 | PPPPP | PPPPA | PPPPA | PPPPP |
| 17–20 | PPPAP | PPAPP | PPPPP | PPPPP |

**b)**

| Company | Present |
|---|---|
| 1 | 5 |
| 2 | 5 |
| 3 | 5 |
| 4 | 4 |
| 5 | 5 |
| 6 | 4 |
| 7 | 3 |
| 8 | 4 |
| 9 | 4 |
| 10 | 2 |
| 11 | 4 |
| 12 | 4 |
| 13 | 5 |
| 14 | 4 |
| 15 | 4 |
| 16 | 5 |
| 17 | 4 |
| 18 | 4 |
| 19 | 5 |
| 20 | 5 |

**c) i)** P(all P) $= \frac{8}{20} = \frac{2}{5} = 0.4$

**ii)** mean = 4.25 spaces occupied

**d) i)** $p = 0.4437$

**ii)** $p = 5 \times 0.85^4 \times 0.15$
= 0.3915

**5 a)**

| Eggs | 4 | 5 | 6 | 7 | 8 | 9 | 10 |
|---|---|---|---|---|---|---|---|
| RND | 0–7 | 8–17 | 18–33 | 34–58 | 59–88 | 89–95 | 96–99 |

*Example simulation:*

| Pair | RND | Eggs re clutch | Survivors[1] | Surviving adults[2] | Surviving chicks[3] |
|---|---|---|---|---|---|
| 1 | 90 | 9 | 7 | 2 | 2 |
| 2 | 46 | 7 | 3 | 0 | 1 |
| 3 | 10 | 5 | 3 | 2 | 1 |
| 4 | 43 | 7 | 3 | 2 | 2 |
| 5 | 10 | 5 | 2 | 1 | 0 |
| 6 | 42 | 7 | 5 | 0 | 1 |
| 7 | 69 | 8 | 6 | 2 | 3 |
| 8 | 63 | 8 | 8 | 0 | 6 |
| 9 | 38 | 7 | 5 | 2 | 2 |
| 10 | 33 | 6 | 2 | 2 | 0 |
| 11 | 38 | 7 | 2 | 2 | 2 |
| 12 | 25 | 6 | 3 | 1 | 2 |
| 13 | 59 | 8 | 5 | 1 | 1 |
| 14 | 55 | 7 | 4 | 1 | 2 |
| 15 | 27 | 6 | 1 | 1 | 0 |
| 16 | 70 | 8 | 5 | 2 | 1 |
| 17 | 42 | 7 | 5 | 2 | 2 |
| 18 | 49 | 7 | 5 | 1 | 2 |
| 19 | 57 | 7 | 4 | 2 | 1 |
| 20 | 27 | 6 | 4 | 0 | 1 |

**1** Survivors calculated from 0–39 die,
40–99 survive for each individual.

**2** Surviving adults (40 adults), 4 die, 7 survive.

**3** Surviving chicks calculated from 0–69 die,
70–99 survive, for each individual with this
simulation after one year 56 idividuals alive.

**b)** Start with a new lot of adults
as those alive in spring, assume
equal numbers of male and
female (so breeding pairs). If an
odd number, drop one from
breeding totals.

# Exercises 12.2B *(p.259)*

**2 a)** 7, 64, 75, 28, 47, 84, 35, 08,
87, 04, 95

**b)** 50, 60, 10, 10, 10, ......

# Consolidation exercises for Chapter 12 *(p.263)*

**1 a)** Rule: win 0–3, loss 4–9

| Run | | No. of bets |
|---|---|---|
| 1 | 5 L 4 L 3 L 2 W 3 L 2 L 1 L 0 | 7 |
| 2 | 5 L 4 L 3 L 2 L 1 W 2 L 1 L 0 | 7 |
| 3 | 5 W 6 W 7 L 6 L 5 W 6 W 7 W 8 L 6 | |
| | L 5 L 4 W 5 L 4 W 5 L 4 L 3 L 2 L 1 | |
| | W 2 L 1 W 2 W 3 W 4 L 3 L 2 W 3 | |
| | L 2 L 1 W 2 W 3 L 2 W 3 L 2 L 1 W 2 | |
| | W 3 L 2 W 3 L 2 L 1 L 0 | 41 |
| 4 | 5 W 6 W 7 L 6 L 5 W 6 L 5 L 4 L 3 | |
| | L 2 W 3 L 1 L 0 | 13 |
| 5 | 5 L 4 L 3 W 4 W 5 L 4 L 3 W 4 L 3 | |
| | L 2 W 3 W 4 W W 5 W 6 L 5 L 4 L 3 | |
| | L 2 W 3 L 2 L 1 L 0 | 21 |
| 6 | 5 W 6 W L W 8 L 7 L 6 W 7 L 6 L 5 | |
| | L 4 L 3 L 2 L 1 W 2 L 1 L 0 | 15 |
| 7 | 5 L 4 L 3 L 2 W 3 W 4 W 5 L 4 W 5 | |
| | L 4 W 5 W 6 L 5 W 6 W 7 L 6 W 7 | 16+ |

**b)** Average ≈ 17 (by ignoring
last unfinished run) or average ≈
18 (if it takes 7 bets to lose on
7th go).

**c)** LLL £2, LLW £4, LWW £6, WWW £8

**d)** £4.29

**2 a)** You need 3 numbers to simulate progress of each ball. At each junction, 0–4 left, 5–9 right.

| Ball | RND | Path | Endpoint | Score |
|---|---|---|---|---|
| 1 | 858 | RRR | D | 5 |
| 2 | 834 | RLL | B | 10 |
| 3 | 844 | RLL | B | 10 |
| 4 | 574 | RRL | C | 10 |
| 5 | 683 | RRL | C | 10 |
| 6 | | | C | 10 |
| 7 | | | C | 10 |
| 8 | | | D | 5 |
| 9 | | | C | 10 |
| 10 | | | C | 10 |

| Ball | Distribution | Score |
|---|---|---|
| 11 | B | 10 |
| 12 | C | 10 |
| 13 | B | 10 |
| 14 | B | 10 |
| 15 | C | 10 |
| 16 | A | 5 |
| 17 | B | 10 |
| 18 | D | 5 |
| 19 | B | 10 |
| 20 | C | 10 |
| 21 | D | 5 |
| 22 | B | 10 |
| 23 | B | 10 |
| 24 | C | 10 |

**b) i)** Mean score 8.958

**ii)** Expected mean $\frac{70}{8}$ = 8.75

**3 a)** Rules: A wins point 0–55, B wins point 56–99, [P(A wining game) = $\frac{5}{8}$]

**b)**

| Game | Points won | | A | B |
|---|---|---|---|---|
| 1 | AABABBAA | A wins (5, 3) | 1 | 0 |
| 2 | BBAAABAA | A wins | 2 | 0 |
| 3 | ABBBB | B wins | 2 | 1 |
| 4 | ABBAABABBB | B wins | 2 | 2 |
| 5 | AABAA | A wins | 3 | 2 |
| 6 | BBBB | B wins | 3 | 3 |
| 7 | ABABAA | A wins | 4 | 3 |
| 8 | BABBAB | B wins | 4 | 4 |
| 9 | BAABBB | B wins | 4 | 5 |
| 10 | AABAA | A wins | 5 | 5 |
| 11 | AABBAA | A wins | 6 | 5 |
| 12 | ABBAAA | A wins | 7 | 5 |

A wins 7–5.

**c)** A wins 0–27, B wins 28–49, C wins 50–99

Points B, B, C, B, C, C, A, B, C, A, A, A, C, C

C wins 6 – 4 – 4

**4 a)** Inter-arrival times 3, 2, 2, 4, 2
Rule: 2 minutes 0–24,

3 minutes 25–74,
4 minutes 75–99

**b)** Service times 2, 5, 2, 11, 1
Rule: 1 minute 0–31,
2 minutes 32–79,
5 minutes 80–95
Ignore 96–99

**c)**

| Customer | Arrival time | Inter-arrival | Start of service | Service time | End of service | Q |
|---|---|---|---|---|---|---|
| 1 | 0 | 3 | 0 | 2 | 2 | 0 |
| 2 | 3 | 2 | 3 | 5 | 8 | 0 |
| 3 | 5 | 2 | 8 | 2 | 10 | 1 |
| 4 | 7 | 4 | 10 | 1 | 11 | 2 |
| 5 | 11 | 2 | 11 | 1 | 12 | 1 |
| 6 | 13 | – | 13 | 1 | 14 | 0 |

**d)** Mean time in queue $\frac{6}{6}$ = 1 minute

Mean queue length (per minute) $\frac{6}{14}$ = 0.43 people

Server utilisation $\frac{12}{14}$ × 100 = 86%

**e)** Needs a much longer run.

**5 a)** Rule: $\frac{100}{6}$ = 6 A: 0–53, B: 54–71, C: 72–89, D: 90–95
Ignore 96–99

**b)**

```
C  A  A   B  C  B   A  A  C
A  B  C   A  C  A   D  A  A
A  A  A   A  A  A   A  C  D
A  B  C   A  A  B   A  B  A
A  C  A   A  A  C   B  C  C
D  B  A   A  C  D   D  A  A
A  A  A   A  D  C
```

Marked in plots of 3.
The proportion of plots with 3As is $\frac{3}{20}$ = 0.15

**c)** 0  1111  1111  11
12 groups of 3As by this method.
Proportion $\frac{12}{58}$ = 0.21
Out of 58 since 58 of the 60 have 2 following.

**d)** $(\frac{9}{16})^3 \approx 0.18$

**6 a) i)** nut centres 0–14, not nut 15–99

**ii)**

| Carton | 1 | 2 | 3 | 4 | 5 | 6 | 7 | 8 | 9 | 10 |
|---|---|---|---|---|---|---|---|---|---|---|
| Nut centres | 3 | 5 | 3 | 2 | 1 | 4 | 1 | 3 | 1 | 4 |

**iii)** $p$(3 nuts) = 0.3
$p$(< 3 nuts) = 0.4

**iv)** nut centre  0–13
not nut  14–97
ignore  98, 99

**b) i)** $p$ is the number of nut centres in a carton.

**ii)** $f(15) = (1 – 15/100)^{20} + 20(15/100)(1 – 15/100)^{19} + ...$
= 0.405
simulation may be inaccurate increase the number of runs.

**7 a)** If today is wet
dry tomorrow – 0–69
wet tomorrow – 70–97
ignore 98,99

**b)** If today is D, then the following days are D, D, D, W, D, D, W, D, D, D, W, D, D, D

**c)** proportion of wet days 3/14 = 0.214

**d)** $p = \frac{7}{14}$

**e)** Longer run, more runs.

**8 a) i)** Winning centres
1.5-2.5, 5.5–7.5, 10.5–12.5, 15.5–17.5, 20.5–22.5, 25.5–27.5, 30.5–32.5, 35.5–37.5, 40.6–42.5, 45.5–47.5, 50.5–51.5

**ii)** Rules  00000 – 1.5 cm up to 98 000 for 51.5 cm
2000 to 1 cm

| $x$-values: | 34.118 | 41.5385 | 42.7905 |
|---|---|---|---|
| Win | ✗ | ✓ | ✗ |
| | 22.0425 | 45.1365 | 4.7835 |
| | ✗ | ✓ | ✗ |
| 34.052 | 38.4715 | 2.337 | 19.7075 |
| ✗ | ✗ | ✓ | ✗ |

**b)** $p$(win) = 0.3774

**c)** $e$ = 6 cm

**9 a)** Observed p(accepting shipment 1% defective)
Scheme 1 – $\frac{14}{20}$ = 0.7

Scheme 2 – $\frac{14}{20}$ = 0.7

**b)**

| % defective | 1 | 5 | 8 | 10 | 12 | 15 | 20 | 25 | 30 |
|---|---|---|---|---|---|---|---|---|---|
| Scheme 1 | 1 | 0.8 | 0.75 | 0.7 | 0.7 | 0.6 | 0.35 | 0.25 | 0.1 |
| Scheme 2 | 1 | 0.9 | 0.8 | 0.7 | 0.5 | 0.35 | 0.2 | 0.1 | 0.05 |

**c), d)**

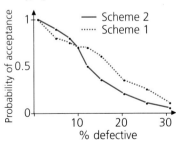

**e)** Scheme 2 does better but it is more expensive.

**10** Break the time interval for 10–20 minutes into 1-minute intervals.
For extractions + fillings <u>assume</u> that time of treatment are equally distributed.

Check-up 10 minutes 33% of appointments  Fillings 10 5% Extractions 15 2%

| | |
|---|---|
| 11 5% | 16 2% |
| 12 5% | 17 2% |
| 13 5% | 18 2% |
| 14 5% | 19 2% |
| 15 5% | 20 2% |
| 16 5% | |
| 17 5% | |
| 18 5% | |
| 19 5% | |
| 20 5% | |

| Treatment time | 10 | 11 | 12 | 13 | 14 | 15 | 16 | 17 | 18 | 19 | 20 |
|---|---|---|---|---|---|---|---|---|---|---|---|
| % frequency | 38 | 5 | 5 | 5 | 5 | 7 | 7 | 7 | 7 | 7 | 7 |
| RND | 0–37 | 38–42 | 43–47 | 48–52 | 53–57 | 58–64 | 65–71 | 72–78 | 79–85 | 86–92 | 93–99 |

Other assumptions – assume all patients arrive at the surgery when the appointment is due to begin. (Ignore waiting time before appointments)
Set out the table as follows:

| Appointment time | Treatment starts | RND | Length | Treatment ends | Waiting |
|---|---|---|---|---|---|
| 8.30 | 8.30 | 39 | 11 | 8.41 | 0 |
| 8.45 | 8.45 | 56 | 14 | 8.59 | 0 |
| 9.00 | 9.00 | 98 | 20 | 9.20 | 0 |
| 9.15 | 9.20 | 17 | 10 | 9.30 | 5 |
| 9.30 | 9.30 | 62 | 15 | 9.45 | 0 |
| 9.45 | 9.45 | 83 | 18 | 10.03 | 0 |
| 10.00 | 10.03 | 40 | 12 | 10.15 | 3 |
| 10.15 | 10.15 | 57 | 14 | 10.29 | 0 |
| 10.30 | 10.30 | 32 | 10 | 10.40 | 0 |
| 10.45 | 10.45 | 30 | 10 | 10.55 | 0 |
| 11.00 | 11.00 | 14 | 10 | 11.10 | 0 |
| 11.15 | 11.15 | 78 | 17 | 11.32 | 0 |
| 11.30 | 11.32 | 64 | 15 | 11.47 | 2 |
| 11.45 | 11.47 | 63 | 15 | 12.04 | 2 |
| 12.00 | 12.04 | 5 | 10 | 12.14 | 4 |
| 12.15 | 12.15 | 13 | 10 | 12.15 | 0 |

The system seems to work OK.

## CHAPTER 13
## Decision-making

### Exercises **13.1A** (p.278)
**1** Total number of different combinations is 232.
**2** 20 ways of choosing one soap and one shampoo for each guest room.
60 ways of choosing one soap, one shampoo, one toothpaste.

---

**3 a) i)** 343 **ii)** 210
**b)** 35
**4 a)** 792 **b)** 86 (45 ways of 3 women and 2 men + 31 ways of 2 women and three men)
**5** 2 118 760 (= $^{50}C_5$ ways)
$Pr$ (at least one defective) = 0.09 (= 9%)

### Exercises **13.1B** (p.278)
**1 a)** 35 **b)** 12 **c)** 72
**2** $\frac{1}{4}$, $\frac{1}{6}$
**3 a)** $10^6$ (including 000000)
**b)** $9 \times 10^5$
**c)** $10^3(10^3 - 1)$
**d)** $\frac{10!}{4} = 151\,200$
**e)** $9 \times 9 \times 8 \times 7 \times 6 \times 5 = 136\,080$
**4 a) i)** $^{49}C_6 = 13\,983\,816$, $\frac{1}{13983816}$
**ii)** $\frac{6 \times 43}{13983816}$
**b)** $\frac{6 \times 43}{42 \times 13983816}$
**5 a)** $^{10}C_2 = 45$
**b)** 210
**c) i)** 21!
**ii)** $10! \times 7! \times 4!$
**ii)** $6 \times 10! \times 7! \times 4!$

### Exercises **13.2A** (p.282)
**1** EMV = £81 750
It would be in the manufacturer's interest to develop the product.

**2**
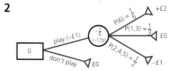

**3 a)** 6 (three ways)
11 (two ways)   16 (one way)
**b)**
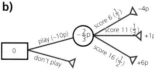

**4**
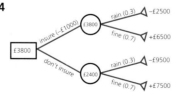

The organiser would be well-advised to take out insurance.

---

**5**
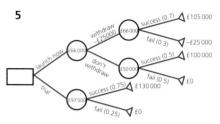

The company could spend up to £31 500 on a trial in order to keep the increase viable.

### Exercises **13.2B** (p. 284)
**1 a)** Do not enter.
**b)** Purchase book and enter.
**2 a)** Do not sink pit.
**b)** Survey; if probable result then sink, if improbable do not sink.
**3** Do not gamble.
**4 a)** Purchase cheaper item,
**b)** $p > 0.5025$ **c)** Cost ≥ £9.62
**5** The company should market the product, if there is no competition then set a high price, if there is competition then set a medium price.

### Exercises **13.3A** (p.290)
**1 a)** Ali studies Maths and German.
**b)** Ali studies Maths and not Art.
**c)** Ali studies Maths and German and not Art.
**d)** Ali studies Maths and not German or Art.

**2**

| $p$ | $q$ | $p \Rightarrow \sim q$ |
|---|---|---|
| T | T | F |
| T | F | T |
| F | T | T |
| F | F | F |

**3**

| $p$ | $q$ | $p \vee q$ | $\sim p \wedge q$ | $(p \vee q) \Leftrightarrow (\sim p \wedge q)$ |
|---|---|---|---|---|
| T | T | T | F | F |
| T | F | T | F | F |
| F | T | T | T | T |
| F | F | F | F | T |

**4**

| $p$ | $q$ | Output |
|---|---|---|
| 0 | 0 | 1 |
| 0 | 1 | 1 |
| 1 | 0 | 0 |
| 1 | 1 | 1 |

**5**

| $p$ | $q$ | Output |
|---|---|---|
| 0 | 1 | 1 |
| 1 | 1 | 1 |
| 1 | 0 | 0 |
| 0 | 0 | 0 |

## Exercises **13.3B** *(p.291)*

**1 a)** Only $r$ is sufficient for $s$ to be true.

**b)** Neither statement is true.

**3 b)** $b \wedge (c \vee b') \wedge ((c' \wedge a') \vee (a \vee (b \wedge c)))$

**4 a)** $c = x_1 \wedge x_2$  $s = (x_1 \vee x_2) \wedge (x_1 \wedge x_2)'$

**b)** $s$ is the right hand digit on the sum of two single digits, $c$ is a single carry digit.

## Consolidation exercises for Chapter 13 *(p.293)*

**1 a)** $10^3 = 1000$

**b)** $26^3 = 17\,576$

**c)** $26 \times 1000 \times 17\,576 = 456\,976\,000$

**2 a)** $8! = 40\,320$

**b)** $4 \times 6! = 2880$

**3 a)** 15 ways  **b)** 18 ways

**4 a)** 3654  **b)** 84  **c)** 5456

**5 a)** Advise her not to drill.

**b) c)**

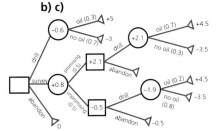

EMV = £800 000
Advise her to have a survey carried out. If it is promising then drill; if not promising then abandon.

**6 a)**

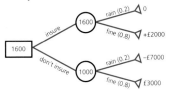

She should insure.

**b)**

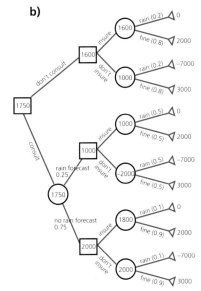

She could pay up to £150 for the consultation.

**7** The best option is not to play! If the player does play and picks an ace he should opt not to throw the dice.

**8 a)** $p \Rightarrow q$

**b)** I don't work hard and I pass my exams.

**c)** I work hard and I don't take regular exercise so I will not pass my exams.

**9 a)** $\sim (p \Rightarrow (q \Rightarrow p))$

| $p$ | $q$ | $p \Rightarrow q$ | $p \Rightarrow (q \Rightarrow p)$ | $\sim (p \Rightarrow (q \Rightarrow p))$ |
|---|---|---|---|---|
| T | T | T | T | F |
| T | F | T | T | F |
| F | T | F | T | F |
| F | F | T | T | F |

**b)** $p \Rightarrow (p \wedge \sim (q \vee r))$

| $p$ | $q$ | $r$ | $\sim (q \vee r)$ | $p \wedge \sim (q \vee r)$ | $p \Rightarrow (p \wedge \sim (q \vee r))$ |
|---|---|---|---|---|---|
| T | T | T | F | F | F |
| T | T | F | F | F | F |
| T | F | T | F | F | F |
| T | F | F | T | T | T |
| F | T | T | F | F | T |
| F | T | F | F | F | T |
| F | F | T | F | F | T |
| F | F | F | T | F | T |

**10 a)** $p \Rightarrow q$    T F T T

| $p$ | $q$ | $\sim (p \wedge q)$ | $\sim (\sim (p \wedge q) \wedge p)$ |
|---|---|---|---|
| T | T | F | T |
| T | F | T | F |
| F | T | T | T |
| F | F | T | T |

**b)**

| $x$ | $y$ | $x \Rightarrow y$ | $x$ | $y$ | A | B |
|---|---|---|---|---|---|---|
| 0 | 0 | 1 | 0 | 0 | 1 | 1 |
| 0 | 1 | 1 | 0 | 1 | 1 | 1 |
| 1 | 0 | 0 | 1 | 0 | 1 | 0 |
| 1 | 1 | 1 | 1 | 1 | 0 | 1 |

# Glossary

**activity network**
see precedence network

**activity on arc network**
precedence network in which the arcs represent the activities, while the vertices represent events

**activity on vertex network**
precedence network in which the activities are represented by the vertices of a network, and edges show the order of precedence

**adjacency matrix**
matrix representing the vertices and edges of a graph

**algorithm**
systematic process for finding a solution to a problem

**alternating path**
path in a bipartite graph that joins vertices from one subset to vertices in the other, in such a way that alternate edges only are in the initial matching, and initial and final vertices are not incident with an edge in the matching

**analytical solution**
the solution of a recurrence relation given as a formula as opposed to a sequence of numbers

**arc (or edge)**
line connecting two vertices in a graph in a directed network, an edge that is directed

**artificial backward capacity**
record of flow being sent down an arc in a network, made with a reversed arrow along the arc

**ascending bubble sort**
see bubble sort

**auxiliary equation**
the equation obtained for the base $m$ when $x_n = m^n$ is substituted into a homogeneous recurrence relation

**back substitution stage**
this stage of the Gaussian elimination method is the process of working back through the equations solving for the variables

**bin**
'space' forming the containers for bin-packing

**bin-packing**
sorting methods, see page 14

**binary codes**
code that uses two digits 0 and 1 known as binary digits or bits

**binary codeword**
a sequence of 0s and 1s

**bipartite graph**
graph which can be divided into two subsets in such a way that each edge of the graph joins a vertex from one subset to a vertex in the other

**bit**
see binary code

**bubble sort**
a sorting method, see page 6

**cascade activity number**
numbers allocated to activities in a project to ensure a logical sequence through the activities

**cascade chart**
diagram showing precedence relations, in the form of bars drawn against a time scale, that helps to make best use of all resources

**combinations**
unordered selections without repetition $^nC_r$

**complete bipartite graph**
bipartite graph in which each vertex in one subset is joined to every vertex in the other by an edge

**complete graph**
a graph where each vertex is connected to every other vertex

**connected graph**
a graph with at least one route between any pair of vertices

**constraints**
limits on the values of variables in linear programming

**cost function**
relation linking the cost to the variables that are included in the cost

**cost path**
cheapest, (shortest) path through a cost network

**counting problem**
we need to know the number of objects or solutions

**critical activities**
those activities that must be started on time to avoid delaying the whole project

**critical path**
minimum time needed to complete a project, following the longest path through the precedence network and including all the critical activities

**cut**
a line dividing a network into two parts, one containing the source and the other containing the sink

**cycle**
a path that completes a loop and returns to its starting point

**decoding**
process of transferring a collection of codewords backwards into a readable message

**degree of a vertex**
the number of edges incident at the vertex

**descending bubble sort**
see bubble sort

**deterministic model**
model that assumes chance events do not occur

**dimension of code**
if a code has a maximum of $2^k$ codewords then the number $k$ is the dimension of the code

**directed network**
network in which there is a set direction of movement

**discrete situation**
a situation modelled by a discrete sequence of numbers

**distance matrix**
matrix representing a network of distances

**dummy arc**
arc introduced when an activity has to be preceded by more than one other activity; they are marked as dotted lines and have zero duration

**dynamic programming**
an algorithm for solving network problem by working backwards through the network

**edge (or arc)**
line connecting two vertices in a graph

**edge-disjoint**
paths between two vertices are edge-disjoint if they have no edges in common

**Elimination stage**
this stage of the Gaussian elimination method is the process of subtracting or adding a multiple of one row from another

**Encoding**
process of transferring a message into codewords

**enumerations problems**
counting problems and listing problems are types of enumeration problems

**Eulerian cycle**
a cycle that includes every edge of a graph exactly once

**event**
finish of one activity and the start of another, in a precedence network

**exhaustive search**
finding the shortest path by looking for every possible path

**face**
region of the plane bounded by edges of a planar graph

**feasible flow**
a flow through a network that satisfies given capacities

**feasible region**
region within a set of straight-line graphs showing linear programming constraints, within which the solution lies

**first-fit algorithm**
a sorting algorithm that places item in the first available bin

**float**
maximum time by which the start of an activity can be delayed without delaying the whole project

**flow augmenting paths**
paths that consist entirely of unsaturated arcs

**full-bin algorithm**
a sorting algorithm that fills bins one at a time

**Gaussian elimination method**
method of solving simultaneous equations by eliminating one variable and solving for the other – it can also be used for systems of more that two equations, using a similar principle

**general solution**
analytical solution of a recurrence relation that contains unknown constants

**generator matrix**
matrix in which the rows are the generator set

**generator set**
minimal set of codewords that can generate an entire linear code

**graph**
diagram of vertices and edges representing how objects are related to each other

**greedy algorithm**
an algorithm that chooses the best option at each stage

**Hamiltonian cycle**
a cycle that passes through every vertex of a graph once, and only once, and returns to the starting point

**Hamming distance**
distance between two codewords i.e. the number of places in which their bits differ

**heuristic algorithm**
algorithm which produces a good solution to a problem using logical and intuitive steps

**homogeneous**
a second order recurrence relation in the form $x_n = ax_{n-1} + bx_{n-2} + p$ is homogeneous if $p = 0$

**independent float**
when the activities with float are independent of each other

**infinite face**
area outside the edges of a planar graph

**initial condition**
the starting value that generates the sequence from a recurrence relation

**interfering float**
when the start time for one activity with a float affects the float for other activities

**isomorphic**
a graph having the same number of vertices as another graph, with the degrees of corresponding pairs of vertices being the same

**iteration**
the step-by-step process of solving recurrence relations and other problems

**$k$th order recurrence relation**
a recurrence relation is of $k$th order if the difference between the highest and lowest subscript is $k$

**length of codeword**
number of bits in a codeword

**line**
arc or edge

**linear code**
if $x$ and $y$ are codewords a given code and if $x + y$ is a codeword then the code is a linear code

**linear function**
relation between two variables that can be expressed as an equation and drawn as a straight line

**linear programming problem**
problem made up of a linear objective function and constraints formed by linear inequalities

**listing problem**
list all the possible objects or solutions

**matching**
a graph that links some of the vertices in one subset of a bipartite graph to some of the vertices in the other subset

**matrix (matrices)**
an array of numbers that can be used to convey information in a condensed form; also used in the solution of simultaneous equations

**maximum capacity**
maximum number of 'items' that can pass along an arc in a network

**maximum matching**
the optimum solution linking as many pairs of vertices as possible in a bipartite graph

**minimum cut**
the cut with the minimum capacity

**minimum distance of a code**
the smallest Hamming distance between two distinct codewords of the code

**minimum-connector problem**
spanning tree of minimum length

**multiplier**
number used to multiply one (or more) equation, to enable the solution of simultaneous equations

**Nearest neighbour algorithm**
*see page 60*

**network**
a graph in which every edge has a value called its weight

**node**
*see* vertex

**non-linear**
an equation, inequality or recurrence relation in which the powers of the variables are not unity

**objective function**
main function in a linear programming problem that has to be minimised or maximised

**optimal**
best solution according to given conditions

**parity check digit**
an extra digit added to a codeword to ensure that the number of 1s in the codeword is even

**pass**
one iteration of an algorithm

**path**
a route through a graph which does not go along any edge more than once or visit any vertex more than once.

**permutations**
ordered selections without repetition, $^nP_r$

**Petersen graph**
*see page 52*

**pivot column**
column with the smallest (i.e. 'most negative') value in the bottom row of the simplex tableau

**pivot element**
element in the simplex tableau which falls in both the pivot row and pivot column

**pivot row**
when the entries in the right hand column of the tableau are divided by the entries in the pivot column, the small value obtained gives the pivot row

**planar graph**
graph that can be drawn in the plane with no edges crossing

**platonic solid**
three-dimensional solid with plane faces

**point**
vertex

**precedence network**
diagram showing the precedence relations

**precedence relations**
the relationship showing the order in which activities must be done

**predecessor**
immediate preceding activity

**principle of inclusion – exclusion**
a method of counting the total number when the sets are not disjoint

**quicksort**
a sorting method, *see page 10*

**rate or efficiency of a code**
the ratio $\frac{\text{dimension}}{\text{length}}$

**recurrence relation**
an equation whose solution is a sequence of numbers and models a discrete situation; the terms in the sequence are called iterations

**redundancy**
defined by 1 – rate of a code

**repetition code**
code made up of repeated digits

**resource histogram**
diagram to show the number of people needed at any stage of a project

**resource levelling**
system of managing a project to balance the amount of work to be done and the resources available

**risk path**
safest path through a network

**saturated**
carrying the maximum capacity

**simplex method**
process of using Gaussian elimination steps to solve linear programming problems

**simplex tableau**
equations of a linear programming problem written in tabular form

**simulation**
process that can be used to test what might happen in situations where an experiment with real subjects may be too long or too dangerous

**sink**
finishing point of a directed network

**slack variables**
variables introduced in linear programming problems which change the constraints in the form of given inequalities into equations; the solution of the problem occurs at one of the corners of the intersection of the planes

**source**
starting point of a directed network

**spanning tree**
a subgraph that includes all the vertices in the original graph and is also a tree

**star graph**
complete bipartite graph in which the number of vertices in the first subset is 1

**stochastic model**
model that allows for an element of chance in the outcome

**subdivision of a graph**
formed by inserting vertices of degree 2 into a graph

**subgraph**
part of a graph which is itself a graph

**supersink**
artificial sink introduced into a network and fed from all the sinks

**supersource**
artificial source introduced into a network feeding all the sources

**tour**
route which visits every vertex in the network

**tree**
a connected graph that contains no cycles

**trivial constraints**
limits on values of variables that give little information as to how they are related, such as $x \geq 0, y \geq 0$

**vertex**
point on a graph representing a point where edges meet; also called a node

**weight**
value on the edge of a network

**weight of a codeword**
number of 1s in a codeword